AF411437

Thermal and Electro-thermal System Simulation 2020

Thermal and Electro-thermal System Simulation 2020

Editors

Márta Rencz
Lorenzo Codecasa
Andras Poppe

MDPI • Basel • Beijing • Wuhan • Barcelona • Belgrade • Manchester • Tokyo • Cluj • Tianjin

Editors
Márta Rencz
Budapest University of Technology and Economics
Hungary

Lorenzo Codecasa
Politecnico di Milano
Italy

Andras Poppe
Budapest University of Technology and Economics
Hungary

Editorial Office
MDPI
St. Alban-Anlage 66
4052 Basel, Switzerland

This is a reprint of articles from the Special Issue published online in the open access journal *Energies* (ISSN 1996-1073) (available at: https://www.mdpi.com/journal/energies/special_issues/thermal_electro_thermal_system_2020).

For citation purposes, cite each article independently as indicated on the article page online and as indicated below:

LastName, A.A.; LastName, B.B.; LastName, C.C. Article Title. *Journal Name* **Year**, *Volume Number*, Page Range.

ISBN 978-3-03943-831-0 (Hbk)
ISBN 978-3-03943-832-7 (PDF)

About the Editors

Márta Rencz received her Electrical Engineering degree and PhD from the Budapest University of Technology and Economics. She was a PI in numerous international research projects, mostly in the field of investigating, measuring and modeling multi-physical effects in electronics. She has published her theoretical and practical results in more than 350 technical papers. She was a co-founder and CEO of MicRed Ltd, that is now part of Siemens DI, where she still holds a research director position. She holds various awards of excellence, among others the Harvey Rosten award (2001) and the Allan Krauss thermal management award of ASME (2015). In 2013, she received a Doctor Honoris Causa degree from the Tallinn University of Technology in Estonia. In 2019, she received the Thermal Hall of Fame, Lifetime achievement award at Semitherm in San Jose, CA, USA.

Lorenzo Codecasa received his PhD in Electronic Engineering from Politecnico di Milano in 2001. From 2002 to 2010, he worked as an Assistant Professor of Electrical Engineering with the Department of Electronics, Information, and Bioengineering of Politecnico di Milano. His main research contributions are in the theoretical analysis and the computational investigation of electric circuits and electromagnetic fields. In his research on heat transfer and thermal management of electronic components, he has introduced original industrial strength approaches to the extraction of compact thermal models, currently also available in commercial software. For these activities, in 2016, he received the Harvey Rosten Award for Excellence. He currently serves as an Associate Editor for the IEEE Transactions of Components, Packaging and Manufacturing Technology. In his research areas, he has authored or co-authored over 200 papers in refereed international journals and conference proceedings.

Andras Poppe obtained his MSc in electrical engineering and his PhD from the Technical University of Budapest (BME), Faculty of Electrical Engineering, where he now serves as a professor, and the head of the Department of Electron Devices of BME. His recent fields of research include electro-thermal simulation, thermal simulation and issues of thermal transient testing of IC-s, MEMS and electrical components, combined thermal and radiometric/photometric characterization of power LEDs and the characterization of OLEDs. He is an internationally recognized expert of multi-physics modelling and the characterization of electronics components. For his work on LED multiphysics modelling, he has already received the Harvey Rosten Award twice. He was one of the co-founders of MicReD LtD, now part of Siemens DI. He has published his research results in more than 300 scientific papers.

Preface to "Thermal and Electro-thermal System Simulation 2020"

Microelectronics thermal experts from four continents came together in the fall of 2019 at the 25th THERMINIC Workshop at Lake Como in Italy, to discuss the latest issues in the design, characterization and simulation of the thermal and reliability problems in electronic devices and systems. The event, as usual, was sponsored by the IEEE Society, by the IEEE Components, Packaging, and Manufacturing Technology Society, and by numerous companies. The workshop is largely application-oriented and shows a rare balance of contributions from both academy and industry, often in great synergism. This issue had participants from 25 countries. The proceedings for the THERMINIC formal workshop are not usually published, but it is a tradition that the most valuable papers from the workshop appear in Special Issues or special sections in leading international journals. These journals are selected according to the nature and scope of the workshop in that particular year. At THERMINIC 2019, significant results were presented about thermal and electro-thermal simulations. For this reason, it was thought to organize a Special Issue of Energies entitled "Thermal and Electro-Thermal System Simulation". In this Special Issue, papers have been accepted whether they are derived from THERMINIC 2019 contributions or not. At the end of a very rigorous revision process, 14 papers are selected. Several papers, here proposed to the thermal community, have turned out to be extended versions of papers presented at THERMINIC 2019, thus confirming the fact that THERMINIC 2019 was a stage of choice for presenting outstanding contributions on thermal and electro-thermal simulation in electronic systems. The papers selected for this Special Issue testify in particular great activity in parametric thermal and electro-thermal modeling, multi-physics simulation of LEDs, electro-thermal simulation of power electronics applications. For the first time, the thermal modeling of batteries and modeling the thermal influenced motion of magnetic nanoparticles in micro-channels are also among the subjects featured. We hope that all the selected papers will provide useful information to the readers who are certainly interested in these recent important questions of microelectronics thermal issues.

Márta Rencz, Lorenzo Codecasa, Andras Poppe
Editors

Article

TRIC: A Thermal Resistance and Impedance Calculator for Electronic Packages [†]

Lorenzo Codecasa [1], Francesca De Viti [2], Vincenzo d'Alessandro [3,*], Donata Gualandris [2], Arianna Morelli [2] and Claudio Maria Villa [2]

[1] Department of Electronics, Information and Bioengineering, Politecnico di Milano, 20133 Milan, Italy; lorenzo.codecasa@polimi.it

[2] STMicroelectronics, 20864 Agrate Brianza, Italy; francesca.deviti@st.com (F.D.V.); donata.gualandris@st.com (D.G.); arianna.morelli@st.com (A.M.); claudio-maria.villa@st.com (C.M.V.)

[3] Department of Electrical Engineering and Information Technology, University Federico II, 80125 Naples, Italy

* Correspondence: vindales@unina.it

[†] This manuscript is based on the conference paper "Thermal Resistance and Impedance Calculator (TRIC)" included in the Proceedings of the 25th International Workshop on Thermal Investigations of ICs and Systems, Lecco, Italy, 25–27 September 2019.

Received: 21 February 2020; Accepted: 21 April 2020; Published: 4 May 2020

Abstract: This paper presents the Thermal Resistance and Impedance Calculator (TRIC) tool devised for the automatic extraction of thermal metrics of package families of electronic components in both static and transient conditions. TRIC relies on a solution algorithm based on a novel projection-based approach, which—unlike previous techniques—allows (i) dealing with parametric detailed thermal models (pDTMs) of package families that exhibit generic non-Manhattan variations of geometries and meshes, and (ii) transforming such pDTMs into compact thermal models that can be solved in short times. Thermal models of several package families are available, and dies with multiple active areas can be handled. It is shown that transient thermal responses of chosen packages can be obtained in a CPU (central processing unit) time much shorter than that required by a widely used software relying on the finite-volume method without sacrificing accuracy.

Keywords: electronic packages; detailed thermal model; Joint Electron Device Engineering Council (JEDEC) metrics; thermal impedance; thermal simulation

---

1. Introduction

The thermal analysis of electronic devices, circuits, and systems has always been an activity of utmost relevance in the semiconductor industry and academia. A thermally aware design can be achieved with the aid of numerical simulations, which are very challenging in terms of CPU time and memory storage, if a high level of accuracy is desired. This has stimulated many research groups to develop tools relying on suitable algorithms to accelerate the solution process (e.g., [1–3]).

For the specific case of packaged components, the preferred approach is to build boundary condition independent (BCI) compact thermal models (CTMs) to alleviate the computational burden without perceptible accuracy loss [4–7]; an interesting review of BCI CTMs for electronic parts is offered in [8].

Standardized procedures have been introduced to allow a fair comparison among packaged components in terms of thermal performances. More specifically, Joint Electron Device Engineering Council (JEDEC) metrics [9] are evaluated and included in the product datasheets. In order to satisfy the request for JEDEC thermal metrics of chosen package families, the authors developed the Thermal Resistance Advanced Calculator (TRAC) tool, the features of which were initially sketched in [10] and then fully described in [11]. TRAC was conceived (i) to allow a straightforward definition of a parametric detailed thermal model (pDTM) of a package with Manhattan geometry/mesh (as well as Manhattan geometry/mesh variations) and (ii) to automatically determine the thermal metrics of the package from the simulated (static) temperature field. TRAC makes use of a model-order reduction (MOR) technique (e.g., [12–23]) to transform the pDTM into a CTM, which can be solved in a short time. Various package families were covered, namely, exposed-pad (epad) low-profile (thick) and thin quad flat packages (eLQFPs and eTQFPs, respectively) as well as exposed-pad quad-flat no-leads (eQFN) packages. However, only one heat source (HS) with arbitrary size and position could be activated within the semiconductor die.

TRAC was believed to fulfill the requests of vendors in the semiconductor industry. It has the potential to allow people not necessarily endowed with expertise in the thermal field (such as system engineers, marketing people, etc.) to easily get the thermal metrics of the packages of interest. Additionally, it is suited to free thermal experts from these standard and repetitive tasks, giving them the possibility to focus on crucial thermal issues.

However, as additional package families were considered, soon emerged the need to deal with variations in geometries/meshes much more complex than those manageable by MOR methods like the one implemented in TRAC. For this reason, a new simulation tool referred to as Thermal Resistance and Impedance Calculator (TRIC) [24] has been realized, which enriches the functionalities of TRAC as follows: (i) TRIC relies on a novel advanced projection-based approach that allows deriving CTMs from the pDTMs of package families with Manhattan geometry/mesh and generic non-Manhattan geometry/mesh transformations without efficiency loss; (ii) in addition to the package families available in TRAC, also full-plastic LQFPs (pLQFPs) are included; (iii) the temperature field can also be evaluated under transient conditions; (iv) the pDTMs and the related CTMs of components with *multiple* HSs can be generated and simulated. It must be remarked that for both pLQFPs and multi-source packages the nature of the geometry/mesh variations is inherently non-Manhattan.

After the contribution [24] was presented, a new TRIC version was released, which also covers eQFN packages with multiple rows of pins (eQFN-mr), as well as PowerSSO packages. Consequently, such a release can handle pDTMs corresponding to a massive amount of package families, which is destined to further increase in the near future.

The aim of this paper is to extend [24] by

- providing an exhaustive picture of the TRIC features;
- reporting and discussing a larger number of results, including those obtained for the newly included package families (eQFN-mr and PowerSSO);
- adding a detailed comparison (only touched upon in [24]) in terms of accuracy and CPU time with the commercial finite-volume (FV) software FloTHERM [25];
- showing a simulated temperature map at a chosen time instant for a multi-source case of practical relevance.

The remainder of the paper is articulated as follows. In Section 2, TRIC is described and the details concerning all the thermally-modeled packages are provided. Section 3 probes into the solution algorithm. The numerical results and the main findings, as well as the comparison with FloTHERM, are shown and discussed in Section 4. Conclusions are then given in Section 5.

2. TRIC Features

TRIC, like the former release TRAC, is suited to automatically extract the JEDEC metrics ϑ_{JA}, Ψ_{JB}, Ψ_{JCtop}, ϑ_{JB}, ϑ_{JCtop}, and $\vartheta_{JCbottom}$ in four ambients [9] for each specimen in a package family. The ambients mainly differ in terms of thermal path followed by the heat generated within the HS and emerging from the die; more specifically, the ambient to evaluate $\vartheta_{JCbottom}$ requires a cold plate in intimate contact with the package backside; the plate is located over the top surface when aiming to compute ϑ_{JCtop}; in the ambient for determining ϑ_{JB}, a cold ring surrounds the package; no cooling systems are adopted in the ambient common to ϑ_{JA}, Ψ_{JB}, and Ψ_{JCtop}. To this end, a purely conductive pDTM is defined for each family of packages immersed in a specific ambient, within which the boundary conditions (BCs) are calibrated following the JEDEC environment specifications. For instance, for the evaluation of $\vartheta_{JCbottom}$, an extremely high heat transfer coefficient was applied to the bottom surface to describe the thermal path from the die to the heat sink. The geometry, assumed to be Manhattan (whereas its variation can also be non-Manhattan), is modeled by subdividing the domain into rectangular parallelepipeds (also referred to as *cells* or, more picturesquely, as *bricks*) with edges parallel to the x, y, and z axes, for which dimensions, material properties, and heat generation are provided. All data on geometry and properties are stored in a parameter vector **p** varying in a set P. The pDTM includes information to automatically generate a Cartesian mesh for all specimens belonging to a family of packages in the chosen ambient.

It is worth noting that the pDTMs are created by resorting to reasonable simplifications (coming from the vendor's experience) that allow a marked reduction in computational burden while negligibly affecting the simulation accuracy. More specifically, in all ambients, the board over which the package is mounted is modeled with a single finely-meshed parallelepiped with a thermal conductivity adjusted to account for the aggregate effect of metal traces and vias, the detailed representation of which would have led to a far too complex problem. In a similar fashion, the pins in the eQFN and eQFN-mr package families, as well as the leads in the pLQFP and PowerSSO families, were thermally represented by a rectangular parallelepiped, the thermal conductivity of which was determined through a weighted averaging procedure.

The thermal metrics can be automatically evaluated through a graphical user interface, in which a chosen specimen in a package family is determined by selecting the corresponding set of parameters. Package size and thickness, pad size, lead count, die size, and thickness are examples of the data that the user can input.

As mentioned in [11], a preliminary convergence analysis of the 3-D mesh discretization of the constructed pDTMs was performed for selected packages; in particular, the calculated thermal metrics were monitored by increasing the degrees of freedom (DoF) until a negligible mesh sensitivity was observed. Then the discretization leading to about 0.1% inaccuracy was chosen to avoid unnecessarily onerous too-fine meshes.

So far, the pDTMs of many package families are available. The list, along with the main geometrical features, is reported below.

- eQFPs, which are surface mount integrated circuit packages with a flat rectangular body, and leads extending from all the four sides. In particular, both the eLQFP and eTQFP variants are available, which differ in terms of body thickness (1.4 and 1 mm, respectively). The square epad structure, which is a standard lead frame wherein the die pad is depressed down to the package bottom face, represents a valuable solution to ease the heat dissipation from die to board. The horizontal size of the body can be 7×7, 10×10, 14×14, 20×20, 24×24 mm^2, the total number of leads can be 32, 48, 64, 80, 100, 128, 144, 176, 216 for both variants; several sizes for the epad are available, which span from 3.5×3.5 to 9×9 mm^2. Various types of glue for die attach can be selected. The die thickness can amount to 100, 280, 375 (used in the simulations shown in Section 4), and 580 µm (the latter value for the eLQFPs only), while any technologically-reasonable horizontal size can be chosen. It is worth noting that only parameter sets corresponding to *real* packages fabricated by STMicroelectronics can be selected, whereas all other combinations are obviously not possible; examples are: the 10×10 mm^2 eTQFP with 80 leads can be equipped only with an epad with sizes 3.5×3.5, 5.4×5.4, 6.2×6.2 mm^2; many sizes are instead possible for the epad in the 14×14 mm^2 eLQFPs with 100 leads, namely, 3.5×3.5, 4.5×4.5, 6.0×6.0, 7.2×7.2, 7.6×7.6, 8.5×8.5 mm^2; 32, 40, and 48 leads are available only for the 7×7 mm^2 eLQFP.
- pLQFPs, where, contrary to the eLQFP counterparts, the mold covers the entire package surface, so that the metal base of the lead frame is not "exposed" and thus not visible from the package bottom. With a few exceptions, all the parameter sets already reported for the eLQFPs are possible.

- eQFN packages, which are lead-less flat molded structures built with a metal lead-frame manufactured by etching, and represent a popular cost-effective and high-performance packaging solution by virtue of the lower inductance than in leaded packages. Several horizontal sizes of the square package body are available, spanning from 2×2 to 15×15 mm^2, while the thickness can be equal to 0.55, 0.75, and 0.9 mm (the latter being adopted for the simulations in Section 4). The horizontal sizes of epad and die can be arbitrarily chosen in the ranges allowed by the design rules. Various types of die attach can be selected. Specimens of this family can be equipped with a single row of pins (single-row QFN) or with multiple rows of pins (eQFN-mr), this option being not available in the former TRIC version.

- PowerSSO packages, which are derived from the well-known Small Outline (SO) family and benefit from footprint and height 30%–50% smaller than a conventional dual in-line package. More specifically, epad PowerSSO structures are considered, which are conceived to favor the heat removal without extra cost penalty. Many packages belonging to the PowerSSO family have been included, namely (i) PowerSSO-12, PowerSSO-14, and PowerSSO-16, all sharing a $4.9 \times 3.9 \times 1.5$ mm^3 body and equipped with 12 leads (the lead pitch being 0.8 mm), 14 leads (0.65 mm), and 16 leads (0.5 mm), respectively; (ii) PowerSSO-24, PowerSSO-28, and PowerSSO-36, all sharing a $7.5 \times 10.3 \times 2.3$ mm^3 body, and equipped with 24 leads (the lead pitch being 0.8 mm), 28 leads (0.65 mm), and 36 leads (0.5 mm), respectively. PowerSSO packages were not covered by the first TRIC release.

Bottom side views of the above-reported packages are shown in Figure 1, while examples of DTMs are depicted in Figure 2. Thermal models of further electronic components can be created with relatively little effort.

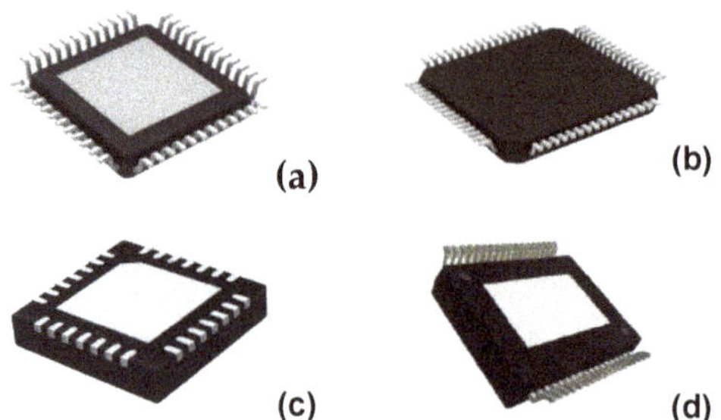

Figure 1. Bottom side views of specimens of the (**a**) eLQFP, (**b**) pLQFP, (**c**) single-row eQFN, and (**d**) PowerSSO-36 package families.

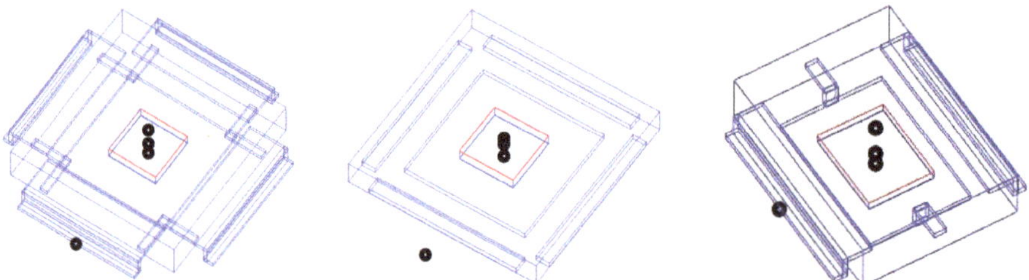

Figure 2. DTMs of the eLQFP (left), single-row eQFN (center), and PowerSSO-36 (right) families. The black circles represent the temperature probes needed to determine the thermal metrics.

Differently from TRAC, in TRIC the evaluation of the temperature field under transient conditions can be enabled for any profile of dissipated power at the HSs. In addition, TRIC allows coping with die layouts integrating circuitries with multiple separate active areas (i.e., HSs), for which the assumption of a uniform power distribution over the whole die surface would have been unacceptably inaccurate. As a result, TRIC can be used with many practical aims. As an example, by emulating the dissipation over the die with only one HS, the thermal impedance can be computed by determining the thermal response to a power step. Moreover, power profiles typically encountered in real applications, like the antilock braking system (ABS) in vehicles, injection, etc., can be taken into account.

3. TRIC Solution Algorithm

The parametric MOR techniques implemented in TRAC are only suited to deal with Manhattan variations of geometry/mesh. A Manhattan transformation is a geometry transformation that converts the x, y, and z coordinates into coordinates X, Y, and Z given by

$$\begin{aligned} X &= f_x(x) \\ Y &= f_y(y) \\ Z &= f_z(z) \end{aligned} \tag{1}$$

with f_x, f_y, and f_z continuous and monotonic functions. Conveniently, contrary to TRAC, TRIC exploits a two-step approach to cope with Manhattan pDTMs associated with non-Manhattan variations of geometry/mesh. This is, for example, the case (i) of pLQFPs, where the relative position of the leads changes by varying the die thickness (Figure 3), and (ii) of multi-source dies, where the relative position of the HSs can be modified (Figure 4).

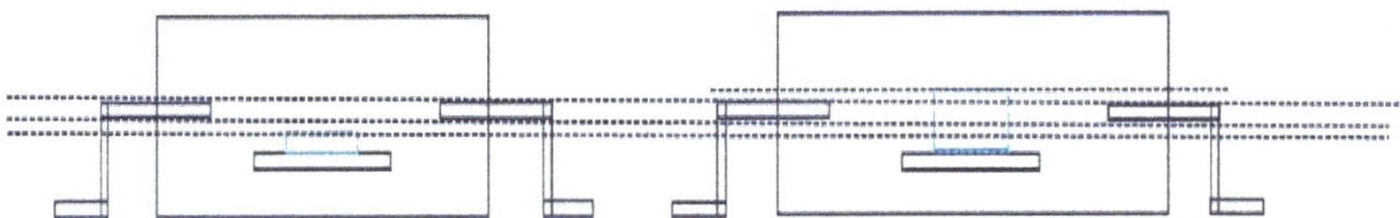

Figure 3. Schematic representation of a non-Manhattan transformation for a pLQFP, where the die (blue rectangle) thickness increases.

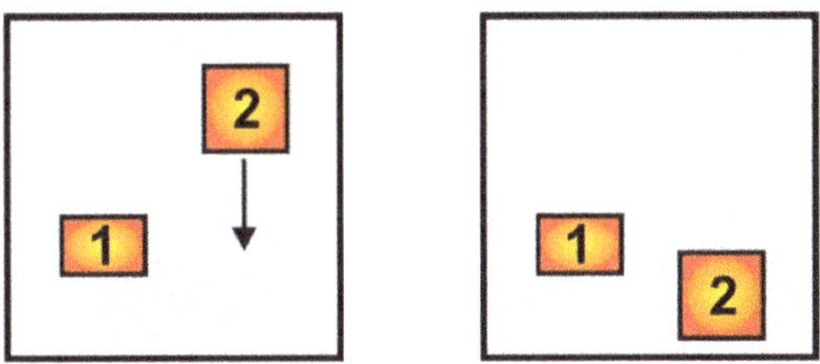

Figure 4. Sketch of a non-Manhattan transformation for a multi-source die.

The functioning principle of TRIC can be described as follows. The DTM of a specimen in a family of packages immersed in a standard environment, automatically extracted by TRIC using the FV method, has the following form:

$$\mathbf{M}(\mathbf{p})\frac{d\vartheta}{dt}(t,\mathbf{p}) + \mathbf{K}(\mathbf{p})\vartheta(t,\mathbf{p}) = \mathbf{G}(\mathbf{p})\mathbf{P}(t) \tag{2}$$

in which $\vartheta(t,\mathbf{p})$ is the $N(\mathbf{p})$ rows vector with the DoF of temperature rise at each time instant t, $\mathbf{M}(\mathbf{p})$ is the $N(\mathbf{p})$-order mass matrix, $\mathbf{K}(\mathbf{p})$ is the $N(\mathbf{p})$-order stiffness matrix, $\mathbf{G}(\mathbf{p})$ is the $N(\mathbf{p}) \times M$ power density rectangular matrix, and $\mathbf{P}(t)$ is the M rows vector of source powers.

In the first step, a basis for the projection of (2) is determined by Algorithm 1 exploiting Algorithm 2. In the second step, using this projection basis, a fast solution of (2) is provided for a chosen value of $\mathbf{p}$ by Algorithm 3, which again exploits Algorithm 2.

Algorithm 1: Extraction of Projection Basis.

---

1   **for** $m = 1, \dots, M$ **do**
     for each σ **do**
     pick a random value $\mathbf{p}$ in $\mathbf{P}$
     set $\hat{\boldsymbol{\Theta}}_m(\sigma, \mathbf{p}) = 0$
     set space $S_m(\sigma){:} = \varnothing$
     set $\rho{:} = +\infty$ (norm of the residual)
     while $\rho > \varepsilon$ **do**
2   solve (2) for $\boldsymbol{\Theta}_m(\sigma, \mathbf{p})$ using $\hat{\boldsymbol{\Theta}}_m(\sigma, \mathbf{p})$ as initial guess
3   add $(\mathbf{p}, \boldsymbol{\Theta}_m(\sigma, \mathbf{p}))$ to space $S_m(\sigma)$
     for Ξ times **do**
     pick a random value $\mathbf{p}$ in $\mathbf{P}$
     apply Algorithm 2
4   compute residual ρ of (2) for $\hat{\boldsymbol{\Theta}}_m(\sigma, \mathbf{p})$
     if $\rho > \varepsilon$ **do**
     break

In Algorithm 1, at line 1, a set of complex frequency values, proper for characterizing the thermal behavior of the family of packages, is chosen following [13]. At line 2, the detailed thermal problem in the complex frequency domain

$$[\sigma\mathbf{M}(\mathbf{p}){+}\mathbf{K}(\mathbf{p})]\boldsymbol{\Theta}_m(\sigma, \mathbf{p}) = \mathbf{g}_m, \tag{3}$$

in which $\mathbf{g}_m$ is the m-th column of $\mathbf{G}$ and is numerically solved by an iterative solver. A multigrid solver is used, and the number of iterations is reduced by assuming as initial guess the $\hat{\boldsymbol{\Theta}}_m(\sigma, \mathbf{p})$ estimation. At line 3, the solutions $\boldsymbol{\Theta}_m(\sigma, \mathbf{p})$ are added to $S(\sigma)$. At line 4, the residual ρ is determined substituting $\boldsymbol{\Theta}_m(\sigma, \mathbf{p})$ with $\hat{\boldsymbol{\Theta}}_m(\sigma, \mathbf{p})$ in (3).

Algorithm 2: Performing Projection.

---

1   set $\mathbf{V}{:} = \varnothing$
     for each *element* $(\mathbf{q}, \boldsymbol{\Theta})$ of $S_m(\sigma)$ **do**
     transform $\boldsymbol{\Theta}$ into $\hat{\boldsymbol{\Theta}}$
     set $\mathbf{V}{:} = [\mathbf{V}, \hat{\boldsymbol{\Theta}}]$
2   project (3) onto the space spanned by columns of $\mathbf{V}$
3   solve (4) determining $\hat{\boldsymbol{\Theta}}_m(\sigma, \mathbf{p})$ as an approximation of $\boldsymbol{\Theta}_m(\sigma, \mathbf{p})$

In Algorithm 2, at line 1, the spatial geometry of the DTM for value $\mathbf{p}$ of the parameter vector is expressed as a map of the spatial geometry of the DTM for value $\mathbf{q}$ of the parameters vector. This map is thus applied to vector $\boldsymbol{\Theta}$, getting vector $\hat{\boldsymbol{\Theta}}$. At line 2, the projected equations are

$$\left(\sigma\hat{\mathbf{M}} + \hat{\mathbf{K}}\right)\hat{\boldsymbol{\xi}}_m(\sigma) = \hat{\mathbf{g}}_m \tag{4}$$

where $\hat{\mathbf{M}}$ and $\hat{\mathbf{K}}$ are projected matrices given by $\hat{\mathbf{M}} = \hat{\mathbf{V}}^T\mathbf{M}(\mathbf{p})\hat{\mathbf{V}}$ and $\hat{\mathbf{K}} = \hat{\mathbf{V}}^T\mathbf{K}(\mathbf{p})\hat{\mathbf{V}}$, while $\hat{\boldsymbol{\xi}}_m$ is the DoF column vector and $\hat{\mathbf{g}}_m = \hat{\mathbf{V}}^T\mathbf{g}_m$. At line 3, vector $\boldsymbol{\Theta}_m(\sigma, \mathbf{p})$ is approximated by

$$\hat{\boldsymbol{\Theta}}_m(\sigma, \mathbf{p}) = \hat{\mathbf{V}}\hat{\boldsymbol{\xi}}_m(\sigma) \tag{5}$$

In Algorithm 3, for a chosen value of $\mathbf{p}$, a space $S(\mathbf{p})$ is determined. Projecting (2) onto such a space, the CTM ensues, the response of which to any power profile is computed for approximating $\vartheta(t, \mathbf{p})$.

Algorithm 3: Performing Projection

set space $S(\mathbf{p}):=\varnothing$
for $m = 1, \dots , M$ **do**
for each σ **do**
apply Algorithm 2
add $\hat{\mathbf{\Theta}}_m(\sigma, \mathbf{p})$ to space $S(\mathbf{p})$

It must be remarked that in commercial numerical codes the complexity for the transient solutions is proportional to $(n_x \times n_y \times n_z)^{\alpha \times nt}$, where α is in the range $1 \div 1.5$ (depending on the iterative method adopted); n_x, n_y, and n_z are the numbers of grid points along x, y, and z, and n_t is the number of time instants in which the problem has to be solved. Conveniently, the corresponding complexity in TRIC is proportional to n_t and independent of n_x, n_y, and n_z.

4. Numerical Results

Except for the multi-source analysis (Section 4.4), the intrinsic symmetry of the packages under test allowed meshing and simulating only a quarter of each structure, thus mitigating the computational burden; the missing portions were virtually restored by applying adiabatic BCs (i.e., zero heat flux) over the planes of symmetry.

4.1. Full-Plastic LQFP vs. Epad LQFP

Figure 5 shows the static thermal metrics ϑ_{JA} and ϑ_{JCtop} corresponding to eLQFPs and pLQFPs for two horizontal package sizes, namely (a) 10×10 and (b) 14×14 mm^2, the epad size being 6×6 and 7.2×7.2 mm^2 for cases (a) and (b), respectively. The metrics are determined by TRIC for various sizes of the square die, i.e., $2 \times 2, 3 \times 3, 4 \times 4, 5 \times 5, 6 \times 6$ mm^2, the latter only for case (b). It can be inferred that the presence of the pad, which eases the heat removal, yields a sizable beneficial impact on ϑ_{JA}, whereas the closer proximity of the die to the top of the package in pLQFPs prevails over the cooling action of the epad in terms of ϑ_{JCtop}.

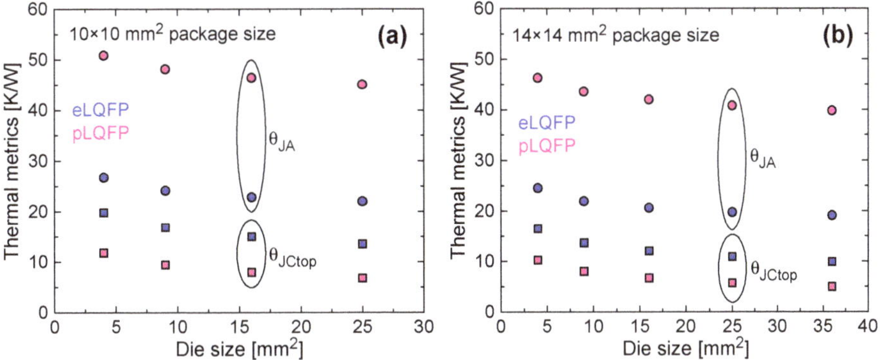

Figure 5. Thermal metrics ϑ_{JA} (circles) and ϑ_{JCtop} (squares) against die size: comparison between eLQFP (blue) and pLQFP (magenta) for (**a**) 10×10 and (**b**) 14×14 mm^2 package size with (**a**) 6×6 mm^2 and (**b**) 7.2×7.2 mm^2 die sizes.

4.2. Thermal Impedances

One of the main features of TRIC not available in TRAC is the possibility to determine the transient thermal responses of a specimen of a package family for any profile of the power sources. In particular, the thermal impedances—often used to characterize the dynamic thermal behavior of components—can be evaluated as the thermal responses (temperature rises normalized to power) to a power step applied at t = 0 to a die modeled with only one HS. Such a capability is witnessed through the analysis of a large set of cases.

Figure 6 shows the junction-to-ambient thermal impedance $Z_{THJA} = \vartheta_{JA}(t)$ of a 10×10 mm^2 eTQFP equipped with a 6×6 mm^2 epad for four different die sizes. The simulations allow quantifying the favorable influence of a large die, which benefits from a lower power density. Static conditions are reached at about 400 s, regardless of die size. Figure 7 illustrates the Z_{THJA} of a 10×10 mm^2 pLQFP with a 4×4 mm^2 die for three different pad sizes. All curves coincide for short times (<0.1 s), where the heat emerging from the die has not hit the pad yet. For this case, (i) a complex evolution with an inflection takes place due to the involved package geometry: the heat propagates through the pad, the mold, and then reaches the leads, which are in direct contact with the board, and (ii) the impedances flatten at 1200 s, as induced by the absence of the cooling epad action. Figure 8 reports the Z_{THJA} of 9×9 mm^2 single-row eQFN packages for three combinations of epad and die sizes. Here the positive influence of a bigger epad (7×7 mm^2 instead of 5.7×5.7 mm^2) for the same die size is evident; again, the impedances overlap for short times. Similar to the study conducted for the eTQFP family, also for eQFN packages the thermal impedance is reduced and delayed for bigger dies. Figure 9 confirms the cooling impact for medium/long times of a larger package body and/or a larger epad for eQFN-mr packages with two rows of pins. Lastly, the dynamic thermal behavior of a 10.3×7.5 mm^2 PowerSSO-36 with a 4.09×3.17 mm^2 die is determined for four different epad sizes in Figure 10.

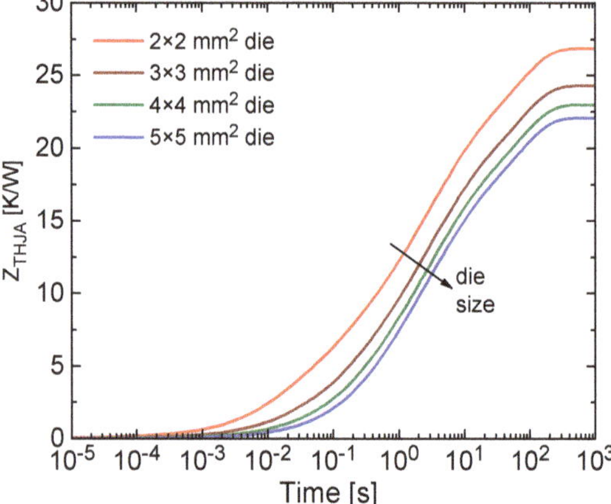

Figure 6. Thermal impedance Z_{THJA} vs. time of a 10×10 mm^2 eTQFP with a 6×6 mm^2 epad for four different die sizes.

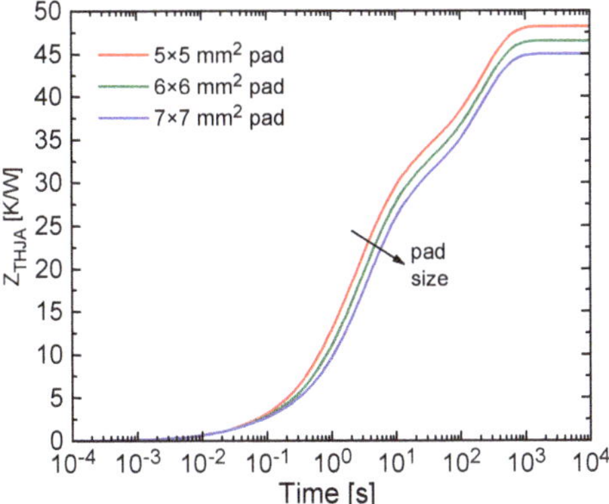

Figure 7. Thermal impedance Z_{THJA} against time of a 10×10 mm² pLQFP with a 4×4 mm² die for three different pad sizes.

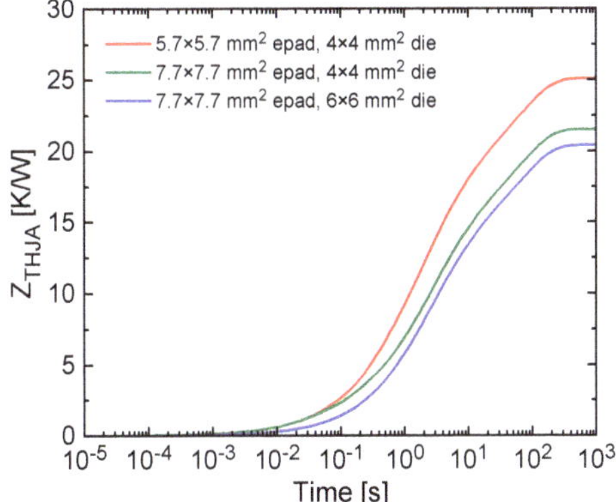

Figure 8. Thermal impedance Z_{THJA} vs. time of a 9×9 mm² single-row eQFN package for three combinations of epad and die sizes.

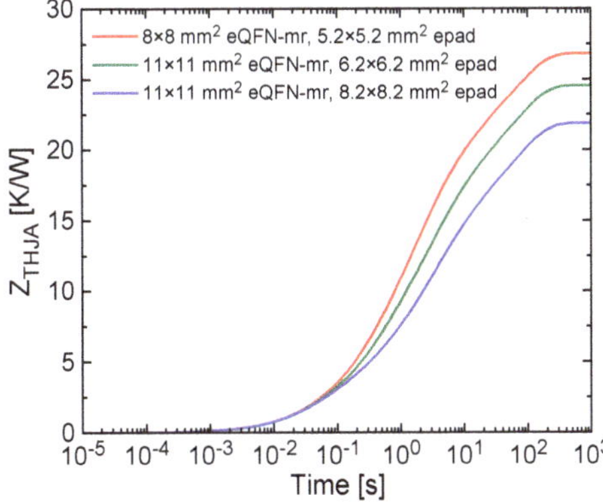

Figure 9. Thermal impedance Z_{THJA} vs. time of various dual-row eQFN-mr packages sharing a die size of 3×3 mm².

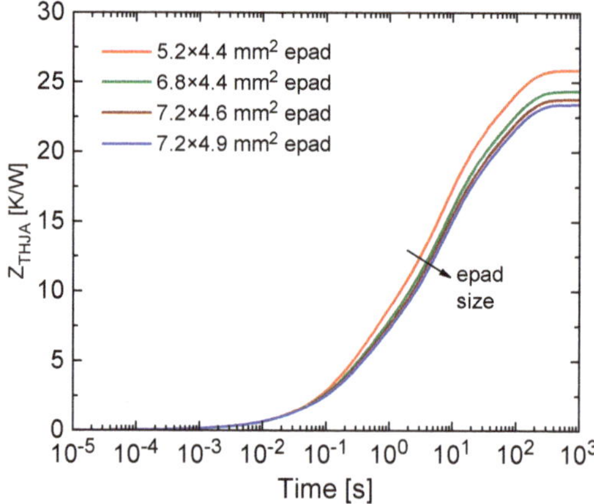

Figure 10. Thermal impedance Z_{THJA} of a 10.3×7.5 mm² PowerSSO-36 L with a 4.09×3.17 mm² die for four different epad sizes.

4.3. Comparison with FloTHERM

First and foremost, it must be underlined that the user-friendly graphical interface of TRIC allows avoiding the long, painstaking, and prone-to-error geometry/mesh construction process often required by conventional numerical tools, thus markedly lowering the pre-processing effort and time. As far as the accuracy and efficiency of TRIC are concerned, they were estimated by comparison with the widely used commercial software FloTHERM. The tools share the same geometry simplifications and BCs. The favorable matching between the thermal metrics determined by TRAC and the FV software was already shown in [11] for eLQFPs, eTQFPs, and eQFN packages; the slight discrepancy (typically <2%, the maximum value being around 3%) was mainly attributed to the different mesh styles of the simulators. Evidence of the good agreement is also provided in Figure 11, which shows the (static) maps of temperature rise over ambient ($T_{amb} = 20\,°C$) determined for three eTQFPs with dies dissipating 1 W in the ϑ_{JA}-related ambient; in particular, (a) corresponds to a 14×14 mm² package size with a 9×9 mm² epad and a 3×3 mm² die; (b) to a 10×10 mm² package with a 6×6 mm² epad and a 4×4 mm² die; (c) to a 10×10 mm² package with a 6×6 mm² epad and a 2×2 mm² die.

The accuracy ensured by TRIC under transient conditions can be inferred from Figure 12, which shows the favorable matching with data computed by FloTHERM for three packages belonging to the eTQFP family. Again, the discrepancy is below 3% within the whole time range. It is worth noting that the CPU time required to obtain an impedance by TRIC for a typical number of 2×10^6 grid points is about 1 min on a workstation with an Intel Xeon E5-2630 v4 @ 2.2 GHz equipped with a 64 GB RAM, whereas more than 15–20 minutes are needed when using FloTHERM. This notable gain in terms of efficiency is expected to take place with respect to most popular simulators based on numerical methods.

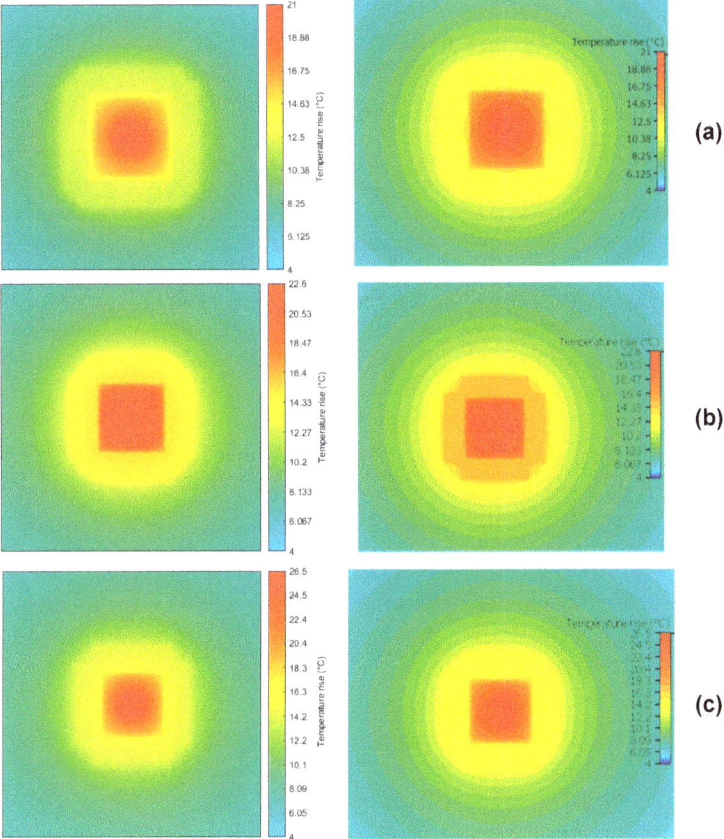

Figure 11. Temperature rise maps for three eTQFPs dissipating 1 W, as calculated by TRIC (left) and FloTHERM (right): (**a**) 14×14 mm^2 package size with a 9×9 mm^2 epad and a 3×3 mm^2 die; (**b**) 10×10 mm^2 package with a 6×6 mm^2 epad and a 4×4 mm^2 die; (**c**) 10×10 mm^2 package with a 6×6 mm^2 epad and a 2×2 mm^2 die.

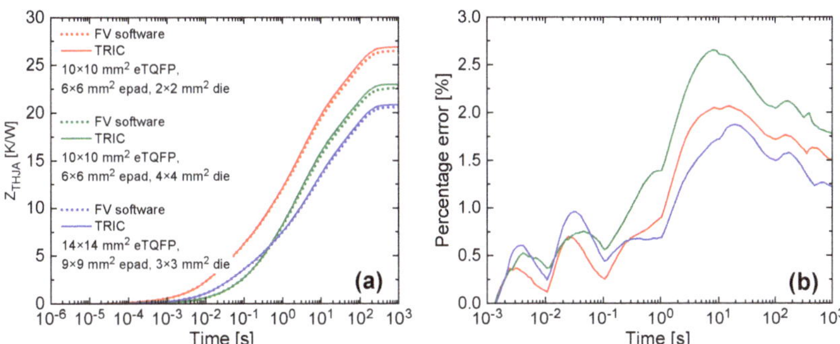

Figure 12. (**a**) Comparison between TRIC and FloTHERM for various eTQFP cases; (**b**) corresponding percentage error vs. time.

4.4. Multi-Source Analysis

Unlike TRAC, TRIC allows also simulating realistic packages that integrate multiple HSs, as evidenced through the following illustrative examples.

First, a 9×9 mm^2 eQFN package is considered, with a 5×5 mm^2 epad and a 4×4 mm^2 die, the latter including four 1×1 mm^2 active areas (i.e., HSs), each dissipating 0.5 W. As sketched in Figure 13, four positions of the HSs are chosen to describe practical layouts. The static temperature rise over ambient was monitored in five critical points, namely, at the die center (ΔT_{center}) and at the centers of the HSs (ΔT_1, ΔT_2, ΔT_3, ΔT_4). Results corresponding to the four layouts are reported in Table 1, along with the maximum temperature rise over the whole die (ΔT_{max}). Again, a fairly good agreement with the temperature maps determined by FloTHERM (not shown here) was obtained, the discrepancy between the maxima being <3%. The data plainly illustrate how the temperature field over the die modifies depending on the specific layout. The main findings are (i) all the HSs share the same temperature in cases #1 and #3 for symmetry reasons; (ii) as expected, in layout #1 ΔT_{center} = ΔT_{max} due to the concurrent influence of all the HSs, while (iii) ΔT_{max} is reached near the die side in layout #2. This simple analysis shows that TRIC can be effectively exploited to identify the most thermally efficient layout. In addition, the information gained on the temperature field over the die is also important to properly place temperature sensors.

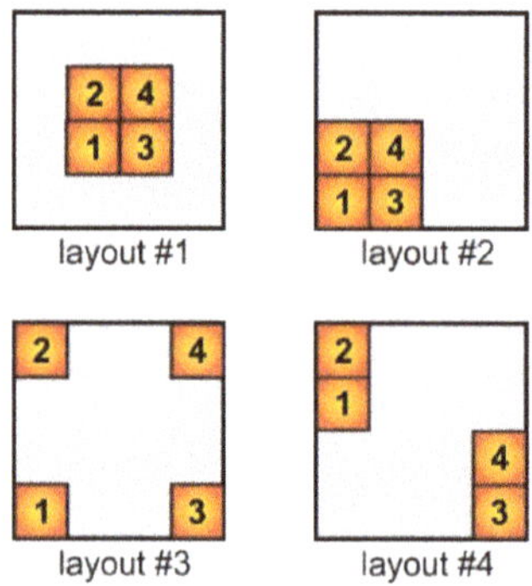

Figure 13. Representation of four layouts corresponding to a four-source die. The temperature-sensing points are located at the center of the die and at the centers of the HSs.

Table 1. Temperature rises over ambient (K) computed by TRIC.

	ΔT_1	ΔT_2	ΔT_3	ΔT_4	ΔT_{center}	ΔT_{max}
Layout #1	56.31	56.31	56.31	56.31	57.06	57.06
Layout #2	59.56	58.20	58.20	57.05	54.13	59.58
Layout #3	54.89	54.89	54.89	54.89	52.25	55.15
Layout #4	55.77	56.12	56.12	55.77	52.70	56.35

4.5. ABS Source Profile

The TRIC capability to cover complex die layouts and transient power profiles is demonstrated by simulating the die temperature dictated by a realistic ABS power profile. The examined package is a 14×14 mm^2 eTQFP with an 8×8 mm^2 epad and a 30 mm^2 die, the circuitry over which presents eight active areas (HSs), as depicted in Figure 14a. The geometry of the system and the external BCs were adapted to this specific application. Figure 14b illustrates the TRIC interface with the probes (placed at the centers of the HSs) where the temperatures are taken. Figure 14c shows the evolution of the temperature rises over ambient from 0 to 20 s. Figure 14d reports the temperature rise field at the most thermally critical time instant, i.e., 2.5 s.

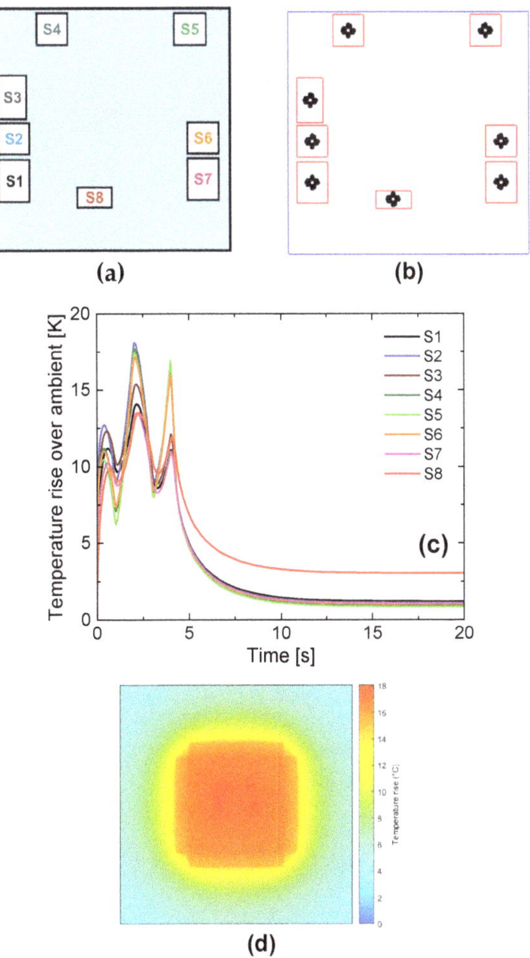

Figure 14. (**a**) Schematic top-view of the die floorplan; (**b**) corresponding TRIC interface; (**c**) evolution of the temperature rises vs. time; (**d**) temperature rise map at the time instant t = 2.5 s.

5. Conclusions

In this paper, a tool denoted as Thermal Resistance and Impedance Calculator (TRIC) has been presented. TRIC allows the automatic extraction of thermal metrics of package families of electronic components under both static and transient conditions. It exploits a solution algorithm based on a novel projection-based approach, which allows dealing with non-Manhattan geometry and mesh variations in the parametric detailed thermal model (pDTM) of a package family. The pDTMs of many relevant package families have been included, and dies with multiple active areas can be handled. An extensive simulation campaign, focused on cases of practical interest, has been performed. A comparison between TRIC and the FV program FloTHERM has been carried out, with the aim of validating the accuracy and assessing the efficiency of the proposed tool; the main findings can be summarized as follows: (i) the discrepancy in terms of thermal metrics calculated by the simulators amounts at most to 3% and is mainly ascribable to the different mesh styles; (ii) thanks to its advanced solution algorithm, TRIC allows obtaining a reduction in CPU time by a factor of 15–20 with respect to FloTHERM when simulating a transient thermal response. Owing to the above reasons, TRIC can be considered particularly helpful for industry specialists who are involved in designing packaged devices and have to cope with thermal flow problems.

Author Contributions: Methodology, L.C.; Software, L.C., F.D.V., D.G., A.M., and C.M.V.; Validation, F.D.V. and L.C.; Writing—Original Draft Preparation, V.d.; Writing—Review & Editing, V.d.; Supervision, L.C. and C.M.V. All authors have read and agreed to the published version of the manuscript.

Funding: This research received no external funding.

Conflicts of Interest: The authors declare no conflict of interest.

Nomenclature

TRIC	Thermal Resistance and Impedance Calculator
TRAC	Thermal Resistance Advanced Calculator
JEDEC	Joint Electron Device Engineering Council
thermal resistance (K/W)	temperature increase over a reference temperature taken over a point (or a region) of interest of the component under test, and normalized to the dissipated power; it is a property depending upon geometry and material parameters, and can be reviewed as an indicator of the *heat dissipation inaptitude* of the component
thermal impedance (K/W)	thermal resistance vs. time resulting from the application of a constant power step
ϑ_{JA} (K/W)	junction-to-ambient thermal resistance
Ψ_{JB} (K/W)	thermal characterization parameter to report the difference between junction temperature and the temperature of the board measured at the top surface of the board
Ψ_{JCtop} (K/W)	thermal characterization parameter to report the difference between junction temperature and the temperature at the top center of the outside surface of the component package
ϑ_{JB} (K/W)	junction-to-board thermal resistance
ϑ_{JCtop} (K/W)	junction-to-case top thermal resistance
$\vartheta_{JCbottom}$ (K/W)	junction-to-case bottom thermal resistance
Z_{THJA} (K/W)	junction-to-ambient thermal impedance
BC	boundary condition
BCI	boundary condition independent
MOR	model-order reduction
CTM	compact thermal model
DTM	detailed thermal model
pDTM	parametric DTM
FV	finite volume
DoF	degree of freedom
HS	heat source
CPU	central processing unit
epad	exposed pad
QFP	quad flat package
eLQFP	exposed-pad low-profile (thick) QFP
eTQFP	exposed-pad thin QFP
pLQFP	full-plastic low-profile (thick) QFP
QFN	quad flat no-leads package
eQFN	exposed-pad QFN
eQFN-mr	multi-row eQFN
PowerSSO	package belonging to the Small Outline family
ABS	antilock braking system

References

1. Smy, T.; Walkey, D.; Dew, S.K. A 3D thermal simulation tool for integrated devices-Atar. *IEEE Trans. Comput. Aided Des. Integr. Circuits Syst.* **2001**, *20*, 105–115. [CrossRef]
2. Huang, W.T.; Ghosh, S.; Sankaranarayanan, K.; Skadron, K.; Stan, M.R. Hotspot: Thermal modeling for CMOS VLSI systems. *IEEE Trans. Large Scale Integr. VLSI Syst.* **2006**, *14*, 501–513. [CrossRef]
3. Ziabari, A.; Park, J.H.; Ardestani, E.K.; Renau, J.; Kang, S.M.; Shakouri, A. Power blurring: Fast static and transient thermal analysis method for packaged integrated circuits and power devices. *IEEE Trans. Large Scale Integr. VLSI Syst.* **2014**, *22*, 2366–2379. [CrossRef]
4. Bar-Cohen, A.; Elperin, T.; Eliasi, R. ϑ_{JC} characterization of chip packages – justification, limitations, and future. *IEEE Trans. Compon. Hybrids Manuf. Technol.* **1989**, *12*, 724–731. [CrossRef]
5. Lasance, C.J.M.; Vinke, H.; Rosten, H. Thermal characterization of electronic devices with boundary condition independent compact models. *IEEE Trans. Compon. Packag. Manuf. Technol. Part A* **1995**, *18*, 723–731. [CrossRef]
6. Pape, H.; Schweitzer, D.; Janssen, J.H.J.; Morelli, A.; Villa, C.M. Thermal transient modeling and experimental validation in the European project PROFIT. *IEEE Trans. Compon. Packag. Technol.* **2004**, *27*, 530–538. [CrossRef]
7. Sabry, M.N. Flexible profile compact thermal models for practical geometries. *J. Electr. Packag.* **2007**, *129*, 256–259. [CrossRef]
8. Lasance, C.J.M. Ten years of boundary-condition-independent compact thermal modeling of electronic parts: A review. *Heat Transf. Eng.* **2008**, *29*, 149–169. [CrossRef]
9. JESD51-12. *Guidelines for Reporting and Using Electronic Package Thermal Information*; JEDEC: Arlington, VA, USA, May 2005.
10. Codecasa, L.; Race, S.; d'Alessandro, V.; Gualandris, D.; Morelli, A.; Villa, C.M. Thermal resistance advanced calculator (TRAC). In Proceedings of the International Workshop on THERMal INvestigation of ICs and Systems (THERMINIC), Stockholm, Sweden, 26–28 September 2018.
11. Codecasa, L.; Race, S.; d'Alessandro, V.; Gualandris, D.; Morelli, A.; Villa, C.M. TRAC: A thermal resistance advanced calculator for electronic packages. *Energies* **2019**, *12*, 1050. [CrossRef]
12. Codecasa, L.; D'Amore, D.; Maffezzoni, P. Modeling the thermal response of semiconductor devices through equivalent electrical networks. *IEEE Trans. Circuits Syst. I Fundam. Theory Appl.* **2002**, *49*, 1187–1197. [CrossRef]
13. Codecasa, L.; D'Amore, D.; Maffezzoni, P. Compact modeling of electrical devices for electrothermal analysis. *IEEE Trans. Circuits Syst. I Fundam. Theory Appl.* **2003**, *50*, 465–476. [CrossRef]
14. Codecasa, L.; D'Amore, D.; Maffezzoni, P. Multipoint moment matching reduction from port responses of dynamic thermal networks. *IEEE Trans. Compon. Packag. Technol.* **2005**, *28*, 605–614. [CrossRef]
15. Codecasa, L.; d'Alessandro, V.; Magnani, A.; Rinaldi, N.; Zampardi, P.J. FAst novel thermal analysis simulation tool for integrated circuits (FANTASTIC). In Proceedings of the International Workshop on THERMal INvestigation of ICs and systems (THERMINIC), London, UK, 24–26 September 2014.
16. Codecasa, L.; d'Alessandro, V.; Magnani, A.; Rinaldi, N. Parametric compact thermal models by moment matching for variable geometry. In Proceedings of the International Workshop on THERMal INvestigation of ICs and systems (THERMINIC), London, UK, 24–26 September 2014.
17. Magnani, A.; d'Alessandro, V.; Codecasa, L.; Zampardi, P.J.; Moser, B.; Rinaldi, N. Analysis of the influence of layout and technology parameters on the thermal impedance of GaAs HBT/BiFET using a highly-efficient tool. In Proceedings of the IEEE Compound Semiconductor Integrated Circuit Symposium (CSICS), La Jolla, CA, USA, 19–22 October 2014.
18. Codecasa, L.; d'Alessandro, V.; Magnani, A.; Rinaldi, N. Matrix reduction tool for creating boundary condition independent dynamic compact thermal models. In Proceedings of the International Workshop on THERMal INvestigation of ICs and Systems (THERMINIC), Paris, France, 30 September–2 October 2015.
19. Janssen, J.H.J.; Codecasa, L. Why matrix reduction is better than objective function based optimization in compact thermal model creation. In Proceedings of the International Workshop on THERMal Investigation of ICs and Systems (THERMINIC), Paris, France, 30 September–2 October 2015.
20. Codecasa, L.; d'Alessandro, V.; Magnani, A.; Rinaldi, N. Structure preserving approach to parametric dynamic compact thermal models of nonlinear heat conduction. In Proceedings of the International Workshop on THERMal INvestigation of ICs and Systems (THERMINIC), Paris, France, 30 September–2 October 2015.

21. Codecasa, L.; d'Alessandro, V.; Magnani, A.; Irace, A. Circuit-based electrothermal simulation of power devices by an ultrafast nonlinear MOS approach. *IEEE Trans. Power Electron.* **2016**, *31*, 5906–5916. [CrossRef]

22. Rogié, B.; Codecasa, L.; Monier-Vinard, E.; Bissuel, V.; Laraqi, N.; Daniel, O.; D'Amore, D.; Magnani, A.; d'Alessandro, V.; Rinaldi, N. Delphi-like dynamical compact thermal models using model order reduction. In Proceedings of the International Workshop on THERMal INvestigation of ICs and Systems (THERMINIC), Amsterdam, The Netherlands, 27–29 September 2017. (best paper award).

23. Codecasa, L.; d'Alessandro, V.; Magnani, A.; Rinaldi, N. Novel approach for the extraction of nonlinear compact thermal models. In Proceedings of the International Workshop on THERMal INvestigation of ICs and systems (THERMINIC), Amsterdam, The Netherlands, 27–29 September 2017.

24. Codecasa, L.; De Viti, F.; Race, S.; d'Alessandro, V.; Gualandris, D.; Morelli, A.; Villa, C.M. Thermal Resistance and Impedance Calculator (TRIC). In Proceedings of the International Workshop on THERMal INvestigation of ICs and Systems (THERMINIC), Lecco, Italy, 25–27 September 2019.

25. FloTHERM®v12.2. *User's Guide*; Mentor Graphics: Wilsonville, OR, USA, 2018.

Article

Compact Thermal Modelling Tool for Fast Design Space Exploration of 3D ICs with Integrated Microchannels

Piotr Zając

Department of Microelectronics and Computer Science, Lodz University of Technology, 90-924 Łódź, Poland; pzajac@dmcs.pl

Received: 30 March 2020; Accepted: 23 April 2020; Published: 2 May 2020

Abstract: Integrated microchannel cooling is a very promising concept for thermal management of 3D ICs, because it offers much higher cooling performance than conventional forced-air convection. The thermo-fluidic simulations of such chips are usually performed using a computational fluid dynamics (CFD) approach. However, due to the complexity of the fluid flow modelling, such simulations are typically very long and faster models are therefore considered. This paper demonstrates the advantages of TIMiTIC—a compact thermal simulator for chips with liquid cooling—and shows its practical usefulness in design space exploration of 3D ICs with integrated microchannels. Moreover, thermal simulations of a 3D processor model using the proposed tool are used to estimate the optimal power dissipation profile in the chip and to prove that such an optimal profile allows for a very significant (more than 10 °C) peak temperature reduction. Finally, a custom correlation metric is introduced which allows the comparison of the power distribution profiles in terms of the peak chip temperature that they produce. Statistical analysis of the simulation results demonstrates that this metric is very accurate and can be used for example in thermal-aware task scheduling or dynamic voltage and frequency scaling (DVFS) algorithms.

Keywords: 3D IC; microchannels; liquid cooling; compact thermal model; thermal simulation; hotspot; thermal-aware task scheduling; DVFS; statistical analysis

---

1. Introduction

Vertically stacked 3D integrated circuits [1–3] have many important advantages, like smaller footprint, lower delay, higher operating speed, potentially higher yield, combining different technologies on a single die, etc. With the first 3D memory chips already on the market, it seems inevitable that this trend will continue, and soon the first processors with multiple silicon layers will be produced. While going 3D is certainly a major step towards maintaining Moore's Law, there are still many challenges which need to be addressed. Most importantly, the technology of vertical interconnections using through-silicon vias (TSV) has to be mastered. Recent breakthroughs in this field [4–7] indicate that chip manufacturers are close to achieving this goal. The bonding technology and mechanical stability is another crucial issue [8,9].

From a thermal perspective, 3D stacking poses significant challenges due to quite obvious reasons: each stacked layer increases the power dissipated in almost the same area. As a consequence, the power density and the resulting temperatures increase rapidly as more and more layers are stacked. Another issue is the difficulty of removing heat from the bottom layers. In a conventional cooling system, most heat is dissipated to ambient through the heat spreader and heat sink, which are put on the top layer. In a 3D chip, as the layers are usually bonded with a material with very low thermal conductivity (typically around several W/(mK)), the vertical thermal resistance between the layers is high (compared to the thermal resistance of the silicon layer itself). With multiple layers present, this results in a very high thermal resistance between the bottom layer and the heat sink and, therefore, higher peak temperatures are produced.

To tackle these two problems, integrated microchannel cooling [10,11] has been suggested. In fact, it was even considered before as a potential cooling solution for 2D chips, but this particular idea is especially promising when considering 3D ICs. The reason is that it perfectly addresses the two issues mentioned in the previous paragraph. First, liquid cooling offers much higher cooling performance: it is estimated that the heat transfer coefficient may be several orders of magnitude higher than in the case of the classic approach based on forced air convection [12]. Second, if microchannels are implemented in all chip layers, the problem of high vertical thermal resistance is eliminated, and the heat is removed almost equally from all chip layers. In other words, the performance of microchannel cooling is scalable: stacking more layers increases power density, but also proportionally increases cooling, so the resulting peak temperature should remain almost the same.

The most valuable research on microchannel cooling is based on real measurements using specially designed test chips [13,14]. However, often, measurements of 3D ICs with liquid cooling are not possible for various reasons. Sometimes, the manufacturing technology is not yet available, or the costs may be too high. Therefore, researchers use simulations as a means to estimate the influence of chip and cooling parameters on chip temperatures. The most commonly used approach is based on the finite element method (FEM), using commercial CFD tools like Ansys [15], Comsol [16] or FloTherm [17]. The main advantage of this method is its relatively high accuracy [18,19]. However, this approach uses a coupled thermo-fluidic simulation and, as modelling fluid dynamics requires a very fine mesh, the resulting number of model nodes is very high (hundreds of thousands) and simulation times are very long [20,21]. In an attempt to shorten the simulation time, many authors have suggested simpler models. In [22], a compact thermal modelling tool 3D-ICE for 3D ICs with microchannels was proposed. By comparing its results with measurements, it was shown that the average error introduced by the model is below 10%. In [23], the authors suggested using a time-variant resistor model to describe the convective heat transfer in heated microchannels. The authors validated their model against Ansys Fluent and obtained an error of 5%. The authors of [24] developed an algorithm allowing the simulation of microchannel-cooled 3D ICs on GPUs. These authors reported errors below 0.5 °C with respect to the traditional CPU-based approach while also achieving considerable speedup with their method. The model based on thermal wake function was introduced in [25]. Numerical pre-simulation was used to extract the function and build a thermal wake aware resistance network model. The authors compared their model with a commercial CFD tool and reported an error of less than 2.0% and a 400× speedup. In [26], the authors extended the existing compact thermal simulator with an equivalent 3D resistive network to model liquid cooling. The model was then used to validate a dynamic thermal management scheme, which allowed up to 95% reduction in hotspots on average. The authors of [27] describe an analytically derived T-equivalent analytical thermal model of a microchannel. Validation of the proposed approach was based on the simulation of one channel in SPICE. Authors reported a good agreement of their model (errors below 2%) with the results obtained with a full CFD simulation.

This paper demonstrates the advantages of TIMiTIC—a compact thermal simulator for chips with liquid cooling—and shows its practical usefulness in design space exploration of 3D ICs with integrated microchannels. In comparison with other approaches, more focus is given on the modelling of convection resistance between the solid and the fluid, especially the thermal entry effects and the variation of the Nusselt number along the channel. The accuracy of the simulator was already analyzed in [28], where its results were extensively compared with the ones obtained using CFD simulation in Comsol. The validation with respect to full CFD simulation showed a very good agreement (the average absolute error of 1.3 °C and the average relative error of 3.7% for the worst analyzed case).

Moreover, thermal simulations of a 3D processor model using the proposed tool are used to estimate the optimal power dissipation profile in the chip and to prove that such an optimal profile allows for a very significant (more than 10 °C) peak temperature reduction. Finally, a custom correlation metric is proposed which allows comparing the power distribution profiles in terms of the peak temperature that they produce. Statistical analysis of the simulation results demonstrates that this metric is very accurate and can be used, for example, in task scheduling or dynamic voltage and frequency scaling (DVFS) algorithms.

2. Characteristics of Microchannel Cooling

A sample 3D chip with microchannels etched in each silicon layer is shown in Figure 1. The inlets/outlets to the chip can be implemented as tubes connected on top of the chip [29] or as channels in the interposer (see Figure 1). Then, the liquid is distributed among the layers using microchannels. The exact description of the manufacturing process of such liquid-cooled chips is beyond the scope of this paper. However, understanding how the heat is removed in such a liquid cooling system is crucial, therefore it will be described in detail.

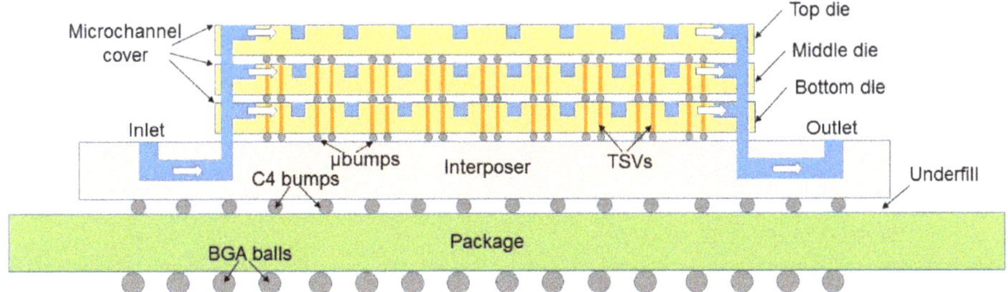

Figure 1. Sample 3D IC with three dies cooled by microchannels. The fluid is inserted into the dies through the interposer.

Figure 2 shows the three phases of heat removal. In the first phase, the heat generated in the active layer of silicon travels to the border of the microchannel by means of conduction (1). The conduction thermal resistance is inversely proportional to the thermal conductivity of silicon k and is comparably low. In the second phase, the heat from the solid is absorbed by the fluid by means of convection. The thermal resistance across the solid–fluid border is inversely proportional to the channel surface area A_{wall} and to the heat transfer coefficient (HTC) h, see (1). It is worth emphasizing that the HTC is not constant: in fact, it is proportional to the Nusselt number Nu, which decreases significantly from the channel inlet to channel outlet [30,31]. Therefore, the convection thermal resistance is always higher at the inlet side of the chip and lower at the outlet side. In the third and final phase, the heat is transported by the fluid by means of advection (1). The important detail to notice is that the temperature of the flowing fluid is increased because of all the heat absorbed upstream.

$$R_{cond} = \frac{l}{kA} \qquad q = \dot{m}c_p T_m \qquad R_{conv} = \frac{1}{hA_{wall}} \qquad h = \frac{kNu}{D_h} \tag{1}$$

where l is the length of the solid block, A is the cross-section of the solid block, $\dot{m}$ is the mass flow, c_p is the specific heat of the fluid, T_m is the mean temperature of the fluid and D_h is the hydraulic diameter of the channel.

Consequently, there are two mechanisms which cause gradually worse cooling performance as we move from inlets to outlets, one related to convection and one related to advection.

- HTC between the solid and the fluid decreases from inlets to outlets.
- The temperature of the cooling fluid increases from inlets to outlets.

Thus, the thermal behavior of microchannel-cooled chips is significantly different than that of chips cooled by a heatsink and a fan. In a 3D chip with a conventional air-cooled system, the top chip layers will be cooled best, and the cooling performance will decrease for each layer farther from the heat sink. However, there is no intra-layer cooling performance variability, so all places located in the same layer are cooled almost equally. Meanwhile, in a 3D IC with microchannel-based system, there is no inter-layer cooling performance variability (all layers are cooled equally, provided that channels are implemented in all layers). However, within the same layer, the cooling performance varies: it is the highest near the inlets and the lowest at the outlets.

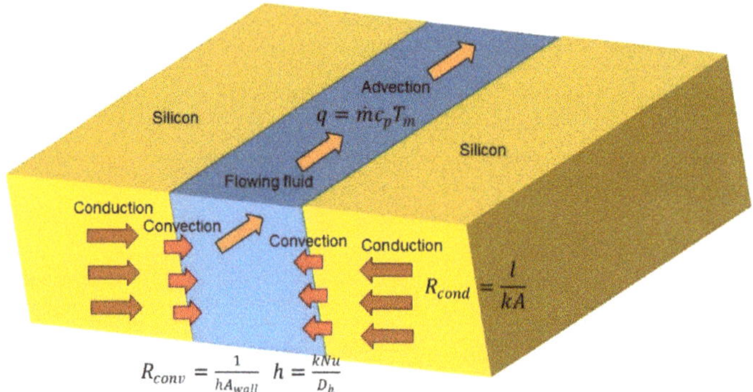

Figure 2. Illustration of the three heat-transport mechanisms in microchannel-cooled chips. See the text for the explanations of the equations.

This leads to two important repercussions:

- The floorplanning strategies which worked the best for chips cooled by forced air convection are not going to work well for IC cooled by microchannels.
- The same applies for all thermal-aware task scheduling mechanisms and DVFS algorithms.

There have been a number of works which tackled the problem of thermal-aware floorplanning in 3D ICs and proposed algorithms to find the optimal floorplan for a given power distribution (power dissipation of every chip unit) [32]. In this paper, this issue will not be analyzed. The reason is that chip floorplans (and especially high-end processors) are primarily optimized for performance (minimizing interconnect delay) and chip manufacturers are unlikely to change this based on the thermal performance. Therefore, in what follows, we will concentrate on demonstrating the usefulness of the proposed model for:

- Thermal-aware task scheduling for multi-core processors, which aims at reducing the peak chip temperature by optimally distributing tasks among processor cores;
- DVFS algorithms which minimize the peak temperature by appropriately changing the voltage and frequency of the cores.

3. TIMiTIC Simulator

The TIMiTIC tool was already extensively described in the previous paper [28]; therefore, here, only its most unique and important aspects will be presented. In general, TIMiTIC follows a widely accepted approach to discretize the chip model into a large number of cells (or nodes). Using the equivalence between the thermal and electrical domain, each cell is then represented by a lumped-capacitance model. The resulting circuit can be then described with Equation (2).

$$C\frac{dT}{dt} = -GT(t) + Q \tag{2}$$

where G is the conductance matrix, C is the capacitance matrix and Q is the heat source matrix. This differential equation can be then solved using standard methods.

The first idea which differentiates TIMiTIC from other published compact models is the improved and flexible approach to convection modeling. While both conduction and advection are quite accurately described by their respective analytical descriptions (1), convection modeling is the weakest link in every compact thermal model. The reason is the dependence of the HTC on the Nusselt number (Nu), which varies depending on many factors:

- The type of flow. The fluid flow can be both hydrodynamically and thermally developing or hydrodynamically developed and thermally developing; see [30] for more detailed information.
- Channel cross-sectional shape. Nu differs not only depending on the channel shape, but also varies for different aspect ratios for rectangular channels [30].
- The distance from the inlet. As already stated in the previous section, Nu decreases quickly from the inlet [30] until the flow becomes thermally developed and Nu becomes constant. The exact shape of the curve Nu(x), which describes Nu in the entrance region depending on the distance from the inlet x, is difficult to estimate.
- The boundary conditions. In the literature, Nu(x) is usually given for two very specific cases (UWF - uniform heat flux or UWT - uniform wall temperature). However, for any other boundary conditions, Nu(x) will be different. It is also worth noting that this Nu variation is quite high. Let us give the simplest example. For a developed flow in a circular channel, Nu equals 4.36 for the UWF boundary condition and 3.66 for the UWT boundary condition, which means that it changes by a factor of 1.19. Thus, significant error when modelling convection thermal resistance can be produced if it is not taken into consideration.

Therefore, it is very difficult to find a Nusselt number formula which would encompass all above-mentioned cases. Most models presented in the literature just use one Nu(x) formula and assume that it should work in all cases. However, it can be estimated that errors up to dozens of percent may be produced when following this approach. In TIMiTIC, the user can choose from many Nusselt number correlations described in literature [33,34], but, more importantly, he can specify his own correlation. In other words, when the user knows the flow type, channel dimensions and boundary conditions, he can easily implement in the simulator the Nusselt number correlation Nu(x) which most accurately describes his case.

Another important feature of the proposed tool is the use of well-established and thoroughly optimized mathematical libraries, like EigenSparseLU [35] or odeint [36], which allow for fast simulation. Especially in steady-state cases, TIMiTIC visibly outperforms FEM simulators, even by a factor of 100,000. The speedup is possible thanks to the use of the C++ language and the use of the solver dedicated to operate on sparse matrices. It was shown in [28] that the simulation time which took 20 min in COMSOL produced almost the same results in just 10 milliseconds using TIMiTIC.

Moreover, as 3D ICs contain a large number of very thin layers, TIMiTIC allows reducing the simulation time by treating these layers as resistances and neglecting their thermal capacity. Thus, the number of nodes in the model can be greatly reduced, and the resulting error should be negligible because the thermal capacity of such thin layers is negligible compared to others.

Lastly, the TIMiTIC simulator is available for everyone to use through the Internet website [37]. The user just has to prepare two files: the first file describes the 3D stack, material properties, channel parameters, etc., and the second file contains the power trace for all chip units. Using the website, the user can then upload these two files and run the simulation. There is no need to download or compile any code, the entire simulation takes place in a browser thanks to the use of Webassembly [38] technology.

4. Simulation Model

4.1. 3D Processor Model

The 3D processor model considered in this paper is based on a modern Intel octa-core processor, namely i9-9900k (see Figure 3). The original design consists of eight cores, each surrounded by two banks of L3 cache, four ring interconnection agents (RIA), the system agent unit (SA) and the GPU. To create a virtual design of a 3D processor based on this design, the following assumptions were made:

- The footprint is reduced so that one layer consists of 8 cores, 16 L3 cache banks and 4 RIAs.
- Two such layers are stacked on top of each other to create a 16-core processor.
- GPU and SA units are moved to the third stacked layer.

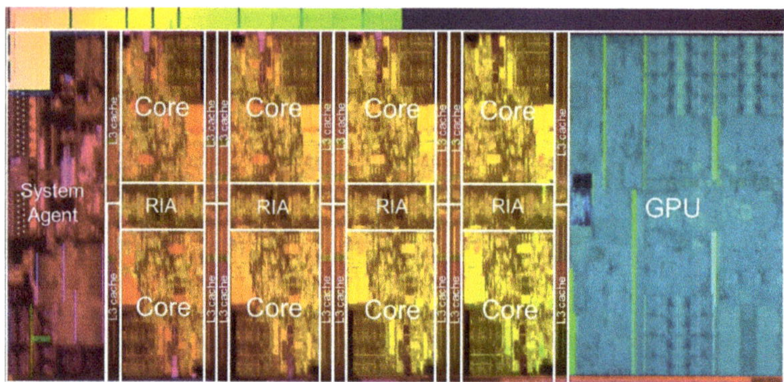

Figure 3. The original floorplan of an Intel i9-9900K processor (source: Intel). GPU–graphics processing unit, RIA–ring interconnection agent. The total die area is about 178 mm^2.

The final 16-core 3D processor design is shown in Figures 4 and 5 whereas its parameters are listed in Table 1. Note that the total power dissipation was increased from 95 W (the original TDP of the Intel processor) to 185 W. Since the 3D processor model also has a smaller footprint, the resulting power density is so high that conventional air-based cooling would not be able to maintain acceptable temperatures. Therefore, the chip is cooled by implementing microchannels in two processor layers. The parameters of the cooling system are also listed in Table 1. Except for the die, the model also includes the thermal interface material (TIM) and the heat spreader. Although the heat dissipated through the spreader to the air is negligible compared to the heat removed by the fluid, it was decided to include it in the model. The use of the spreader may be helpful for the mechanical stability of the chip and it may also make the temperature distribution more uniform, thus reducing the peak temperature.

Table 1. 3D processor design parameters.

Parameter	Value
Die area	9.13 mm × 11.69 mm (106.7 mm^2)
Chip layers	5 (two layers with microchannels, three layers without microchannels)
Cores	16
L3 caches	32
Microchannel layers	2
Number of microchannels in a layer	28
Microchannel cross-sectional size	100 μm × 100 μm (unless otherwise specified)
Silicon layer thickness	200 μm
TIM thickness	60 μm
Heat spreader thickness	1000 μm
Microchannel cover/ILD/underfill equivalent thermal conductance coefficient	1,000,000 W/(m^{2}K)
Convection to air	10 W/(m^{2}K)
Secondary path convection (through underfill/C4 bumps and PCB)	1 W/(m^{2}K)
Number of solid nodes per layer	28 × 96
Number of fluid nodes per layer	28 × 96

Table 1. *Cont.*

Parameter	Value
Total number of nodes	18,816
Inlet fluid temperature	300 K
Inlet fluid velocity	1.5 m/s (unless otherwise specified)
Ambient air temperature	300 K
Solver	EigenSparseLU
Chip layers	silicon, $k = 130$ W/(mK), $\rho = 2330$ kg/m^3, $c_P = 700$ J/(kgK)
Thermal interface material	$k = 2$ W/mK, $\rho = 2000$ kg/m^3, $c_P = 700$ (J/kgK)
Heat spreader	copper, $k = 400$ W/(mK), $\rho = 8960$ kg/m^3, $c_P = 385$ J/(kgK)
Cooling fluid	water, $k = 0.591$ W/(mK), $\rho = 1000$ kg/m^3, $c_P = 4184$ J/(kgK), $\mu = 0.000653$ Ns/m^2, $Pr = 3.56$

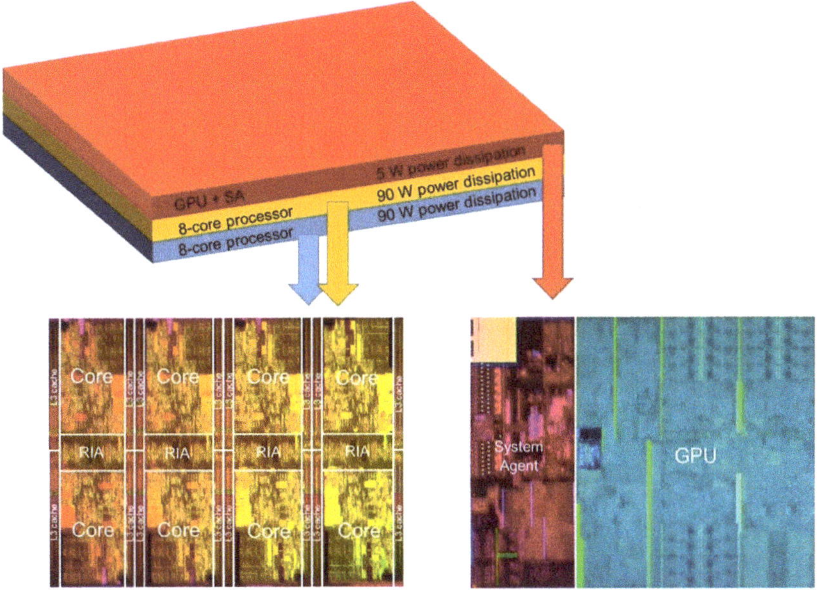

Figure 4. The model of the 3D processor based on the floorplan of Intel i9-9900K. The 3D chip contains two octa-core processors in two layers. The system agent and the GPU occupy the third layer. The total die area is about 107 mm^2. The power density is about 3.2× higher than in the original Intel processor. A more detailed view of all layers is shown in Figure 5.

4.2. Smulation Parameters

Based on the processor model described in the previous section, a simulation model in TIMiTIC was created. Each solid layer was divided into multiple cells: 96 cells in the direction along the channels and 28 cells in the direction perpendicular to the channels. Additionally, two processor layers contain channels which add additional fluid nodes to the model. Thus, the total number of nodes in the model is equal to 18,816 (two layers with channels, each consisting of 5376 nodes, and three layers without channels, each consisting of 2688 nodes). Bonding layers, inter-layer dielectric, and microchannel covers are considered very thin layers and are modeled using the thermal contact approach: their thermal resistance is added to the model, but their thermal capacity is neglected. All other simulation parameters are listed in Table 1.

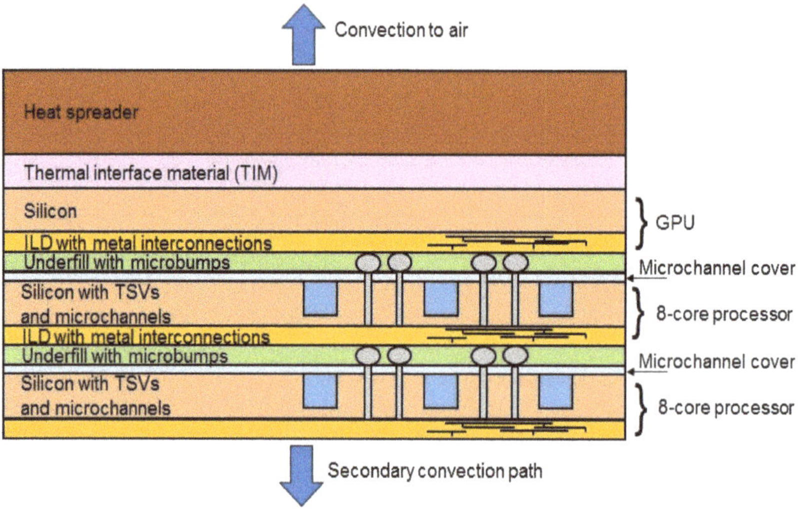

Figure 5. The 3D processor layers included in the 3D chip simulation model. ILD–inter-layer dielectric, TSV–through-silicon via.

4.3. Power Data

The default power distribution among chip units is shown in Table 2. Note that it is based on the thermal design power (TDP) of the original Intel processor, which was equal to 95 W. It was assumed that for a compute-intensive application which does not use graphic processing the power dissipated in the GPU and SA units is low and equal to 5 W, while the rest of the power (90 W) is dissipated by other units (8 cores, 16 L3 caches and 4 RIAs). Therefore, in the 3D chip model, the total power dissipation equals 185 W which, considering that the die area was reduced compared to the original Intel processor, results in a 3.2× increase in the power density. The power breakdown into particular units was assumed based on the idea that cores should have the highest power density followed by L3 caches and RIAs.

Table 2. Power distribution among chip units for the default case.

Processor Unit	Dissipater Power
Cores	16×9 W
L3 caches	32×1 W
RIAs	8×0.5 W
GPU and SA	5 W
Total	185 W

5. Results

5.1. Peak Temperature Variability Simulations

The simulations presented in this section have two goals: first, to demonstrate the usefulness of the simulator and, second, to prove that the appropriate DVFS algorithm or efficient scheduling of tasks to cores can reduce the maximum temperature by more than 10 °C in 3D ICs cooled by microchannels.

The methodology adopted in this section assumes that each core dissipates different power (for example, it can be seen as equivalent to cores executing different tasks, which vary in terms of how computationally intensive they are), but the total power dissipation is always constant. Therefore, 1000 power distributions were generated randomly assuming that the maximum power dissipated in a core can be equal to 13.5 W and the minimum to 4.5 W. Next, 1000 simulations were run using these power data (the default power distribution shown in Table 1 was modified). Then, the distribution of maximum temperatures found in the chip for each case was reported. Note that running 1000 simulations using any FEM-based tool would not be feasible, as one simulation of such a chip can typically take up to several hours. With the TIMiTIC simulator, one simulation took only about two seconds, so the total simulation time needed for 1000 iterations was around 40 min. The same set of simulations was repeated for different values of fluid velocity (mass flow) and channel cross-sectional area. The results are presented in Figures 6 and 7.

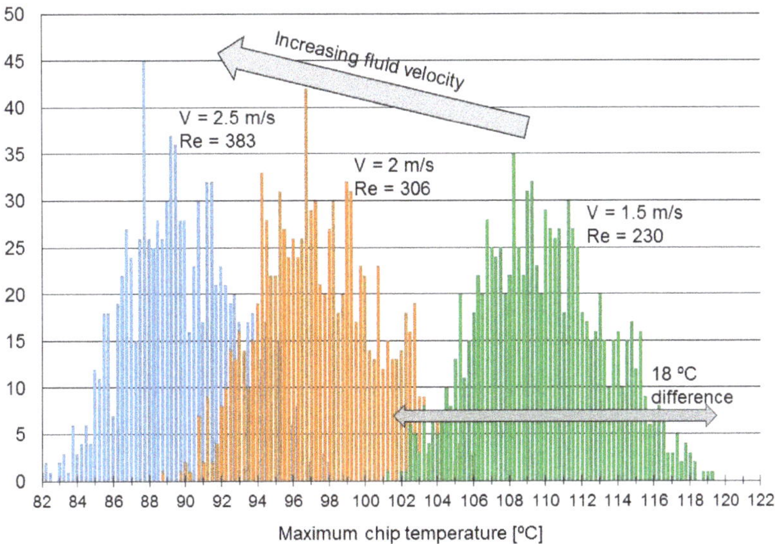

Figure 6. Peak temperature distribution (1000 cases in total) in the simulated 3D processor for various fluid velocities. Power consumption in processor cores was generated randomly once, and only power-to-core mapping was varied between cases. Re indicates the Reynolds number.

As expected, when increasing the fluid velocity, the temperatures decrease for two reasons: first, the advection thermal resistance is lower and, second, the thermal entry length is higher, which results in a higher average HTC between the solid and the fluid. Similarly, when increasing channel height, the convection resistance is decreased and temperatures goes down. These two observations are quite obvious, but other more interesting conclusions can be also drawn.

The obtained results resemble a normal distribution. The distribution shape varies only slightly when the fluid velocity or the channel height are changed, which indicates that, regardless of the parameters of the cooling system, the difference in maximum and minimum temperatures for different power distributions remain almost constant. In other words, we can conclude that peak temperature variability is independent of the cooling system.

It is important to emphasize that the power values were the same in all 1000 cases; just power-to-core-mapping was changed. This leads us to an important observation: the peak temperature variability has to be a direct result of the power distribution variation combined with the inherent property of microchannel cooling (gradually worsening cooling performance from inlets to outlets). To be more precise, when high-power tasks are executed near the outlets and low-power tasks near the inlets, a higher peak temperature is observed than in the opposite case.

The temperature difference between the worst and the best case is quite significant. For example, for the case with 1.5 m/s fluid velocity (right-most distribution in Figure 6), the peak temperature can be as high as 119 °C or as low as 101 °C. Therefore, just by appropriately assigning tasks to cores, a potential 18 °C reduction in chip maximum temperature can be achieved. In relative terms, this means that the peak temperature rise could be reduced by 22% (where temperature rise is calculated with respect to the inlet fluid temperature). Even considering the average temperature (which in this case was about 110 °C) the reduction of hotspot temperature from 110 °C to 101 °C can definitely be seen as a worthwhile goal for thermal designers.

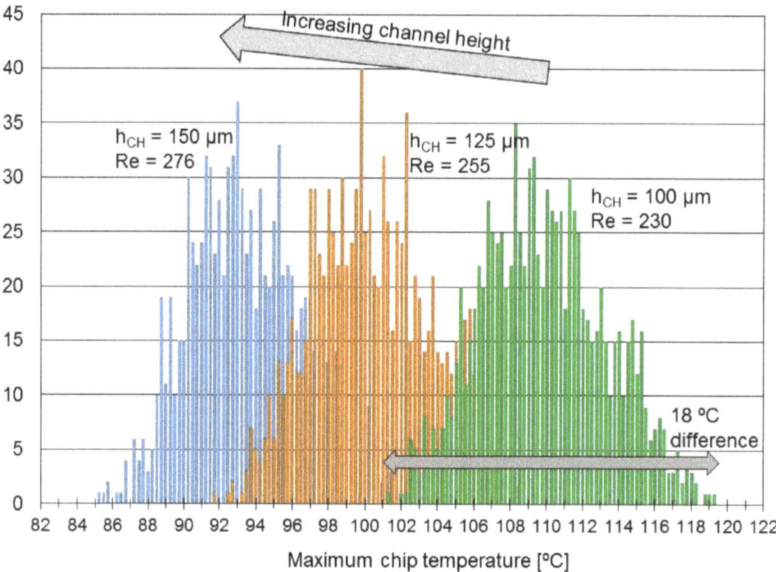

Figure 7. Peak temperature distribution (1000 cases in total) in the simulated 3D processor for various channel heights. Power consumption in processor cores was generated randomly once, and only power-to-core mapping was varied between cases. Re indicates the Reynolds number.

5.2. Finding the Optimal Power Dissipation Profile

In the previous section, it was proven that appropriate task scheduling or DVFS can considerably minimize the peak temperature in liquid-cooled 3D ICs. In case of task-to-core mapping, one can imagine that the algorithm should take into consideration the cooling performance, which varies depending on the core's location in the chip. In case of DVFS, one may consider running cores located near the inlets at a higher frequency than those located near the outlets. In other words, by using a non-uniform power dissipation profile, a more uniform temperature distribution can be achieved resulting in a lower peak temperature. However, the important question is: how uneven should this power profile be? What should be its exact shape? Clearly, for any chip with a microchannel cooling system, there must exist a power dissipation profile which is optimal, i.e., results in the lowest peak temperature.

Therefore, an optimization problem can be formulated: given a microchannel-cooled 3D processor with multiple cores and a set of tasks with different power dissipation, find the task-to-core mapping (or, in case of DVFS, core voltage and frequency settings) which minimizes the peak temperature. To design such an optimization algorithm, one should know how exactly the cooling performance of microchannels decreases from inlet to outlet. Consequently, once it is known how well each location in the chip is cooled, a cooling performance coefficient (CPC) to each processor core can be assigned. Of course, cores with high CPC should be preferred when assigning tasks (or, in case of DVFS, they should run at higher frequencies). One may think that the optimization problem can be then reduced to one simple rule: the higher power dissipation of a task, the closer to inlets it should be executed. However, the problem becomes more complex if we consider processors with multiple layers. Then, the power dissipated in cores located one over the other adds up. Therefore, this simple rule does not work, and a more complex approach has to be implemented. In this section, using the processor model described in Section 4, the process of finding the optimal power dissipation profile is described.

In [18], the authors analytically calculated the optimal (producing the lowest peak temperature) power dissipation profile along the channel (3).

$$\frac{dq}{dx} = q\frac{hP}{\dot{m}c_p}\frac{1}{1-\exp\left(-\frac{hPL}{\dot{m}c_p}\right)}\exp\left(-\frac{hP}{\dot{m}c_p}x\right) \tag{3}$$

where dq/dx is the linear power density, x is the distance from the inlet, h is the average heat transfer coefficient, q is the total dissipated power, $\dot{m}$ is the mass flow in kg/s, c_p is the specific heat, L is the channel length and P is the channel perimeter.

Based on this profile, the optimal power dissipation in all sixteen cores of the processor model (see Section 4) can be found using the steps listed below.

- Find the position of the center of the core area along the channel;
- Using the above-mentioned optimal power profile, calculate the optimal power dissipation in the core area using Equation (3) (note that this area includes all three layers);
- Subtract the power dissipated in the GPU layer in the area directly above cores;
- The remaining power can be divided by two to obtain the optimal power dissipated in both cores located at this position.

Table 3 shows the power values obtained using the above method for the analyzed 3D processor. However, the formula from (3) was derived for a constant HTC along the channel and does not take into consideration the thermal entry effects (higher HTC near the inlets). It can be safely supposed that the true optimal power distribution will be more uneven, with even more power dissipated in cores near the inlets. Thus, additional simulations were run: taking the previously calculated power distribution as a starting point, the power consumption in cores near the inlets was increased and in cores near the outlets decreased (keeping the total power constant). It was discovered that it was still possible to reduce the peak temperature using this method. Thus, it was proven that the power values calculated analytically are in fact a quasi-optimal solution, and a better solution can be found. Consequently, a Monte Carlo method was employed: 1000 simulations were run, and for each core the power was generated randomly within a certain range around the previously found quasi-optimal values. Again, the total power was kept constant. The simulation results are shown in Figure 8. The lowest maximum temperature was obtained for the power values shown in Table 3 (right column). It was 3.5 °C lower that the one obtained for the quasi-optimal solution.

Table 3. Power dissipated in cores and the respective peak temperatures for three cases. Power dissipation in cores in the second processor layer is exactly the same, so it was not shown.

Core Number	Dissipated Power (Uniform)	Dissipater Power (Analytical, Quasi-optimal Solution)	Dissipater Power (Optimal Solution Based on Monte Carlo Simulations)
Cores 1 and 2 (closest to inlets)	9 W	14.59 W	18.24 W
Cores 2 and 3	9 W	9.97 W	9.43 W
Cores 4 and 5	9 W	6.82 W	5.81 W
Cores 6 and 7 (closest to outlets)	9 W	4.62 W	2.52 W
Peak temperature obtained with simulations	105 °C	97.7 °C	94.2 °C

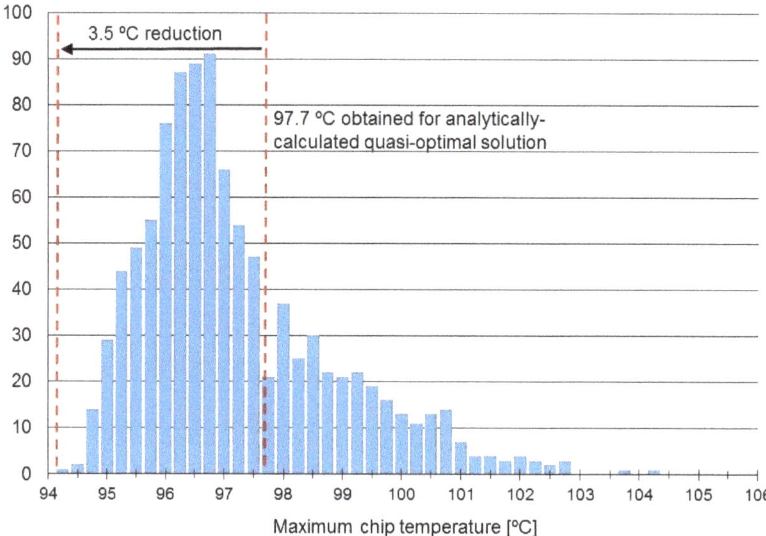

Figure 8. Peak temperature distribution (1000 cases in total) in the simulated 3D processor where power dissipated in cores was generated randomly around the quasi-optimal values.

Additionally, Figures 9–11 show the temperature maps of the middle chip layer for three cases: one with uniform power distribution in cores; one with quasi-optimal power distribution; and one with optimal power distribution, respectively. The temperature maps were superimposed on the chip floorplan for clarity. By comparing the three figures, it can be clearly seen that the more uniform the temperature distribution in all cores, the lower the maximum temperature in the chip. Therefore, it may be argued that the optimal power dissipation in cores should be such that the peak temperature in each core is the same.

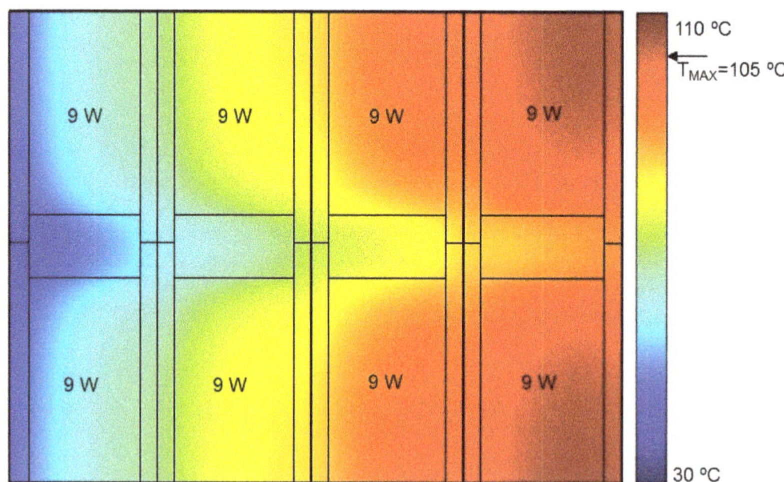

Figure 9. Temperature map of the middle layer of the simulated 3D processor (default case) where powers dissipated in all cores are the same. The powers dissipated in other processor units are listed in Table 2. The inlets are located on the left side of the chip.

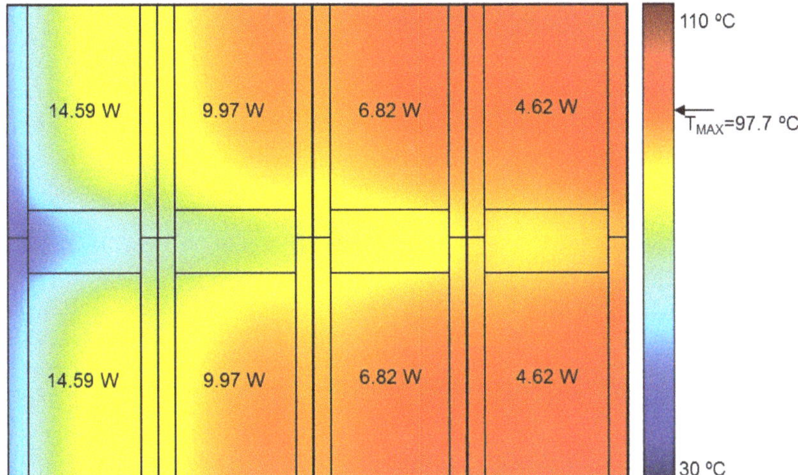

Figure 10. Temperature map of the middle layer of the simulated 3D processor where power dissipated in cores was calculated analytically (the quasi-optimal solution). The powers dissipated in other processor units are listed in Table 2. The inlets are located on the left side of the chip.

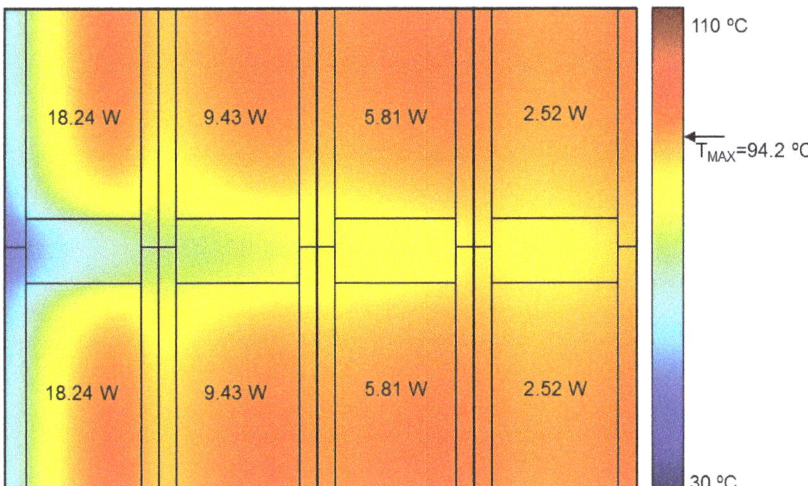

Figure 11. Temperature map of the middle layer of the simulated 3D processor where the power dissipated in cores was found by running Monte Carlo simulations (the optimal solution). The power dissipated in other processor units is listed in Table 2. The inlets are located on the left side of the chip.

There are three main findings that can be derived from the obtained results. First, although the power profile described by Formula (3) allows a considerable reduction in peak temperature, it cannot be seen as the optimal solution. The true optimal power profile contains quite different power values (compare columns two and three in Table 3). Therefore, for best results, the thermal entry effects should not be ignored when developing efficient DVFS or task scheduling schemes. Second, the optimal power profile is extremely difficult to calculate analytically. However, it was possible to find the optimal power values by running a large number of simulations. Note that this Monte Carlo approach would not be possible with CFD simulation, because the simulation time would be prohibitively long. This highlights the advantages and demonstrates the usefulness of the proposed compact thermal modelling tool, which allows exploring the design space in a reasonable time. Third, which may be surprising, the optimal power profile is very non-uniform (18.24 W for cores near the inlets and 2.25 W for cores near the outlets). This means that more than seven times more power should be dissipated in "inlet cores" compared to "outlet cores". Even with very aggressive DVFS, this may not always be possible to achieve. Therefore, for microchannel-cooled processors, the designers should perhaps consider floorplanning schemes which take this effect into consideration, e.g., processing cores should not be put near the outlets, and this outlet area should be reserved for processor units with low power dissipation.

5.3. Power Distribution Correlation Metric

Knowing the optimal power consumption in each core allows introducing a metric which describes how a given power distribution correlates with the optimal one. In other words, every possible power distribution can be quantified in terms of how similar it is to the optimal distribution, which in turn allows choosing the best one, for example when executing thermal-aware task scheduling or calculating the best DVFS settings.

Let $P_0{}^i$ be the optimal power dissipation calculated for core i and P^i the actual power consumption in core i. Let us define T_{max} as the maximum chip temperature. In the first approach, one may think than T_{max} will be strongly correlated with the maximum difference between P^i and $P_0{}^i$ calculated for all cores. To verify this hypothesis, 1000 simulations were run, with random powers generated for all cores while keeping the total power constant. Figure 12 shows the correlation between such a metric and the maximum chip temperature. We can see that this correlation is certainly visible, but the variance is still quite high. For example, let us look at the points A and B in the figure. For the same value of the metric, the difference in peak temperature is around 14 °C, so this metric would not produce satisfactory results and a more accurate approach is needed. Hence, a simple thermal model is proposed here. Consider a one-channel system and the area located at the distance x from the inlet. The temperature of the fluid at location x can be approximated as:

$$T_f(x) = T_i + \frac{Q_U}{\dot{m}c_P} \tag{4}$$

where T_i is the fluid inlet temperature, $\dot{m}$ is the mass flow, c_P is the specific heat of the fluid and Q_U is the power dissipated upstream from the location x. The temperature of the solid at point x can then be calculated as:

$$T(x) = T_f(x) + \frac{Q}{hA} = T_i + \frac{Q_U}{\dot{m}c_P} + \frac{Q}{hA} \tag{5}$$

where h is the heat transfer coefficient between the solid and the fluid, A is the channel area through which the convection occurs, and Q is the power consumption in the area. Equation (5) can then be reformulated to describe the overhead in peak temperature ΔT:

$$\Delta T(x) = \frac{\Delta Q_U}{\dot{m}c_P} + \frac{\Delta Q}{hA} \tag{6}$$

where ΔQ_U is the overhead power dissipated upstream from location x and ΔQ is the overhead power consumption in the area. The overhead power is, of course, calculated as the difference between the actual power and the optimal power. ΔT is the overhead peak temperature: the difference between the actual peak temperature and the temperature resulting from the optimal power.

Although Equation (6) was derived for a one-channel system, it can be applied to estimate ΔT for each core in the chip and then to calculate the maximum of these values ΔT_{max}. Note that, in this case, ΔT_{max} does not give any information about the absolute value of the maximum temperature in the chip; it is simply a relative metric which can be used to compare power distributions. For example, a power distribution with a higher value of this metric will always produce higher maximum temperature and vice versa. Achieving the linearity of the metric is also important. To make this metric dimensionless, it was further divided by inlet temperature T_i. Again, 1000 random-power simulations were run to test the correlation of the metric with the chip peak temperature. Figure 13 shows the obtained results: despite the fact that the used model is relatively simple, we can see that the linearity of the model is surprisingly good and that this metric can be effectively used to predict which power distribution is better, i.e., which power distribution results in lower maximum chip temperature.

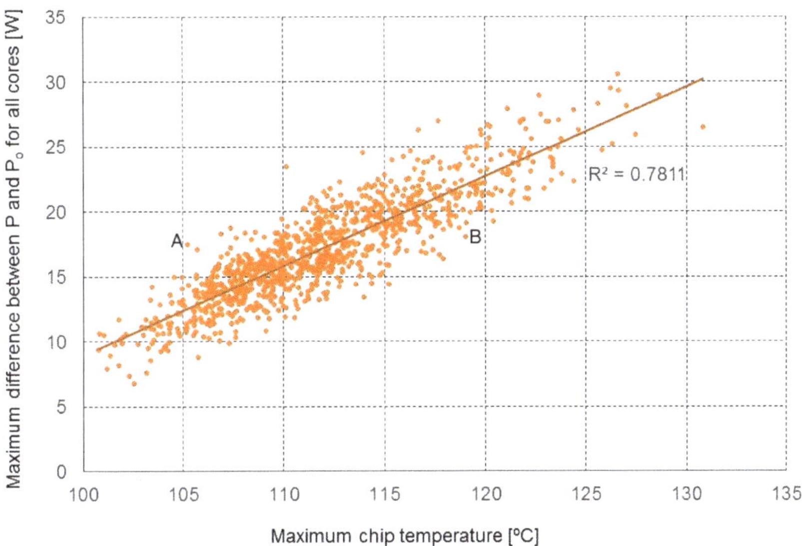

Figure 12. Simulation results of 1000 cases, showing the correlation between the maximum power overhead (P-P_o) calculated for all cores and the maximum chip temperature.

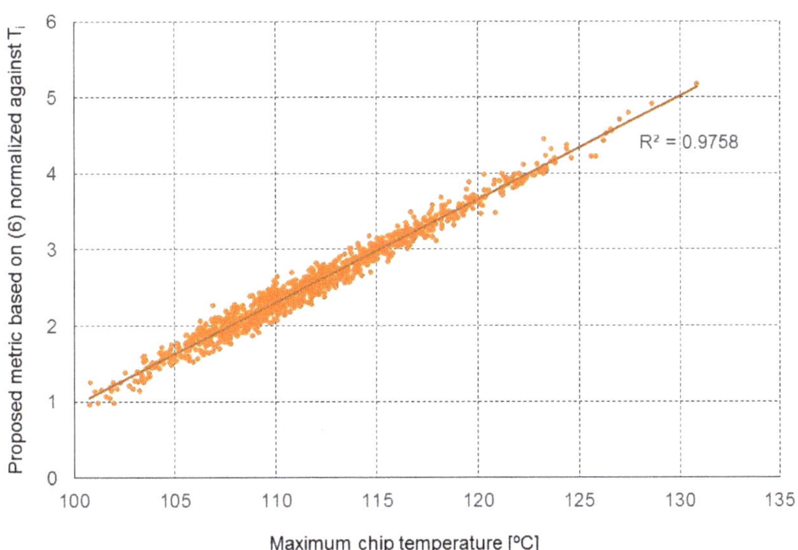

Figure 13. Simulation results of 1000 cases, showing the correlation between the metric proposed in this paper and the maximum chip temperature.

In what follows, the utility of the proposed metric will be discussed. The main advantage is that it is very simple: the optimal power distribution has to be found only once for a given chip. Then, the metric can be easily calculated analytically, so the cost (for example in terms of time or energy) of calculating this metric is negligible. Therefore, it can be used not only in static, but also in dynamic task scheduling algorithms. Such an algorithm could use, for example, the cooling performance coefficient CPC for each processor core, calculated as the inverse of the proposed metric. Then, assuming that a set of tasks is to be scheduled on cores, the algorithm can utilize the calculated CPC values to find the optimal mapping, thus guaranteeing the lowest possible peak temperature in the chip during task execution. It can be also used to for efficient task swapping: if new tasks appear, a new optimal mapping can be found on the fly, and all tasks (including those already being executed) can be rescheduled to different cores. Similarly, researchers who develop DVFS algorithms can make use of the proposed metric to dynamically adapt the voltage and the frequency of each core. For example, taking as input the estimated power of tasks to be executed, the dependencies between them and their potential deadlines, the algorithm can find the optimal frequencies and voltages for the cores which result in the minimization of the peak temperature during task execution.

6. Conclusions and Future Work

The main contributions provided by this research are the following:

- In Section 2, the characteristics of microchannel cooling are explained, with special emphasis on how the cooling performance of microchannels decreases from inlet to outlet. The impact of this phenomenon on other research fields like task scheduling or DVFS algorithms is also discussed.
- In Section 3, the thermal modelling tool used in this paper is presented and its most unique features are listed. In Section 4, the model of the 3D processor (based on the design of a real modern Intel processor) which is used in this study is thoroughly described.
- Section 5 constitutes the central part of the paper. First, the simulations in Section 5.1 prove that the same power distribution can produce drastically different peak temperatures depending on how powers are mapped to cores. The obtained results also show that this effect is independent of the cooling system parameters.
- Second, the analysis provided in Section 5.2 calculates the optimal power dissipation profile for the analyzed chip. It is also shown that the analytical calculation which assumes a constant HTC along the channel (neglects thermal entry effects) can only be used to obtain a quasi-optimal solution, which performs considerably worse that the true optimal solution.
- Third, Section 5.3 introduces a simple analytical metric which effectively compares a given power distribution with the optimal one. It is demonstrated that such a metric can be used to compare the power distribution profiles in a 3D IC in terms of their thermal efficiency. It is also argued that this metric can be used by researchers to develop optimal task scheduling or DVFS strategies for 3D ICs with integrated microchannels.

Future work will include the verification of the proposed metric for temperature-aware task scheduling. The scheduling algorithm using this metric will be implemented in the TIMiTIC simulator and transient thermal simulations will be run to determine its effectiveness.

Funding: This research received no external funding.

Conflicts of Interest: The authors declare no conflict of interest.

References

1. Beyne, E. The 3-D Interconnect Technology Landscape. *IEEE Des. Test* **2016**, *33*, 8–20. [CrossRef]
2. Pavlidis, V.F.; Friedman, E.G. *Three-Dimensional Integrated Circuit Design;* Morgan Kaufmann Publishers Inc.: San Francisco, CA, USA, 2008; ISBN 978-0-08-092186-0.

3.   Sharma, R. *Design of 3D Integrated Circuits and Systems*; CRC Press: Boca Raton, FL, USA, 2018; ISBN 978-1-315-21570-9.

4.   Shen, W.-W.; Chen, K.-N. Three-Dimensional Integrated Circuit (3D IC) Key Technology: Through-Silicon Via (TSV). *Nanoscale Res. Lett.* **2017**, *12*, 56. [CrossRef] [PubMed]

5.   Oh, H.; Gu, J.M.; Hong, S.J.; May, G.S.; Bakir, M.S. High-aspect ratio through-silicon vias for the integration of microfluidic cooling with 3D microsystems. *Microelectron. Eng.* **2015**, *142*, 30–35. [CrossRef]

6.   Bakir, M.S.; Meindl, J.D. *Integrated Interconnect Technologies for 3D Nanoelectronic Systems*; Artech House: Norwood, MA, USA, 2008; ISBN 978-1-59693-247-0.

7.   Kim, S.-W.; Detalle, M.; Peng, L.; Nolmans, P.; Heylen, N.; Velenis, D.; Miller, A.; Beyer, G.; Beyne, E. Ultra-Fine Pitch 3D Integration Using Face-to-Face Hybrid Wafer Bonding Combined with a Via-Middle Through-Silicon-Via Process. In Proceedings of the 2016 IEEE 66th Electronic Components and Technology Conference (ECTC), Las Vegas, NV, USA, 31 May–3 June 2016; pp. 1179–1185.

8.   De Vos, J.; Peng, L.; Phommahaxay, A.; Van Ongeval, J.; Miller, A.; Beyne, E.; Kurz, F.; Wagenleiter, T.; Wimplinger, M.; Uhrmann, T. Importance of alignment control during permanent bonding and its impact on via-last alignment for high density 3D interconnects. In Proceedings of the 2016 IEEE International 3D Systems Integration Conference (3DIC), San Francisco, CA, USA, 8–11 November 2016; pp. 1–5.

9.   Zhang, L.; Liu, Z.; Chen, S.-W.; Wang, Y.; Long, W.-M.; Guo, Y.; Wang, S.; Ye, G.; Liu, W. Materials, processing and reliability of low temperature bonding in 3D chip stacking. *J. Alloy. Compd.* **2018**, *750*, 980–995. [CrossRef]

10. Alfieri, F.; Tiwari, M.K.; Zinovik, I.; Poulikakos, D.; Brunschwiler, T.; Michel, B. 3D Integrated Water Cooling of a Composite Multilayer Stack of Chips. *J. Heat Transf.* **2010**, *132*, 121402. [CrossRef]

11. Koo, J.-M.; Im, S.; Jiang, L.; Goodson, K.E. Integrated microchannel cooling for three-dimensional electronic circuit architectures. *J. Heat Transf.* **2005**, *127*, 49–58. [CrossRef]

12. Mudawar, I.; Bharathan, D.; Kelly, K.; Narumanchi, S. Two-Phase Spray Cooling of Hybrid Vehicle Electronics. *IEEE Trans. Compon. Packag. Technol.* **2009**, *32*, 501–512. [CrossRef]

13. Brunschwiler, T.; Michel, B.; Rothuizen, H.; Kloter, U.; Wunderle, B.; Oppermann, H.; Reichl, H. Forced convective interlayer cooling in vertically integrated packages. In Proceedings of the 2008 11th Intersociety Conference on Thermal and Thermomechanical Phenomena in Electronic Systems, Orlando, FL, USA, 28–31 May 2008; pp. 1114–1125.

14. Takács, G.; Szabó, P.G.; Plesz, B.; Bognár, G. Improved thermal characterization method of integrated microscale heat sinks. *Microelectron. J.* **2014**, *45*, 1740–1745. [CrossRef]

15. Ansys: Engineering Simulation & 3D Design Software. Available online: www.ansys.com (accessed on 23 March 2020).

16. COMSOL Multiphysics® v. 5.2. Available online: www.comsol.com (accessed on 23 March 2020).

17. FloTHERM, CFD Modelling Software. Available online: https://www.mentor.com/products/ (accessed on 23 March 2020).

18. Zając, P.; Maj, C.; Napieralski, A. Peak temperature reduction by optimizing power density distribution in 3D ICs with microchannel cooling. *Microelectron. Reliab.* **2017**, *79*, 488–498. [CrossRef]

19. Lau, J.H.; Yue, T.G. Thermal management of 3D IC integration with TSV (through silicon via). In Proceedings of the 2009 59th Electronic Components and Technology Conference, San Diego, CA, USA, 26–29 May 2009; pp. 635–640.

20. Zając, P.; Napieralski, A. Novel thermal model of microchannel cooling system designed for fast simulation of liquid-cooled ICs. *Microelectron. Reliab.* **2018**, *87*, 245–258. [CrossRef]

21. Zając, P.; Janicki, M.; Napieralski, A. On the applicability of single-layer integrated microchannel cooling in 3D ICs. In Proceedings of the 2018 19th International Conference on Thermal, Mechanical and Multi-Physics Simulation and Experiments in Microelectronics and Microsystems (EuroSimE), Toulouse, France, 15–18 April 2018; pp. 1–6.

22. Sridhar, A.; Vincenzi, A.; Atienza, D.; Brunschwiler, T. 3D-ICE: A Compact Thermal Model for Early-Stage Design of Liquid-Cooled ICs. *IEEE Trans. Comput.* **2014**, *63*, 2576–2589. [CrossRef]

23. Németh, M.; Takács, G.; Jani, L.; Poppe, A. Compact modeling approach for microchannel cooling and its validation. *Microsyst. Technol.* **2018**, *24*, 419–431.

24. Mizunuma, H.; Yang, C.-L.; Lu, Y.-C. Thermal modeling for 3D-ICs with integrated microchannel cooling. In Proceedings of the 2009 IEEE/ACM International Conference on Computer-Aided Design—Digest of Technical Papers, San Jose, CA, USA, 2–5 November 2009; pp. 256–263.

25. Coskun, A.K.; Ayala, J.L.; Atienza, D.; Rosing, T.S. Modeling and dynamic management of 3D multicore systems with liquid cooling. In Proceedings of the 2009 17th IFIP International Conference on Very Large Scale Integration (VLSI-SoC), Florianopolis, Brazil, 12–14 October 2009.

26. Bognár, G.; Takács, G.; Pohl, L.; Szabó, P.G. Thermal modelling of integrated microscale heatsink structures. *Microsyst. Technol.* **2018**, *24*, 433–444. [CrossRef]

27. Feng, Z.; Li, P. Fast Thermal Analysis on GPU for 3D ICs with Integrated Microchannel Cooling. *IEEE Trans. Very Large Scale Integr. (Vlsi) Syst.* **2013**, *21*, 1526–1539. [CrossRef]

28. Zając, P. TIMiTIC: A C++ based Compact Thermal Simulator for 3D ICs with Microchannel Cooling. In Proceedings of the 2019 25th International Workshop on Thermal Investigations of ICs and Systems (THERMINIC), Lecco, Italy, 25–27 September 2019; pp. 1–6.

29. Sekar, D.; King, C.R.; Dang, B.; Spencer, T.; Thacker, H.D.; Joseph, P.K.; Bakir, M.S.; Meindl, J.D. A 3D-IC Technology with Integrated Microchannel Cooling. In Proceedings of the 2008 International Interconnect Technology Conference, Burlingame, CA, USA, 1–4 June 2008.

30. Incropera, F.P.; DeWitt, D.P.; Bergman, T.L.; Lavine, A.S. *Fundamentals of Heat and Mass Transfer*; Wiley: Hoboken, NJ, USA, 2007; ISBN 978-0-471-45728-2.

31. Su, L.; Duan, Z.; He, B.; Ma, H.; Xu, Z. Thermally Developing Flow and Heat Transfer in Elliptical Minichannels with Constant Wall Temperature. *Micromachines* **2019**, *10*, 713. [CrossRef] [PubMed]

32. Cuesta, D.; Risco-Martín, J.L.; Ayala, J.L.; Hidalgo, J.I. Thermal-aware floorplanner for 3D IC, including TSVs, liquid microchannels and thermal domains optimization. *Appl. Soft Comput.* **2015**, *34*, 164–177. [CrossRef]

33. Shah, R.K.; London, A.L. *Laminar Flow Forced Convection in Ducts: A Source Book for Compact Heat Exchanger Analytical Data*; Academic Press: Cambridge, MA, USA, 1978; ISBN 978-0-12-020051-1.

34. Muzychka, Y.S.; Yovanovich, M.M. Laminar Forced Convection Heat Transfer in the Combined Entry Region of Non-Circular Ducts. *J. Heat Transf.* **2004**, *126*, 54–61. [CrossRef]

35. Guennebaud, G.; Jacob, B.; Avery, P.; Bachrach, A.; Barthelemy, S. Eigen v3. 2010. Available online: http://eigen.tuxfamily.org (accessed on 23 March 2020).

36. Ahnert, K.; Mulansky, M. Odeint—Solving ordinary differential equations in C++. *AIP Conf. Proc.* **2011**, *1389*, 1586–1589.

37. TIMiTIC Simulator Website. Available online: timitic.dmcs.pl (accessed on 16 March 2020).

38. Haas, A.; Rossberg, A.; Schuff, D.L.; Titzer, B.L.; Holman, M.; Gohman, D.; Wagner, L.; Zakai, A.; Bastien, J. Bringing the web up to speed with WebAssembly. In Proceedings of the 38th ACM SIGPLAN Conference on Programming Language Design and Implementation, Association for Computing Machinery, Barcelona, Spain, 18–23 June 2017; pp. 185–200.

Article

Thermal Characterization and Modelling of AlGaN-GaN Multilayer Structures for HEMT Applications [†]

Lisa Mitterhuber *, René Hammer, Thomas Dengg and Jürgen Spitaler

Materials Center Leoben Forschung GmbH, Roseggerstrasse 12, 8700 Leoben, Austria;
rene.hammer@mcl.at (R.H.); thomas.dengg@mcl.at (T.D.); juergen.spitaler@mcl.at (J.S.)

* Correspondence: lisa.mitterhuber@mcl.at

† This paper is an extended version of our conference paper "Thermal Investigation of AlGaN-GaN Multilayer Structures", 2019 25th International Workshop on Thermal Investigations of ICs and Systems (THERMINIC), Lecco, Italy, 2019.

Received: 30 March 2020; Accepted: 29 April 2020; Published: 9 May 2020

Abstract: To optimize the thermal design of AlGaN-GaN high-electron-mobility transistors (HEMTs), which incorporate high power densities, an accurate prediction of the underlying thermal transport mechanisms is crucial. Here, a HEMT-structure ($Al_{0.17}Ga_{0.83}N$, GaN, $Al_{0.32}Ga_{0.68}N$ and AlN on a Si substrate) was investigated using a time-domain thermoreflectance (TDTR) setup. The different scattering contributions were investigated in the framework of phonon transport models (Callaway, Holland and Born-von-Karman). The thermal conductivities of all layers were found to decrease with a temperature between 300 K and 773 K, due to Umklapp scattering. The measurement showed that the AlN and GaN thermal conductivities were a magnitude higher than the thermal conductivity of $Al_{0.32}Ga_{0.68}N$ and $Al_{0.17}Ga_{0.83}N$ due to defect scattering. The layer thicknesses of the HEMT structure are in the length scale of the phonon mean free path, causing a reduction of their intrinsic thermal conductivity. The size-effect of the cross-plane thermal conductivity was investigated, which showed that the phonon transport model is a critical factor. At 300 K, we obtained a thermal conductivity of (130 ± 38) $Wm^{-1}K^{-1}$ for the (167 ± 7) nm thick AlN, (220 ± 38) $Wm^{-1}K^{-1}$ for the (1065 ± 7) nm thick GaN, (11.2 ± 0.7) $Wm^{-1}K^{-1}$ for the (423 ± 5) nm thick $Al_{0.32}Ga_{0.68}N$, and (9.7 ± 0.6) $Wm^{-1}K^{-1}$ for the (65 ± 5) nm thick $Al_{0.17}Ga_{0.83}N$. Respectively, these conductivity values were found to be 24%, 90%, 28% and 16% of the bulk values, using the Born-von-Karman model together with the Hua–Minnich suppression function approach. The thermal interface conductance as extracted from the TDTR measurements was compared to results given by the diffuse mismatch model and the phonon radiation limit, suggesting contributions from inelastic phonon-scattering processes at the interface. The knowledge of the individual thermal transport mechanisms is essential for understanding the thermal characteristics of the HEMT, and it is useful for improving the thermal management of HEMTs and their reliability.

Keywords: AlGaN-GaN HEMT; TDTR; thermal conductivity; thermal interface resistance; size effect; phonon transport mechanisms

---

1. Introduction

The ability of gallium nitride (GaN) to form heterojunctions can be used to fabricate high-electron-mobility transistors (HEMTs). Aluminium gallium nitride (AlGaN)-GaN-based HEMTs offer high carrier concentration (~1013 cm^{-2} [1]) and high electron mobility (2000 cm^2/Vs [2]), resulting in a high current density and a low channel resistance. These properties offer great potential in the power amplifier technology due to their high-power and high-frequency performance. Thus, these devices are attractive for communication, radar and space applications [3]. The high output power density in the AlGaN-GaN-HEMT also allows further miniaturization of the device with gate lengths down to 80 nm [4]. However, this high-power density leads to significant self-heating, which decreases the device performance. Thus, to have reliable and long-lasting devices, a thorough thermal management is becoming increasingly relevant [5–7]. Therefore, thermal characterization and modelling are essential in order to guarantee an efficient heat removal from the heterojunction. Thermal bottlenecks that have to be considered in GaN-based HEMTs are: (i) the reduced thermal conductance of transition layers (here: AlGaN) [8], (ii) the heat transport across interfaces [9], and (iii) the reduced thermal conductivity due to phonon-scattering processes [10,11].

The information gained by this study allows thermal engineering of GaN-based HEMTs and, hence, consequent improvement in thermal management to achieve reliable and long-lasting devices. The paper has three major topics:

Firstly, we systematically investigated a material stack of a high-electron-mobility transistor (HEMT) on a silicon (Si) substrate. Both the thermal properties of the constituent materials and their thermal interface conductance were obtained simultaneously by using a time-domain thermoreflectance (TDTR) measurement setup. In literature, the thermal investigations of the individual materials of the HEMT stack can be found. For example, Chen et al. measured the intrinsic thermal conductivity of AlN [12]. Park and Bayram [13] were investigating GaN grown on different substrates to extract the influence of dislocations on the thermal behaviour. Lui and Balandin [14] did thermal investigations on $Al_xGa_{1-x}N$ alloys. Other studies investigated the whole device and reported on the temperature field, such as Chatterjee et al. [15]. Lundh et al. [16] used Raman spectroscopy and thermoreflectance thermal imaging to experimentally obtain the lateral and vertical steady-state operating temperature of an AlGaN-GaN HEMT. In this study, we demonstrate a thermal investigation of the holistic HEMT stack, including the thermal conductivities of each layer and the interfaces.

Secondly, the intrinsic thermal conductivity and the mean free path of the phonons were extracted by TDTR measurements at different temperatures. Three different phonon transport models obtained the different phonon scattering mechanisms: the Callaway, Holland, and Born-von-Karman (BvK) models. We describe the significant differences and assumptions of these models. Recently, it was recognised that the cumulative sum over all phonon contributions from the smallest mean free paths up to infinity constitutes a complete description of the bulk thermal conductivity but is particularly powerful when size effects are being investigated. For instance, Yang and Dames [17] analysed phonon transport models for bulk Si and Si nanowires. Regner et al. [18] were also determining the effect of different phonon transport models in terms of accumulated thermal conductivity. In our work, we evaluated the phonon scattering times within the said phonon transport models on the materials of the HEMT stack, based on our temperature-dependent experimental results. We show that the availability of temperature-dependent data is crucial for the analysis of the different scattering contributions.

Thirdly, we revised the calculation of the size effect of the layers with three different approaches: first, with a suppression function approach as introduced by Hua and Minnich [19]. Second, by a common approach using the accumulation of thermal conductivity over the boundary scattering [20]. Third, we determined the size effect via a phonon hydrodynamic equation as proposed by Gua and Wang [21]. All three approaches are compared for predicting the size-dependent cross-plane thermal conductivity.

2. Materials and Methods

The investigated device was an ungated GaN-based HEMT consisting of an AlGaN-GaN heterostructure. The layers were produced by a metal-organic chemical vapour deposition (MOCVD) process [22]. The AlGaN-GaN heterostructure includes a Si (111) substrate, a 2 nm thick silicon nitride layer, a (167 ± 7) nm thick aluminium nitride (AlN) nucleation layer, a (423 ± 5) nm thick $Al_{0.32}Ga_{0.68}N$ transition layer, a (1065 ± 7) nm thick GaN buffer layer, and a (65 ± 5) nm thick $Al_{0.17}Ga_{0.83}N$ top barrier layer. A cross-sectional scanning electron microscope (SEM) investigation determined the thickness of all layers (see Figure 1—right-hand side). The $Al_xGa_{1-x}N$ layer composition was analysed via X-ray diffraction [22]. At the interface between the buffer and top barrier layer, lies the heterojunction of the HEMT, where the two-dimensional electron gas channel is formed. To gain an in-depth knowledge of the heat transport mechanisms within HEMTs, four samples, encompassing different number of sublayers of the overall heterostructure (see Figure 1 Samples A–D), were produced and thermally characterized.

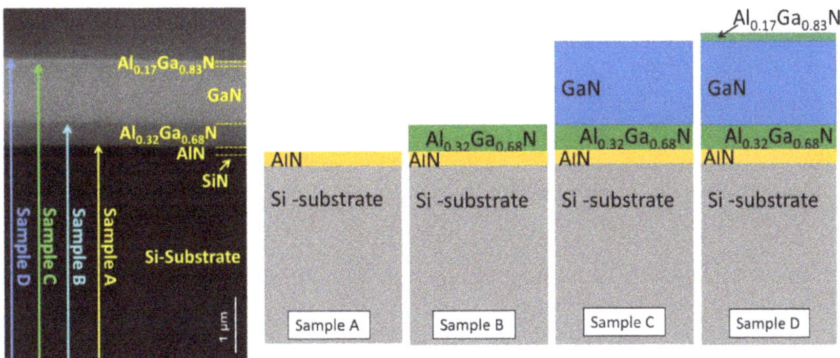

Figure 1. On the left-hand side, the SEM image of the cross-section of the AlGaN-GaN heterostructure (Sample D) is shown. The schematic of the studied AlGaN-GaN-based high-electron-mobility transistor (HEMT) structures is visualised on the right (Samples A–D).

The cross-plane thermal characterization of the four samples A to D was accomplished with a TDTR measurement (PicoTR, Netzsch [23]), which was also used in previous experiments [24]. An integrated oven with a continuous nitrogen flow allowed the measurements to be performed in the range of 300 K to 773 K. The samples A–D were investigated subsequently (see Figure 1). A 100 nm-thick platinum (Pt) layer covered all samples as a transducer. The TDTR setup used was a pump-probe technique, where the pump beam had a wavelength of 1550 nm, a spot radius of 45 µm and a pulse energy of 25 mW. The pump beam and the probe beam had a pulse width of 0.5 ps. For the purpose of lock-in detection, the pump beam was modulated at 200 kHz. The pump beam was focused on the top face of the Pt-layer, and the probe beam was focused on the same position. The temperature change caused by the pump beam was monitored by the probe beam, with a wavelength of 775 nm, by using the two-colour thermoreflectance principle. The probe beam was electrically time delayed to the pump beam with a picosecond time resolution. This time resolution allowed an analysis of both the thermal conductivity of the individual layers and the thermal interface conductance. These thermal properties were determined by fitting the phase delay to an analytical heat conduction model [25]. The phase signal over time showed a max. noise of 1° and the uncertainty of the measurements were calculated according to Yang et al. [26].

3. Results

The thermal conductivity (κ) of the constituent materials and thermal interface conductance were measured by the TDTR. The phase signal of the lock-in amplifier of the TDTR was analysed by using the analytical heat flow model of the multi-layered structures as proposed by Cahill [25]. The thickness (see Figure 1) and the volumetric heat capacity of all materials and κ of the Si substrate were used as input for the calculations. All temperature-dependent thermophysical properties of Si were taken from the literature [27,28]. The heat capacity of AlN and GaN at 300 K was 2.67 $\mathrm{MJm^{-3}K^{-1}}$ and 2.63 $\mathrm{MJm^{-3}K^{-1}}$. These values and their temperature dependency were taken from [29] and [30], respectively. By using the rule of mixture, the heat capacity of 2.65 $\mathrm{MJm^{-3}K^{-1}}$ for $Al_{0.32}Ga_{0.68}N$ was calculated. A 4-point probe measurement captured the electrical sheet resistance of the Pt-transducer layer. The thermal conductivity (κ) of the Pt transducer was determined to be 14 W/mK according to the Wiedemann–Franz law.

Starting from Sample A, κ of the AlN layer (κ_{AlN}) and the thermal interface conductance between Pt and AlN were measured via TDTR. These values were used as input parameters for Sample B, which provided the κ for the alloy $Al_{0.32}Ga_{0.68}N$ and the corresponding thermal interface conductance. Consequently, for the analysis of Sample C we used the results from Sample B. The same procedure was applied for Sample D, using the results obtained from Sample C. [31]

The temperature-dependence of κ_{AlN} for the (167 ± 7) nm thick AlN of Sample A is presented in Figure 2a. κ_{AlN} decreased from (130 ± 38) $Wm^{-1}K^{-1}$ at 300 K to (16 ± 13) $Wm^{-1}K^{-1}$ at 773 K. Our results, as can be seen, are half of the values reported by Slack et al. [32]. Such a difference potentially can be explained by size effects (as investigated in more detail below) and different defect densities, stemming from different growth processes, which might reduce the thermal conductivity. Slack et al. [32] investigated a single AlN crystal with a thickness of 3 mm, grown from high-purity AlN powder by vapour-phase transport.

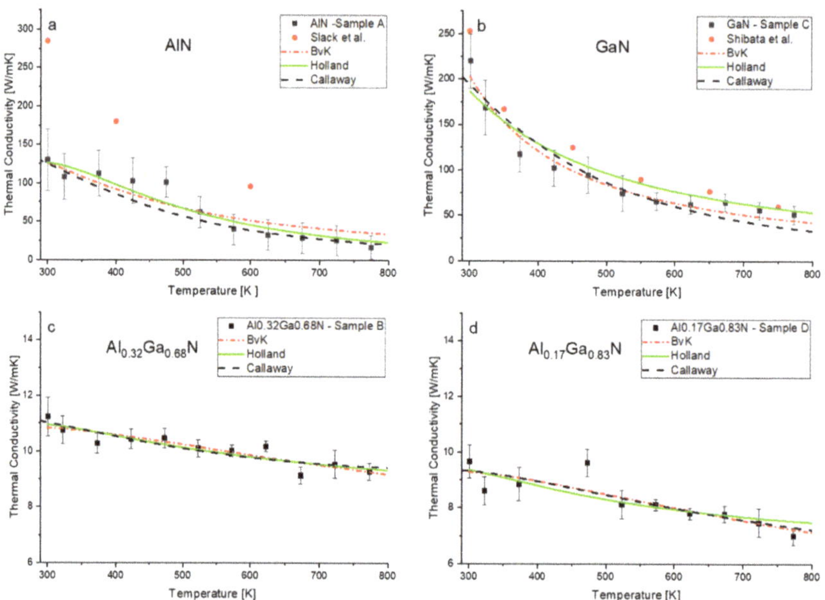

Figure 2. Measured and calculated thermal conductivities as a function of temperature. Measured κ of (**a**) AlN, (**b**) GaN, (**c**) $Al_{0.32}Ga_{0.68}N$, and (**d**) $Al_{0.17}Ga_{0.83}N$ are marked as black squares, while thermal conductivities from literature are shown as red circles. Solid lines depict the models for lattice thermal conductivities, describing the scattering processes in the materials (dotted-dashed red: Born-von Karman, solid green: Holland and dashed black: Callaway model).

Figure 2c shows the cross-plane thermal conductivity of the (423 ± 5) nm $Al_{0.32}Ga_{0.68}N$ (obtained from the measurements on Sample B). The previous results of κ_{AlN} (obtained from Sample A) served as an input parameter. The $Al_{0.32}Ga_{0.68}N$ layer had a thermal conductivity of (11.2 ± 0.7) $Wm^{-1}K^{-1}$ at 300 K. κ of $Al_{0.32}Ga_{0.68}N$ was one order of magnitude smaller than κ_{AlN}, due to the phonon scattering processes from defects/ alloying elements [33]. The thermal conductivity of the (65 ± 7) nm $Al_{0.17}Ga_{0.83}N$ of Sample D was (9.7 ± 0.6) $Wm^{-1}K^{-1}$ at 300 K and decreased to (7.0 ± 0.3) $Wm^{-1}K^{-1}$ at 773 K (see Figure 2d). In our case, the thermal conductivity of $Al_{0.17}Ga_{0.83}N$ was even lower than that of $Al_{0.32}Ga_{0.68}N$, explainable by their difference in layer thickness (enhanced boundary scattering). The temperature-dependence of the thermal conductivity of these alloys was much weaker than that obtained for κ_{AlN}. This behavior was in agreement with the results reported by Daly [34], who investigated $Al_{0.18}Ga_{0.82}N$, $Al_{0.20}Ga_{0.80}N$, and $Al_{0.44}Ga_{0.56}N$.

Figure 2c shows the thermal conductivity of the GaN film (κ_{GaN}) from Sample C. κ_{GaN} decreases from (220 ± 38) Wm^{-1}K^{-1} at 300 K to (51 ± 9) Wm^{-1}K^{-1} at 773 K. This value was similar to the findings reported by Shibata et al. [35] for a GaN film grown by a hydride-vapour phase epitaxy process (here, the MOCVD process).

This work aims to clarify the origin of the reduced thermal conductivity of all those layers, within the HEMT layer stack, by applying different phonon transport and boundary models (see below).

The thermal interface conductance between Pt and AlN (obtained from the TDTR measurements of Sample A) increased with temperature (red squares in Figure 3a). Above the Debye temperature of Pt (240 K [36]), this increase indicates a dominating inelastic scattering of phonons at the interface [37]. The temperature dependence of the thermal boundary conductance is also in line with results obtained from ab-initio molecular dynamics simulations [38], which showed an increase of the thermal conductivity with temperature due to the increase of available spatially localized modes at the interface. In the range of temperatures between 300 K and 523 K, the value increased from (269 ± 54) MWm^{-2}K^{-1} to (380 ± 76) MWm^{-2}K^{-1}. A similar change of the thermal interface conductance was previously reported by [37]. The difference between their results (red triangles in Figure 3a) and our data can be explained by differences in the interface itself, e.g., different levels of roughness, disorder or mixing. Regarding the thermal interface conductance between Pt and GaN of Sample C, it increased with temperature until reaching 623 K, similar to the thermal conductance of the Pt/AlN interface (Figure 3a—grey circles). Above that temperature, the thermal interface conductance showed a constant value of ~3.8 m^{2}K/GW. This constant value is in agreement with the analysis published by Hopkins et al. [39]. There it was reported that above the Debye temperature of both materials at the interface (Pt: 240 K and GaN: 655 K), the inelastic scattering became temperature-independent.

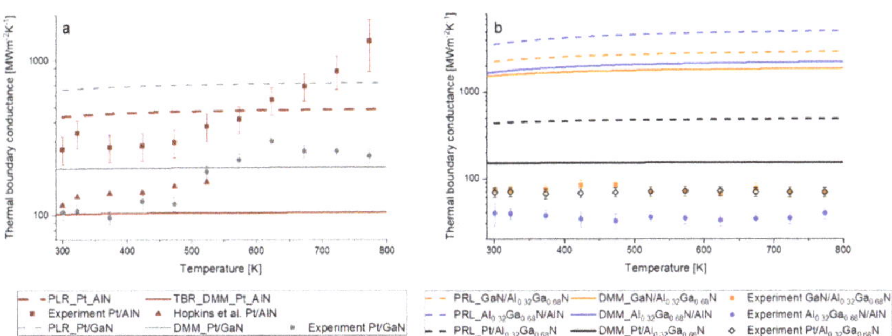

Figure 3. (a) Thermal boundary conductance across a Pt/AlN (red squares) and a Pt/GaN interface from 300 K to 773 K; together with the results for Pt/AlN boundary conductance published by Hopkins et al. [37]. (b) Thermal boundary conductance across the $Al_{0.32}Ga_{0.68}N$/GaN (orange squares), $Al_{0.32}Ga_{0.68}N$/AlN (blue circles) and Pt/$Al_{0.32}Ga_{0.68}N$ (black rhombi). The calculated thermal conductance is shown as solid lines according to the phonon radiation limit (PRL) model and as dashed lines according to the diffuse mismatch model (DMM).

Concerning the thermal interface conductance across Pt/$Al_{0.32}Ga_{0.68}N$ of Sample B (Figure 3b), it was an order of magnitude lower than that across Pt/AlN of Sample A. It did not feature a significant change with temperature in the measured range (300 K–773 K). Similar behaviour was obtained for the thermal interface conductance between $Al_{0.32}Ga_{0.68}N$/AlN and GaN/$Al_{0.32}Ga_{0.68}N$ (Figure 3b), which were comparable to the results of a GaSb/GaAs interface published in [40].

To visualise the relative contribution and the importance of the interface conductance in the HEMT layer stack, we visualised the normalized temperature profile from the GaN to the bottom of the AlN, for the system at 300 K and 773 K (Figure 4). The $Al_{0.32}Ga_{0.68}N$ showed the highest temperature gradient.

However, the temperature jump at $Al_{0.32}Ga_{0.68}N$/AlN interface is two-third of the temperature decrease in the $Al_{0.32}Ga_{0.68}N$ layer and the GaN/$Al_{0.32}Ga_{0.68}N$ one-third. It can also be seen that the temperature gradient in the GaN and AlN layer increases significantly from 300 K to 775 K. Note that we did not visualise the leftmost layer since its interface conductance cannot be resolved.

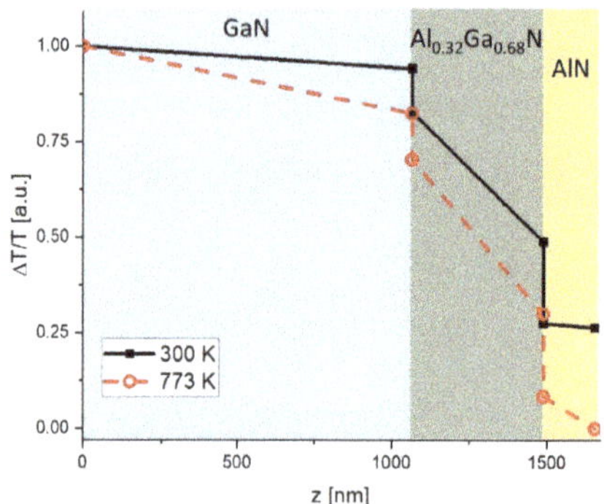

Figure 4. The normalized temperature profile of the HEMT structure: Starting with the GaN, which is underneath the heat source of the HEMT. The solid black line shows the temperature profile at 300 K, while the dashed red line at 773 K.

4. Modelling Methodology and Discussion

4.1. Thermal Interface Conductance

The thermal interface conductance from phonon scattering was modelled with the diffuse mismatch model (DMM—Equation (1)) [41] and the phonon radiation limit (PRL—Equation (2)) [42]. The DMM and the PRL models assume both elastic scattering of the phonons at the interface. Both models predict a constant thermal interface conductance in the high-temperature limit, at temperatures well above the Debye temperature. The DMM uses a transmission coefficient (a), which is a function of the phonon frequency (ω). a was computed from the group velocities and density of states (DOS) of both materials at the interface [42]. The behaviour in the high-temperature limit comes from the fact that in the DMM, the only temperature-dependent quantity is the phonon population on the hot side of the interface. In this model, the phonon population is given by the change of the Bose–Einstein distribution function with temperature ($\frac{\delta f(\omega,T)}{\delta T}$) multiplied by the DOS of the material in the hot side ($D_{hot,i}$). Thus we obtain:

$$h_{BD,DMM} = \frac{1}{4} \sum_i v_{hot,i} \int_0^{\omega_{hot,max,i}} a(\omega) \hbar \omega \, D_{hot,i}(\omega) \frac{\delta f(\omega,T)}{\delta T} \, d\omega, \tag{1}$$

where $\hbar$ is the Planck constant, $v_{hot,i}$ is the phonon velocity of the acoustic phonon mode (i) in the hot side material and $\omega_{hot,max,i}$ the corresponding cut-off frequency.

For the PRL model, the transmission coefficient is assumed to be one, and a cut-off frequency is given by the highest frequency of the material on the hot side. Besides that, the only temperature-dependent quantity in the PRL model is $\frac{\delta f(\omega,T)}{\delta T}$ multiplied by the DOS of the material on the cold ($D_{cold,i}$) side. This gives the equation:

$$h_{BD,PRL} = \frac{1}{4} \sum_i v_{cold,i} \int_0^{\omega_{hot,max,i}} \hbar \omega \, D_{cold,i}(\omega) \frac{\delta f(\omega,T)}{\delta T} \, d\omega. \tag{2}$$

The PRL always gives a higher value than the DMM, since it represents the upper limit of the elastic contribution to the thermal transport [43].

In the high-temperature limit: if the material on the hot side has a much lower Debye temperature than that on the cold side, the phonon population in the integrated frequency window (between zero and $\omega_{hot,max,i}$) does not change with temperature on both sides of the interface. Thus, the high-temperature limit of the interface conductance is always set by the material on the hot side, either by the DOS in the DMM or by introducing a cut-off in the PRL. If the Debye temperature of the material on the hot side is much higher than that of the cold side, the material from the hot side is still responsible for the temperature behaviour of the thermal interface conductance according to the DMM. In the PRL model, conversely, the material in the cold side is determining the temperature dependence of the thermal interface conductance.

The experimental thermal interface conductance of both Pt/AlN and Pt/GaN increased both with temperature (see Figure 3a). Conversely, as the Debye temperature of Pt is 240 K [36], the values calculated from the DMM and PRL do not show a significant increase with temperature. Hopkins et al. [37] also reported an increasing Pt/GaN interface conductance with temperature. They suggested that this temperature-dependent behaviour results from an inelastic phonon-scattering processes at the interface. This inelastic scattering affects the conductance, offering more channels for transport than the DMM and the PRL. In these models, the transmission coefficient is independent of temperature at high temperatures.

Concerning the thermal interface conductance involving AlGaN alloys (Figure 3b), both the DMM and PRL predicted a lower thermal interface conductance for the Pt/$Al_{0.32}Ga_{0.68}$N than for the $Al_{0.32}Ga_{0.68}$N/AlN and the GaN/$Al_{0.32}Ga_{0.68}$N interfaces. This is reasonable, as the Pt/$Al_{0.32}Ga_{0.68}$N has the highest mismatch in their sound velocities, which determines the transmission coefficient. For all interfaces, the PRL and the DMM overestimate the thermal interface conductance. These models, however, do not account for scattering mechanisms from defects at the interface, changes in the interatomic bonds near the interface or interfacial disorder (especially in well-matched materials) [44], that results in a reduction of the thermal conductance [43].

4.2. Models for Phonon Scattering

To provide insight into the thermal transport in a HEMT device, κ of the constituent materials were also modelled by using analytical scattering models for phonons: Callaway [45], Holland [46] and Born-von-Karman (BvK) [47,48]. All analytical models have as a starting point the formula for the lattice thermal conductivity (κ), having its origin in kinetic theory and being derived from the Peierls–Boltzmann transport equation. κ can be expressed as a function of the phonon frequency (ω) and the temperature (T), including the summation over the phonon modes (i):

$$\kappa = \frac{1}{3} \sum_i \int C(\omega,T) v_g^{\,2}(\omega)\, \tau_{eff}(\omega,T) d\omega, \tag{3}$$

where C is the volumetric specific heat capacity, v_g is the phonon group velocity and τ_{eff} is the total relaxation time. τ_{eff} depends on the phonon scattering mechanisms. $v_g = \delta\omega/\delta k$ and $C(\omega,T) = \hbar\omega D(\omega)\left(\frac{\delta f_{BE}}{\delta T}\right) = \left(\frac{\hbar\omega}{T}\right)^2 D(\omega)\, e^{\frac{\hbar\omega}{k_B T}} / \left[k_B\left(e^{\frac{\hbar\omega}{k_B T}} - 1\right)^2 \right]$ are determined by the phonon dispersion relation.

Here, k is the wavenumber, $\hbar$ is the Planck constant, k_B the Boltzmann constant, $D(\omega)$ the phonon DOS and f_{BE} is the Bose–Einstein distribution.

The Callaway model is based on a linear dispersion relation (Debye model), while the BvK model uses a sine-type dispersion, capturing the increased DOS near the Brillouin zone edge. Details about the analytical models used are given in Appendix A. The models only consider heat transport by acoustic phonons (i.e., the heat transport by optical phonons were neglected).

Following Matthiessens's rule, τ_{eff} can be expressed in terms of the relaxation times of different scattering processes as:

$$\tau_{eff}^{-1} = \tau_U^{-1} + \tau_{PD}^{-1} + \tau_B^{-1} \tag{4}$$

$\tau_U, \tau_{PD}, \tau_B$ are associated with Umklapp scattering [49], point-defect scattering [50], and the boundary scattering [51], respectively. These relaxation times have different dependencies on temperature and frequency [52].

$$\tau_U^{-1}(\omega,\ T) = \frac{2\ k_B \delta\ \gamma^2}{(6\pi^2)^{1/3}\ M v_p^2 v_g}\omega^2 T\ e^{-\frac{\theta}{bT}} \tag{5}$$

$$\tau_{PD}^{-1}(\omega,\ T) = \frac{\delta^3}{4\ \pi\ v_p^2 v_g}\omega^4\ \Gamma \tag{6}$$

$$\tau_B^{-1} = \frac{v_g}{\alpha L} \tag{7}$$

Here, γ denotes the Grüneisen parameter and δ the characteristic length scale of the lattice (i.e., cubic root of the atomic volume). The atomic volume was calculated by dividing the atomic molar mass (M_{Mol}) by the density. M is the average mass of an atom in the crystal (M_{Mol} divided by the number of atoms in the unit cell). b is a constant characteristic of the vibrational spectrum of the material, α is a specularity factor, v_P is the phonon phase velocity and L is the characteristic size of the material. In the case of the Debye model, v_g and v_p were both approximated as the speed of sound v_s.

In Equation (6), Γ denotes the phonon-scattering parameter. For a single element, Γ describes the scattering by point defects ($\Gamma = \Gamma_{imp}$), which is influenced by the doping density and the growth method [14]. For alloys, the scattering parameter also includes the scattering caused by alloying (Γ_{alloy}), $\Gamma = \Gamma_{imp} + \Gamma_{alloy}$ [33]. Γ_{alloy} is related to the difference in mass and strain-field (the lattice constants) between two constituents of an alloy [53].

The material parameters used in our calculations are summarized in Table 1; γ, v, n and ρ were obtained from ab-initio calculations, using ThermElpy [54] and Vienna Ab-initio Simulation Package (VASP) [55]. Here, γ is the high-temperature limit of the Grüneisen parameter, obtained by fitting Birch-Murnaghan equation of states and extracting the pressure derivative of the bulk modulus. For the AlGaN alloys, these values were extracted from an $Al_{0.5}Ga_{0.5}N$ alloy. The values M and ρ of the alloys were calculated by the relation $\sum_i f_i \times z$, where f_i is the mass fraction of the component and z stands for the parameter either M or ρ.

Table 1. Material parameters used for the theoretical models of the lattice thermal conductivity.

Materials	γ	v_L [m/s]	v_{T1} [m/s]	v_{T2} [m/s]	n [Å^{-3}]	M [kg]	ρ [kg/m^3]
AlN	1.77	10,751	6027	6406	0.0941	3.48×10^{-26}	3201
GaN	2.05	7538	4022	4566	0.0852	6.95×10^{-26}	5923
$Al_{0.32}Ga_{0.68}N$	1.91	8962	4783	5322	0.0916	5.82×10^{-26}	5136
$Al_{0.17}Ga_{0.83}N$	1.91	8962	4783	5322	0.0916	6.35×10^{-26}	5510

γ Grüneisen parameter; v_L longitudinal acoustic phonon velocities; v_T transverse acoustic phonon velocities; n number of atoms per volume in the unit cell; M average mass of an atom in the crystal; ρ density.

The determination of the thermal conductivity trend was limited by the temperature-dependence of the aforementioned scattering mechanisms. For the three models considered (Equations (5)–(7)), b, Γ and α were determined by fitting the modelled thermal conductivity to the experimental one (Figure 2). The values of the fitting parameters are listed in Table 2.

Table 2. Fitting parameters of several phonon scattering mechanisms used in the three models.

Materials	Callaway			Holland			BvK		
	b	Γ	α	b	Γ	α	b	Γ	α
AlN	3	3×10^{-4}	0.3	3	0.0021	0.87	3	8×10^{-5}	0.3
GaN	3	2×10^{-4}	1	1.6	0.0026	0.29	0.85	9×10^{-5}	1
$Al_{0.32}Ga_{0.68}N$	3	0.056	1	3	0.0988	0.19	3	0.4	1
$Al_{0.17}Ga_{0.83}N$	3	0.030	0.85	3	0.0854	0.11	3	0.2	0.96

The frequency dependence of the different scattering mechanisms (Equations (5)–(7)) is shown in Figure 5. As can be seen in Equations (5)–(7) only τ_U features a temperature dependence. To show the effect of temperature on τ_U and hence τ_{eff}, these values were plotted at 300 K and at 773 K (see Figure 5: solid vs. dashed lines). In all models τ_B dominates the phonon scattering at low ω (e.g., <3 THz at 773 K in the BvK model). At higher frequencies and high temperature (773 K) τ_U dictates the heat transport. For frequencies >10 THz, τ_{PD} is the main scattering mechanism for $Al_{0.32}Ga_{0.68}N$ (Figure 5b,f) in contrast to GaN (Figure 5a,e). This behaviour of the relaxation times explains why the thermal conductivity of $Al_{0.32}Ga_{0.68}N$ is less affected by temperature. In contrast to the Callaway and Holland models, which use the Debye dispersion, the BvK model shows a frequency-dependent τ_B. In the BvK model, τ_B increases with frequency and τ_U dominates at low frequencies in GaN and in $Al_{0.32}Ga_{0.68}N$ at 773 K. The Holland model offers insight into the scattering mechanisms of each phonon branch (Figure 5c,d). The two transverse phonon branches dominate the low-frequency range while the longitudinal phonon branch becomes dominant at higher frequencies (>5 THz).

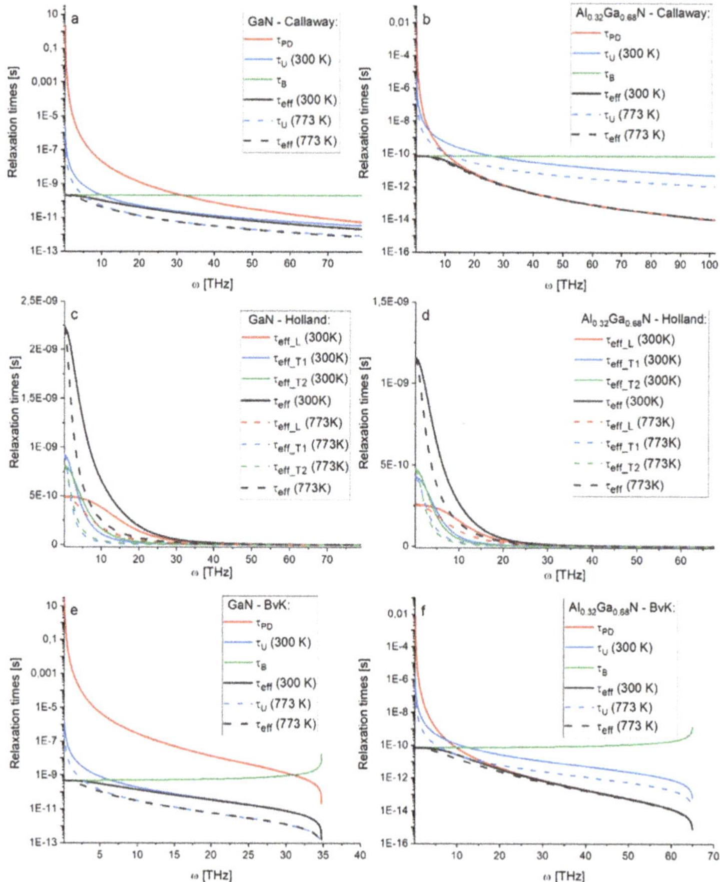

Figure 5. (**a,b**) Frequency dependence of the three different scattering mechanisms according to the Callaway model; (**c,d**) frequency dependence of the relaxation times for the longitudinal and two transverse phonon branches according to the Holland model; (**e,f**) frequency dependence of the three different scattering mechanisms of the BvK model. The left-hand side figures (**a,c,e**) are the relaxation times for GaN, and the right-hand side figures are the relaxation times for $Al_{0.32}Ga_{0.68}N$.

4.3. Models for the Size Effect in Cross-Plane Thermal Transport

We have tested three different approaches to predict the size effect on cross-plane transport given by layer thickness variations:

1. A suppression function-based approach as proposed by Hua and Minnich [19].
2. A model using simply the boundary scattering law as given in Equation (3) [56]. In the following it is referred to as the "simple model".
3. A model derived from the phonon hydrodynamic equations by Guo and Wang [21].

In our work, all three models were compared to the BvK model. In the first approach, we used the thickness-dependent thermal conductivity according to the Hua and Minnich suppression function approach [19], that uses a model analogous to the Fuchs–Sondheimer [57] expression for the thermal conductivity. Accordingly, the calculation of the thickness-dependent thermal conductivity can be written as:

$$\kappa(L) = \int_0^{\omega_{max}} \frac{1}{3} C(\omega)\, v_g\, \Lambda(\omega) S(\text{Kn}(\omega),\, L) d\omega, \tag{8}$$

where $\Lambda(\omega)$ is the bulk phonon mean free path and Kn is the Knudsen number. $S(\text{Kn}(\omega), L) = 1 + 3\text{Kn}(\omega)\cdot\left(E_5\left(\text{Kn}^{-1}(\omega)\right) - 0.25\right)$ is a layer thickness (L) dependent suppression function. $\Lambda(\omega) = v_g \tau_{eff_{fmat}}(\omega)$ was calculated by using the aforementioned fitted models. $\tau_{eff_{fmat}}$ is similar to τ_{eff} in Equation (4) without τ_B. By using the fit parameters in Table 2, $\tau_{eff_{fmat}}$ of all three transport models ((Callaway, Holland and BvK)) was calculated and used in Equation (8). Figure 6 shows the thickness-dependent thermal conductivity for the four materials at 300 K. The Holland model for AlN and GaN predicted lower thermal conductivities at a small thickness ($<10^{-4}$ m) compared with the other models. The size-dependence extracted from the Holland model did not agree well with the experimental values; it predicts thermal conductivity for AlN and GaN that is too low. The thermal conductivity of AlN and GaN showed strong thickness dependence. The BvK model showed the highest bulk thermal conductivity while the Callaway model the smallest. E.g., The 167 nm thick AlN showed a reduction by 41% for the Callaway model and 36% for the BvK model compared the corresponding bulk value. The $\kappa(L)$ of GaN at 1065 nm resulted in 78% of the corresponding bulk value for the Callaway model and 52% for the BvK model, which was also reported by [58]. The values predicted by the BvK model of $\kappa(L)$ for GaN (green dashed-dotted line Figure 6b) were in agreement with the results reported in [59]. Comparing the three transport models for the alloys, there was a difference of 10% for $Al_{0.32}Ga_{0.68}N$ and 18% for $Al_{0.17}Ga_{0.83}N$ in their size-dependencies. For example, the thermal conductivity of the 65 nm thick $Al_{0.17}Ga_{0.83}N$ showed 9% of the bulk thermal conductivity in the BvK model (Figure 6d).

The second approach used to compute the size effect on the cross-plane thermal conductivity takes into account boundary scattering by using the thickness-dependent boundary relaxation time (Equation (7)) to calculate the total relaxation time. The thermal conductivity is computed according to Equation (3) [56]. The results of this model are shown as dashed lines in Figure 7.

The third approach to calculate size effects uses the formula derived from Guo and Wang [21]:

$$\frac{\kappa(L,\omega)}{\kappa_{bulk}(\omega)} = \left(1 + \frac{(1+s)}{(1-s)} \frac{4}{3} \text{Kn}_L(\omega)\right)^{-1} \tag{9}$$

where $\kappa_{bulk}(\omega)$ is the spectral thermal conductivity as given in Equation (3), without integrating over ω. The Knudsen number is given as $\text{Kn}_L(\omega) = \Lambda(\omega)/L$, where $\Lambda(\omega)$ is the mean free path spectrum. The term $(1 + s)/(1 - s)$ accounts for the specularity factor $1/\alpha$ (see Table 2). For instance, for perfectly diffuse boundary scattering $s = 0$ and $\alpha = 1$. For boundaries with specular scattering contribution $s > 0$, approaching 1 for fully specular reflection and $\alpha < 1$, approaching 0, respectively. Thus, the effective thermal conductivity becomes smaller with increasing specular scattering at the boundary for the cross-plane case. Note, this is the opposite of the in-plane case. A detailed discussion of the cross-plane and in-plane effective thermal conductivity is given by Guo and Wang [21]. The effective cross-plane thermal conductivity was calculated by integrating over ω, analogous to Equation (8), by inserting Equation (9) instead of $S(\text{Kn}(\omega), L)$ (see dotted lines in Figure 7).

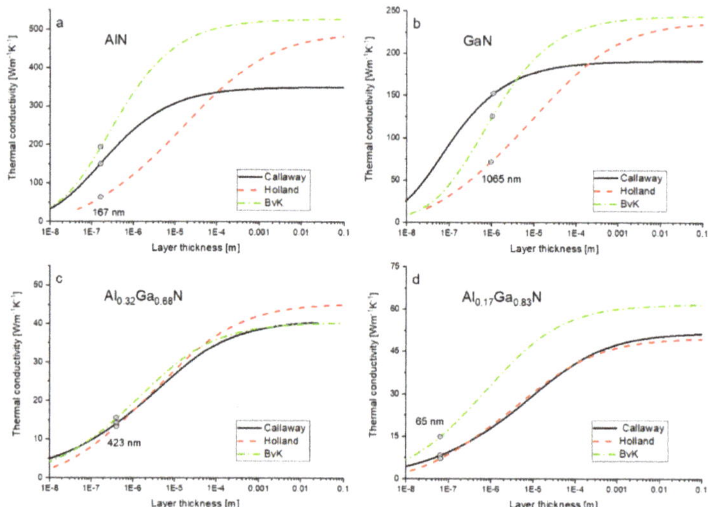

Figure 6. Thickness dependent thermal conductivity as calculated using the Hua–Minnich suppression function approach for (**a**) AlN, (**b**) $Al_{0.32}Ga_{0.68}N$, (**c**) GaN, and (**d**) $Al_{0.17}Ga_{0.83}N$. Calculations were performed at 300 K determined by using Equation (8) with the relaxation times determined by the Callaway, Holland and BvK models.

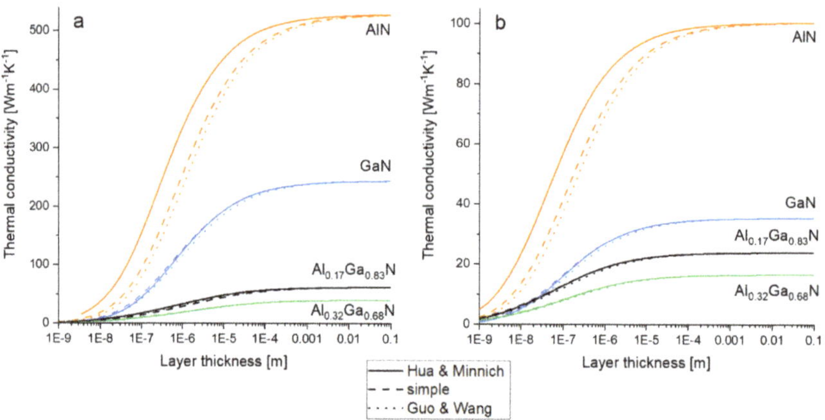

Figure 7. The size-dependent thermal conductivity for AlN (orange), GaN (blue), $Al_{0.32}Ga_{0.68}N$ (green) and $Al_{0.17}Ga_{0.83}N$ (black) at (**a**) 300 K and (**b**) 773 K. The size-dependent thermal conductivity evaluated within the framework of the BvK transport model. The Hua–Minnich approach is drawn as solid lines, the simple boundary scattering approach as dashed lines, and the approach of Guo and Wang as dotted lines.

Figure 7 shows the size-dependent thermal conductivity of the three models at (a) 300 K and (b) 773 K. Comparing the size effect at 300 K and 773 K, the convergence to the bulk thermal conductivity (e.g., 90% of $\kappa_{GaN}(L = \infty)$) happens at smaller thicknesses at 773 K. At higher temperatures, the phonon-phonon scattering increases, and hence the mean free path of the phonons contributing to the thermal conductivity is shorter for 773 K than for 300 K.

All three models for predicting the size effect agree very well when the specularity parameter α is close to one or equal to one (see Table 2). In the AlN, where α is 0.3, the simple model and the model of Gua and Wang showed similar results. For AlN, both models predict a lower thermal conductivity than the Hua and Minnich model for layer thicknesses <0.001 m, the convergence of $\kappa(L)$ is also broader.

5. Conclusions

The constituent materials of an AlGaN-GaN-based HEMT, and the interfaces between them were investigated by a TDTR measurement setup. The measured thermal interface conductance was compared to the values calculated in the framework of the DMM and the PRL models. In contrast to the elastic DMM and PRL models, the measured interface conductance showed a stronger increase with temperature. This observation suggests a contribution of inelastic scattering processes. The Pt/AlN thermal interface conductance increased with temperature from (269 ± 54) MW/m^{2}K at 300 K to (1358 ± 503) MW/m^{2}K at 773 K. The Pt/GaN thermal interface conductance increased from (105 ± 11) MW/m^{2}K at 300 K until reaching a plateau of $3 \cdot 10^8$ W/m^{2}KW at 600 K. The thermal interface conductance with Al$_{0.32}$Ga$_{0.68}$N showed no significant temperature-dependence.

The thermal conductivities of the (167 ± 7) nm AlN and (1065 ± 7) nm GaN at 300 K ($\kappa_{AlN} = (130 \pm 38)$ Wm^{-1}K^{-1} and $\kappa_{GaN} = (220 \pm 38)$ Wm^{-1}K^{-1}) were found to be an order of magnitude higher than that of the alloy. The κ of (423 ± 5) nm Al$_{0.32}$Ga$_{0.68}$N layer was found to be (11.2 ± 0.7) Wm^{-1}K^{-1} and that of the of (65 ± 5) nm Al$_{0.17}$Ga$_{0.83}$N layer was (9.7 ± 0.6) Wm^{-1}K^{-1}. As expected, their thermal conductivities decreased with increasing temperature in the range of 300 to 773 K. From the analysis of the relaxation times, it has been shown that besides boundary scattering, the Umklapp scattering was dominant for the thermal conductivity in pure AlN and pure GaN. Point-defect scattering, conversely, dictated the thermal conductivity of the alloys.

The investigated material layers with nanometer to micrometre thicknesses generally showed a reduced thermal conductivity relative to the bulk. The size effect for the cross-plane phonon transport was calculated by using three different models: the suppression function-based approach, the approach taking into account the boundary scattering, and the approach based on phonon hydrodynamic equations. The last two models include specular phonon scattering effects at the boundary. In contrast, the suppression function-based approach assumes perfectly diffusive scattering. To apply this for the AlGaN-GaN-based HEMT layer stack can be content of future work. We observed an agreement between the three boundary models, in the prediction of size-dependent cross-plane thermal conductivity (see Figure 7). The uncertainty, which is introduced by the transport model (Callaway, Holland, BvK) was much higher than for the boundary model (compare Figures 6 and 7).

The categorical low thermal conductivity of the AlGaN layers in the HEMT structure hinders the heat dissipation from the junction to the substrate and can lead to hot spots. This increases the importance of the GaN-layer as a heat spreader for efficient heat transport from the junction to the substrate. Our work highlights the role of interfaces and size effects; these features lessen the thermal conductance in HEMTs. It also designates that a significant contribution of the phonons that transport heat in GaN have long MFPs in the range of the involved layer thicknesses. This fact should be considered for a better exploitation of the intrinsic thermal conductivity of the materials used in microelectronic devices as HEMTs. Overall, the reported thermal transport results can be used to evaluate self-heating effects in AlGaN-GaN HEMT heterostructure and might serve as a guide for advanced optimization, taking thermal considerations into account.

Author Contributions: Conceptualization, L.M. and R.H.; methodology, L.M.; ab intio calculations, T.D.; formal analysis, R.H. and J.S.; investigation, L.M.; writing—original draft preparation, L.M.; writing—review and editing R.H. and J.S.; project administration, J.S.; All authors have read and agreed to the published version of the manuscript.

Funding: The authors gratefully acknowledge the financial support under the scope of the COMET program within the K2 Center "Integrated Computational Material, Process and Product Engineering (IC-MPPE)" (Project No 859480). This program is supported by the Austrian Federal Ministries for Climate Action, Environment, Energy, Mobility, Innovation and Technology (BMK) and for Digital and Economic Affairs (BMDW), represented by the Austrian research funding association (FFG), and the federal states of Styria, Upper Austria and Tyrol.

Acknowledgments: The authors would like to thank Natalia Bedoya-Martínez for her help.

Conflicts of Interest: The authors declare no conflict of interest.

Appendix A

In the following section, three different analytical phonon transport models are summarized in a detailed way: the Callaway, Holland and BvK models. The models were used to provide insight into the temperature-dependent thermal conductivity. The Callaway model is based on the Debye model that uses a dispersion relation for a single, degenerated phonon branch [45]. The Debye model approximates the dispersion relation as linear, and hence the frequency-dependent phonon group velocity as constant value:

$$\omega = v_s q \tag{A1}$$

where q is the phonon wave vector and v_s is the sound velocity. Here, the v_g and v_p are assumed to be v_s. For v_s the average value over all acoustic phonon branches (longitudinal (v_L) and the two transverse (v_T)), $v_s = 3/\left(v_L^{-1} + v_{T1}^{-1} + v_{T2}^{-1}\right)$ is taken. Note, this simplification, due to the Debye dispersion, causes an overestimation of the group velocity especially of high-frequency phonons in the Debye model $D(\omega) = \frac{3\,\omega^2}{2\pi^2 v_s^3}$. Substituting ω with $x = \frac{\hbar\omega}{k_B T}$, $C(x)$ can be written as:

$$C(x) = \frac{3\,k_B^3\,T^2}{2\pi^2\hbar^2 v_s^3}\,\frac{x^4 e^x}{(e^x - 1)^2} \tag{A2}$$

Here, $\hbar$ denotes the Plank constant, and k_B the Boltzmann constant. By inserting $C(x)$ in Equation (3), the lattice thermal conductivity can be calculated with the following equation:

$$\kappa_{Callaway} = \frac{\kappa_B^4 T^3}{2\pi^2\hbar^3 v_s}\left(\int_0^{\frac{\theta_D}{T}} \tau_c \frac{x^4 e^x}{e^x - 1}dx + \frac{\int_0^{\frac{\theta_D}{T}} \frac{\tau_c}{\tau_N}\frac{x^4 e^x}{(e^x-1)^2}dx}{\int_0^{\frac{\theta_{D,pol}}{T}} \frac{\tau_c}{\tau_N \tau_R}\frac{x^4 e^x}{(e^x-1)^2}dx} \right) \tag{A3}$$

For this model τ_{eff} is the combined scattering relaxation time (τ_C), where $\tau_C^{-1} = \tau_N^{-1} + \tau_R^{-1}$ and $\tau_R^{-1} = \tau_U^{-1} + \tau_{PD}^{-1} + \tau_B^{-1}$. By assuming $\tau_N \gg \tau_R$, $\tau_R \approx \tau_C$ and Equation (A3) reduces to (A4).

$$\kappa_{Callaway} = \frac{k_B^4 T^3}{2\pi^2\hbar^3 v_s} \int_0^{\theta_D/T} \tau_c(x)\,\frac{x^4 e^x}{(e^x - 1)^2}\,dx \tag{A4}$$

where θ_D is the Debye temperature of the material under investigation. The Debye temperature was calculated according to $\theta_D = \frac{\hbar}{k_B}\left(6\pi^2 n\right)^{1/3} v_s$, where n number of atoms per volume in the unit cell.

The second model is the Holland model, which is also based on the Debye model, but taking into account the two types of polarization. The calculation of the thermal conductivity is separated into the contributions of the longitudinal (κ_L) and the two transverse (κ_T) phonon branches. The thermal conductivity (κ_i) of each phonon mode is based on the Equation (A4). For each polarization the dispersion relation is assumed to be linear, so each phonon branch has its phonon velocity of v_L, v_{T1} and v_{T2}.

$$\kappa_{Holland} = \kappa_L + \kappa_{T1} + \kappa_{T2} = \sum_i \kappa_i. \tag{A5}$$

The third model is the BvK model, where the dispersion relation is:

$$\omega = \omega_{max} \sin\left(\frac{\pi q}{2 q_{max}}\right). \tag{A6}$$

Here, the cut-off wave vector $q_{max} = \omega_D/v_s$ and the cut-off frequency $\omega_{max} = 2\omega_D/\pi$ are the same as in the Debye model using $\omega_D = \theta_D k_B/\hbar = \left(6\pi^2 n\right)^{1/3} v_s$. Hence, the phonon group velocity $v(\omega)$ is the same as is used for the models as mentioned above based on the Debye model at low ω. The corresponding temperature can be expressed as $\theta_{max} = \hbar\omega_{max}/k_B$. However, the more realistic dispersion reduces the $v(\omega)$ for high-frequency phonons. The phonon DOS is $D(\omega) = \frac{3}{2\pi^2}\frac{\omega^2}{v_g v_p^2}$, where v_g ($v_g = d\omega/dq$) is the phonon group velocity and v_P ($v_P = \Delta\omega/\Delta q$) is the phonon phase velocity. In the BvK dispersion $v_g = v_s \sqrt{1 - (\omega/\omega_{max})^2} = v_s \sqrt{1 - (Tx/\theta_{max})^2}$ and $v_P = \frac{\omega}{\frac{2}{\pi} q_{max} \sin^{-1}(\omega/\omega_{max})} = \frac{x}{\frac{2}{\pi} q_{max} \sin^{-1}(Tx/\theta_{max})}\frac{k_B T}{\hbar}$, v_s is the average value over all acoustic phonon branches. The spectral heat capacity can be written as:

$$C(\omega) = \frac{6\,\hbar^2 q_{max}^2}{\pi^4 k_B T^2 v_s} \frac{\omega^2 e^{\frac{\hbar\omega}{k_B T}} \left(\sin^{-1}(\omega/\omega_{max})\right)^2}{\left(e^{\frac{\hbar\omega}{k_B T}} - 1\right)^2 \sqrt{1 - (\omega/\omega_{max})^2}} \tag{A7}$$

and

$$C(x) = \frac{6\, q_{max}^2 k_B}{\pi^4 v_s} \frac{x^2 e^x \left(\sin^{-1}(Tx/\theta_{max})\right)^2}{(e^x - 1)^2 \sqrt{1 - (Tx/\theta_{max})^2}} \tag{A8}$$

The thermal conductivity can be calculated as:

$$\kappa_{BvK} = \int_0^{\frac{\theta_{max}}{T}} \left(\frac{k_B T}{\hbar}\right) C(x)\, v_g^2(x)\, \tau_C(x)\, dx. \tag{A9}$$

References

1. Baskaran, S.; Mohanbabu, A.; Anbuselvan, N.; Mohankumar, N.; Godwinraj, D.; Sarkar, C.K. Modeling of 2DEG sheet carrier density and DC characteristics in spacer based AlGaN/AlN/GaN HEMT devices. *Superlattices Microstruct.* **2013**, *64*, 470–482. [CrossRef]
2. Mishra, U.K.; Shen, L.; Kazior, T.E.; Wu, Y.-F. GaN-based RF power devices and amplifiers. *Proc. IEEE* **2008**, *96*, 287–305. [CrossRef]
3. Fitch, R.C.; Walker, D.E.; Green, A.J.; Tetlak, S.E.; Gillespie, J.K.; Gilbert, R.D.; Sutherlin, K.A.; Gouty, W.D.; Theimer, J.P.; Via, G.D.; et al. Implementation of high-power-density X-band AlGaN/GaN high electron mobility transistors in a millimeter-wave monolithic microwave integrated circuit process. *IEEE Electron Device Lett.* **2015**, *36*, 1004–1007. [CrossRef]
4. Luo, X.; Halder, S.; Curtice, W.R.; Hwang, J.C.M.; Chabak, K.D.; Dennis, E.; Dabiran, A.M. Scaling and high-frequency performance of AlN/GaN HEMTs. In Proceedings of the IEEE International Symposium on Radio-Frequency Integration Technology, Beijing, China, 30 November–2 December 2011. [CrossRef]
5. Vallabhaneni, A.K.; Chen, L.; Gupta, M.P.; Kumar, S. Solving nongray boltzmann transport equation in gallium nitride. *J. Heat Transf.* **2017**, *139*. [CrossRef]
6. Yalamarthy, A.S.; So, H.; Muñoz Rojo, M.; Suria, A.J.; Xu, X.; Pop, E.; Senesky, D.G. Tuning electrical and thermal transport in AlGaN/GaN heterostructures via buffer layer engineering. *Adv. Funct. Mater.* **2018**, *28*. [CrossRef]

7. Hao, Q.; Zhao, H.; Xiao, Y.; Kronenfeld, M.B. Electrothermal studies of GaN-based high electron mobility transistors with improved thermal designs. *Int. J. Heat Mass Transf.* **2018**, *116*, 496–506. [CrossRef]

8. Zhang, G.-C.; Feng, S.-W.; Zhou, Z.; Li, J.-W.; Guo, C.-S. Evaluation of thermal resistance constitution for packaged AlGaN/GaN high electron mobility transistors by structure function method. *Chin. Phys. B* **2011**, *20*, 027202. [CrossRef]

9. Sarua, A.; Ji, H.; Hilton, K.P.; Wallis, D.J.; Uren, M.J.; Martin, T.; Kuball, M. Thermal boundary resistance between GaN and substrate in AlGaN/GaN electronic devices. *IEEE Trans. Electron Devices* **2007**, *54*, 3152–3158. [CrossRef]

10. Cho, J.; Bozorg-Grayeli, E.; Altman, D.H.; Asheghi, M.; Goodson, K.E. Low thermal resistances at GaN-SiC interfaces for HEMT technology. *IEEE Electron Device Lett.* **2012**, *33*, 378–380. [CrossRef]

11. Freedman, J.P.; Leach, J.H.; Preble, E.A.; Sitar, Z.; Davis, R.F.; Malen, J.A. Universal phonon mean free path spectra in crystalline semiconductors at high temperature. *Sci. Rep.* **2013**, *3*. [CrossRef]

12. Cheng, Z.; Koh, Y.R.; Mamun, A.; Shi, J.; Bai, T.; Huynh, K.; Yates, L.; Liu, Z.; Li, R.; Lee, E.; et al. Experimental Observation of High Intrinsic Thermal Conductivity of AlN. Available online: https://arxiv.org/ftp/arxiv/papers/1911/1911.01595.pdf (accessed on 10 April 2020).

13. Park, K.; Bayram, C. Impact of dislocations on the thermal conductivity of gallium nitride studied by time-domain thermoreflectance. *J. Appl. Phys.* **2019**, *126*. [CrossRef]

14. Liu, W.; Balandin, A.A. Thermal conduction in AlxGa1-xN alloys and thin films. *J. Appl. Phys.* **2005**, *97*, 1–6. [CrossRef]

15. Chatterjee, B.; Dundar, C.; Beechem, T.E.; Heller, E.; Kendig, D.; Kim, H.; Donmezer, N.; Choi, S. Nanoscale electro-thermal interactions in AlGaN/GaN high electron mobility transistors. *J. Appl. Phys.* **2020**, *127*. [CrossRef]

16. Lundh, J.S.; Chatterjee, B.; Song, Y.; Baca, A.G.; Kaplar, R.J.; Beechem, T.E.; Allerman, A.A.; Armstrong, A.M.; Klein, B.A.; Bansal, A.; et al. Multidimensional thermal analysis of an ultrawide bandgap AlGaN channel high electron mobility transistor. *Appl. Phys. Lett.* **2019**, *115*. [CrossRef]

17. Yang, F.; Dames, C. Mean free path spectra as a tool to understand thermal conductivity in bulk and nanostructures. *Phys. Rev. B Condens. Matter Mater. Phys.* **2013**, *87*, 1–12. [CrossRef]

18. Regner, K.T.; Wei, L.C.; Malen, J.A. Interpretation of thermoreflectance measurements with a two-temperature model including non-surface heat deposition. *J. Appl. Phys.* **2015**, *118*. [CrossRef]

19. Hua, C.; Minnich, A.J. Semi-analytical solution to the frequency-dependent Boltzmann transport equation for cross-plane heat conduction in thin films. *J. Appl. Phys.* **2015**, *117*. [CrossRef]

20. Regner, K.T.; Freedman, J.P.; Malen, J.A. Advances in studying phonon mean free path dependent contributions to thermal conductivity. *Nanoscale Microscale Thermophys. Eng.* **2015**, *19*, 183–205. [CrossRef]

21. Guo, Y.; Wang, M. Phonon hydrodynamics for nanoscale heat transport at ordinary temperatures. *Phys. Rev. B* **2018**, *97*, 1–27. [CrossRef]

22. Reisinger, M.; Tomberger, M.; Zechner, J.; Daumiller, I.; Sartory, B.; Ecker, W.; Keckes, J.; Lechner, R.T.T. Resolving alternating stress gradients and dislocation densities across AlxGa1-xN multilayer structures on Si(111). *Appl. Phys. Lett.* **2017**, *111*. [CrossRef]

23. Netzsch (Ed.) Thermoreflectance by Pulsed Light Heating Thermoreflectance–NanoTR/PicoTR. Available online: https://www.netzsch-thermal-analysis.com/de/produkte-loesungen/waerme-und-temperatur-leitfaehigkeitsbestimmung/nanotrpicotr/ (accessed on 4 May 2020).

24. Fladischer, K.; Leitgeb, V.; Mitterhuber, L.; Keckes, J.; Sagmeister, M.; Carniello, S. Combined thermo-physical investigations of thin layers with Time Domain Thermoreflectance and Scanning Thermal Microscopy on the example of 500 nm thin, CVD grown tungsten. *Thermochim. Acta* **2019**. [CrossRef]

25. Cahill, D.G. Analysis of heat flow in layered structures for time-domain thermoreflectance. *Rev. Sci. Instrum.* **2004**, *75*, 5119–5122. [CrossRef]

26. Yang, J.; Ziade, E.; Schmidt, A.J. Uncertainty analysis of thermoreflectance measurements. *Rev. Sci. Instrum.* **2016**, *87*, 1–11. [CrossRef] [PubMed]

27. Glassbrenner, C.J.; Slack, G.A. Thermal conductivity of silicon and germanium from 3°K to the melting point. *Phys. Rev.* **1964**, *134*, A1058. [CrossRef]

28. Corruccini, R.J.; Gniewek, J.J. *Specific Heats and Enthalpies of Technical Solids at Low Temperatures*; National Bureau of Standards: Washington, DC, USA, 1960. [CrossRef]

29. Palankovski, V.; Quay, R. *Analysis and Simulation of Heterostructure Devices*; Springer: Wien, Austria, 2004. [CrossRef]

30. Danilchenko, B.A.; Paszkiewicz, T.; Wolski, S.; Jezowski, A.; Plackowski, T. Heat capacity and phonon mean free path of wurtzite GaN. *Appl. Phys. Lett.* **2006**, *89*. [CrossRef]

31. Mitterhuber, L.; Hammer, R.; Dengg, T.; Fladischer, K.; Spitaler, J. Thermal investigation of AlGaN-GaN multilayer structures. In Proceedings of the THERMINIC 2019-25th International Workshop on Thermal Investigations of ICs and Systems, Lecco, Italy, 25–27 September 2019. [CrossRef]

32. Slack, G.A.; Tanzilli, R.A.; Pohl, R.O.; Vandersande, J.W. The intrinsic thermal conductivity of AIN. *J. Phys. Chem. Solids* **1987**, *48*, 641–647. [CrossRef]

33. Abeles, B. Lattice thermal conductivity of disordered semiconductor alloys at high temperatures. *Phys. Rev.* **1963**, *131*, 1906–1911. [CrossRef]

34. Daly, B.C.; Maris, H.J.; Nurmikko, A.V.; Kuball, M.; Han, J. Optical pump-and-probe measurement of the thermal conductivity of nitride thin films. *J. Appl. Phys.* **2002**, *92*, 3820–3824. [CrossRef]

35. Shibata, H.; Waseda, Y.; Ohta, H.; Kiyomi, K.; Shimoyama, K.; Fujito, K.; Nagaoka, H.; Kagamitani, Y.; Simura, R.; Fukuda, T. High thermal conductivity of gallium nitride (GaN) crystals grown by HVPE Process. *Mater. Trans.* **2007**, *48*, 2782–2786. [CrossRef]

36. Berg, W.T. The low temperature heat capacity of platinum. *J. Phys. Chem. Solids* **1969**, *30*, 69–72. [CrossRef]

37. Hopkins, P.E.; Norris, P.M.; Stevens, R.J. Influence of inelastic scattering at metal-dielectric interfaces. *J. Heat Transf.* **2008**, *130*, 022401. [CrossRef]

38. Stanley, C.M.; Estreicher, S.K. Phonon dynamics at an oxide layer in silicon: Heat flow and Kapitza resistance. *Phys. Status Solidi Appl. Mater. Sci.* **2019**, *216*, 1–9. [CrossRef]

39. Hopkins, P.E.; Norris, P.M. Relative Contributions of inelastic and elastic diffuse phonon scattering to thermal boundary conductance across solid interfaces. *J. Heat Transf.* **2009**, *131*, 22402. [CrossRef]

40. Hopkins, P.E.; Duda, J.C.; Clark, S.P.; Hains, C.P.; Rotter, T.J.; Phinney, L.M.; Balakrishnan, G. Effect of dislocation density on thermal boundary conductance across GaSb/GaAs interfaces. *Appl. Phys. Lett.* **2011**, *98*. [CrossRef]

41. Swartz, E.T.; Pohl, R.O. Thermal boundary resistance. *Rev. Mod. Phys.* **1989**, *61*, 605–668. [CrossRef]

42. Stoner, R.J.; Maris, H.J. Kapitza conductance and heat-Flow between solids at temperatures from 50 to 300K. *Phys. Rev. B* **1993**, *48*, 16373–16387. [CrossRef]

43. Norris, P.M.; Hopkins, P.E. Examining interfacial diffuse phonon scattering through transient thermoreflectance measurements of thermal boundary conductance. *J. Heat Transf.* **2009**, *131*, 043207. [CrossRef]

44. Keune, W.; Hong, S.; Hu, M.Y.; Zhao, J.; Toellner, T.S.; Alp, E.E.; Sturhahn, W.; Rahman, T.S.; Roldan Cuenya, B. Influence of interfaces on the phonon density of states of nanoscale metallic multilayers: Phonon confinement and localization. *Phys. Rev. B* **2018**, *98*, 1–16. [CrossRef]

45. Callaway, J. Model for lattice thermal conductivity at low temperatures. *Phys. Rev.* **1959**, *113*, 1046–1051. [CrossRef]

46. Morelli, D.T.; Heremans, J.P.; Slack, G.A. Estimation of the isotope effect on the lattice thermal conductivity of group IV and group III-V semiconductors. *Phys. Rev. B Condens. Matter Mater. Phys.* **2002**, *66*, 1953041–1953049. [CrossRef]

47. Jugdersuren, B.; Kearney, B.T.; Queen, D.R.; Metcalf, T.H.; Culbertson, J.C.; Chervin, C.N.; Stroud, R.M.; Nemeth, W.; Wang, Q.; Liu, X. Thermal conductivity of amorphous and nanocrystalline silicon films prepared by hot-wire chemical-vapor deposition. *Phys. Rev. B* **2017**, *96*, 1–8. [CrossRef]

48. Chen, G. Size and interface effects on thermal conductivity of superlattices and periodic thin-film structures. *J. Heat Transf.* **1997**, *119*, 220–229. [CrossRef]

49. Peierls, R. Zur kinetischen Theorie der Wärmeleitung in Kristallen. *Ann. Phys.* **1929**, *395*, 1055–1101. [CrossRef]

50. Klemens, P.G.; White, G.K.; Tainsh, R.J. Scattering of lattice waves by point defects. *Philos. Mag.* **1962**, *7*, 1323–1335. [CrossRef]

51. Wang, Z.; Alaniz, J.E.; Jang, W.; Garay, J.E.; Dames, C. Thermal conductivity of nanocrystalline silicon: Importance of grain size and frequency-dependent mean free paths. *Nano Lett.* **2011**, *11*, 2206–2213. [CrossRef]

52. Toberer, E.S.; Zevalkink, A.; Snyder, G.J. Phonon engineering through crystal chemistry. *J. Mater. Chem.* **2011**, *21*, 15843–15852. [CrossRef]

53. Alekseeva, G.T.; Efimova, B.A.; Ostrovskaya, L.M.; Serebryannikova, O.S.; Tsypin, M.I. Thermal conductivity of solid solutions based on lead telluride. *Sov. Phys. Semicond.* **1971**, *4*, 1122–1125.

54. Dengg, T.; Razumovskiy, V.; Romaner, L.; Kresse, G.; Puschnig, P.; Spitaler, J. Thermal expansion coefficient of WRe alloys from first principles. *Phys. Rev. B* **2017**, *96*. [CrossRef]

55. Kresse, G.; Furthmüller, J. Efficient iterative schemes for ab initio total-energy calculations using a plane-wave basis set. *Phys. Rev. B Condens. Matter Mater. Phys.* **1996**, *54*, 11169–11186. [CrossRef]

56. Fan, Z.; Dong, H.; Harju, A.; Ala-Nissila, T. Homogeneous nonequilibrium molecular dynamics method for heat transport and spectral decomposition with many-body potentials. *Phys. Rev. B* **2019**, *99*, 1–9. [CrossRef]

57. Sondheimer, E.H. The mean free path of electrons in metals. *Adv. Phys.* **1952**, *1*, 1–42. [CrossRef]

58. Beechem, T.E.; McDonald, A.E.; Fuller, E.J.; Talin, A.A.; Rost, C.M.; Maria, J.P.; Gaskins, J.T.; Hopkins, P.E.; Allerman, A.A. Size dictated thermal conductivity of GaN. *J. Appl. Phys.* **2016**, *120*. [CrossRef]

59. Hodges, C.; Anaya Calvo, J.; Stoffels, S.; Marcon, D.; Kuball, M. AlGaN/GaN field effect transistors for power electronics - Effect of finite GaN layer thickness on thermal characteristics. *Appl. Phys. Lett.* **2013**, *103*. [CrossRef]

Article

On the Reproducibility of Thermal Measurements and of Related Thermal Metrics in Static and Transient Tests of Power Devices

Gabor Farkas [1], Dirk Schweitzer [2], Zoltan Sarkany [3] and Marta Rencz [3,*]

[1]   Mentor, a Siemens Business, 1117 Budapest, Hungary; gabor_farkas@mentor.com
[2]   Infineon Technologies AG, 85579 Neubiberg, Germany; dirk.schweitzer@infineon.com
[3]   Department of Electron Devices, Budapest University of Technology and Economics, 1117 Budapest, Hungary; zoltan_sarkany@mentor.com
*   Correspondence: rencz@eet.bme.hu

Received: 5 November 2019; Accepted: 13 January 2020; Published: 23 January 2020

Abstract: Traditionally the thermal behavior of power devices is characterized by temperature measurements at the junction and at accessible external points. In large modules composed of thin chips and materials of high thermal conductivity the shape and distribution of the heat trajectories are influenced by the external boundary represented by the cooling mount. This causes mediocre repeatability of the characteristic R_{thJC} junction to case thermal resistance even in measurements at the same laboratory and causes very poor reproducibility among sites using dissimilar instrumentation. The Transient Dual Interface Methodology (TDIM) is based on the comparison of measured structure functions. With this method high repeatability can be achieved although introducing severe changes into the measurement environment is the essence of this test scheme. There is a systematic difference between thermal data measured with TDIM method and that measured with temperature probes, but we found that this difference was smaller than the scatter of the latter method. For checking production stability, we propose the use of a structure function-based $R_{th@Cth}$ thermal metric, which is the thermal resistance value reached at the thermal capacitance belonging to the mass of the package base. This metric condenses the consistency of internal structural elements into a single number.

Keywords: thermal transient testing; non-destructive testing; thermal testability; accuracy repeatability and reproducibility of thermal measurements; thermal testing standards

---

1. Introduction

The thermal characterization of power devices and assemblies has become more and more important with the growing level of power density. The related measurements may serve different purposes; they can be used in providing data sheet values, for calibrating thermal models of packaged devices, etc.

In *static* tests, steady temperature values are measured at certain locations in an assembly. In *transient* tests, a much larger amount of information can be gained recording the change of the temperature at one or more points over a time period. The two techniques are interrelated; steady state can be reached only through transient events, and transient techniques automatically yield static values when they end.

Transient testing has a deeper theoretical background, presented in References [1–8]. Both static and transient techniques are standardized as treated in related References [9–18]. Some of the tools used for obtaining simulated and measured results presented in this work are referred to in References [19–21].

Power devices and the assemblies composed of them are typically sandwich-like structures. The heat generated in silicon chips flows through a complex structure built of different layers of metals, ceramics, solder, and thermal paste (Figure 1). All layers have different thermal conductance, shear modulus, and other parameters.

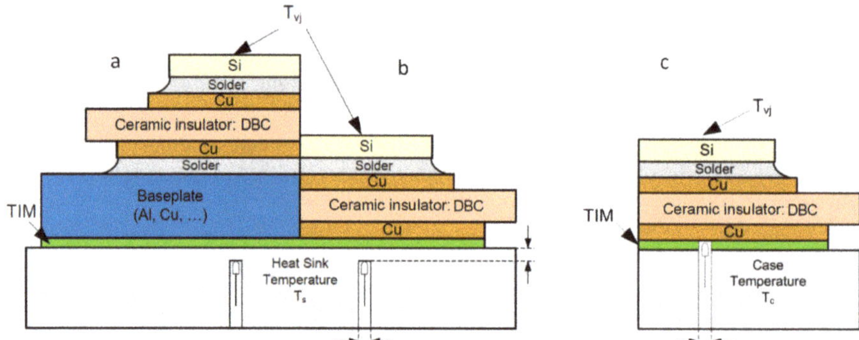

Figure 1. Power device on a cold plate. (**a**) Semiconductor die on direct bonded copper (DBC) in a module with a baseplate. (**b**) The DBC is directly attached to a heat sink. The heat sink temperature is measured. (**c**) The DBC is directly attached to the heat sink. The lower DBC surface temperature is measured. The optional sensor positions are shown as prescribed in Reference [16] (Courtesy of ECPE).

In a thermal test, the temperatures are converted, in most cases, to an electric signal, either measuring the temperature-sensitive electric parameters of the semiconductor chips in the assembly or using dedicated sensors at accessible outer points. A suitable sensitive parameter can be the forward voltage of a pn-type junction in a semiconductor device or the thermal voltage induced by the Seebeck effect in metal–metal junctions (i.e., thermocouples).

The recorded thermal quantities are typically distilled into simpler thermal descriptors, sometimes formulated as charts (e.g., Z_{th} plots, structure functions, pulse thermal resistance diagrams) and sometimes into single numbers (junction to ambient, junction to case thermal resistance, etc.).

Based on theoretical considerations, a transient test can yield partial thermal resistances between internal layers of the assembly. As it is shown in detail in References [4,5], this way a measurement at a single point can provide information on the temperature of structures which are normally not accessible.

In the electric world, measurements are highly repeatable and remain so when they are reproduced at different laboratories with different instrumentation. For example, voltage measurements yield results of 5 to 7 digits, and different instruments provide the same numbers within a fraction of a percent.

For thermal measurements, this is not the case. In electric measurements, the "conductive" and "insulating" parts of the measurement arrangement differ in their conductivity at a ratio of $1:10^{12}$; in thermal tests, this ratio is 1:100. Accordingly, parallel heat flow paths which exist besides the main one can influence the calibration and measurement process. Although it is expected that the thermal tests comply with related standards and actual temperatures can be measured with an accuracy of a few percent, the calculated thermal metrics can be up to 30% different when carried out at a different site with other instruments and thermal environment.

In this study, we first define the thermal quantities which can be measured and the relevant thermal metrics which can be gained from them. Then, we introduce the concept of transient and static thermal tests. Further on, related thermal measurement standards are discussed. Lastly, the reproducibility of thermal parameters measured in different test concepts is examined, and conclusions are drawn.

2. Simple Thermal Metrics: The Junction to Ambient and the Junction to Case Thermal Resistance

Several thermal measurement standards have been defined in order to simplify the description of thermal behavior with single numbers [9–16]. Based on the fact that the thermal conductivity of the typical materials used in power packages is nearly constant in the temperature range of their use, these descriptors are often *partial thermal resistances*. In a strict treatment, such a partial resistance is interpreted between two isothermal surfaces in an assembly, and they express that the temperature drop between such surfaces is proportional to the heat flux flowing between them. These isothermal surfaces are not accessible in most cases for attaching temperature sensors, and the accessible geometries in the assembly are rarely isothermal. This contradiction can be resolved in many cases using transient characterization techniques as demonstrated in References [1–6].

The primary descriptor used for characterizing a full assembly is the R_{thJA} *junction to ambient* thermal resistance, and the one for a power device with a dedicated cooling surface is the R_{thJC} *junction to case* thermal resistance. These already give a general impression on the thermal performance of an assembly or a device and can be used for approximate back-of-the envelope calculations.

The context and the interpretation of these metrics slightly differ in various standards, now we use the most consistent approach defined in the JEDEC JESD51 set of standards [12].

The standard describes the thermal system as a single heat source (junction) where P power is generated, and then a heat flux flows, partly or fully, through reference surfaces which are accessible for temperature probes.

In Figure 2 we cumulated all threads of the heat flow in a usual power device package structure into a thermal network equivalent. The heat is supposed to be generated at the point J. The part of the material through which the heat flux flows from the junction towards an X reference surface is represented by an R_A thermal resistance; the next part where the flux leaves X towards the ambient is denoted by R_B. A portion of the heat does not flow through X, and the corresponding portion of the assembly is cumulated into R_H.

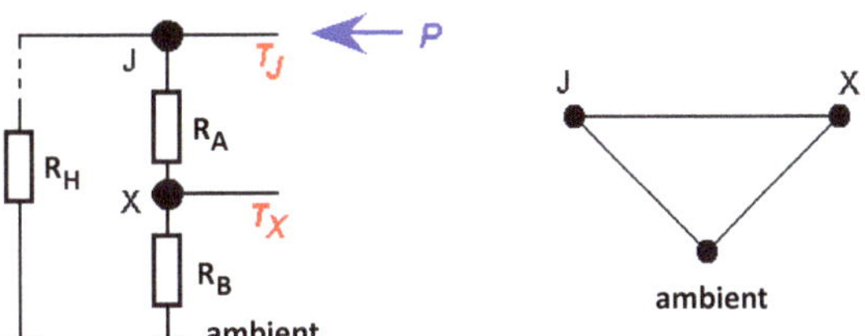

Figure 2. A simple network model for interpreting a partial thermal resistance between a single heat source and a reference point.

For the usual cases, when most of the heat flows through X, the standard defines an R_{thJX} thermal resistance as:

$$R_{thJX} = (T_J - T_X)/P \tag{1}$$

where T_J is the temperature of the junction and T_X is that of the reference surface. We can observe that in this definition it is tacitly supposed that the temperature distribution on such a reference surface is nearly homogeneous; the geometrical surfaces in the system coincide with isothermal surfaces (which is rarely true).

Of course, at the end, all heat flows towards the ambient. The R_{thJA} junction to ambient thermal resistance is defined as:

$$R_{thJA} = (T_J - T_A)/P \tag{2}$$

So far, one might think that the best approach is to measure the junction temperature and the temperature of a point on the X surface. Measuring the junction temperature is a challenge in itself as we show later. What is even worse, as it is shown in Reference [6] and in Section 6, the errors made in measuring T_J and T_A can be added up. A more relevant measurement approach is composing the difference in time, rather than in space.

For example, in a junction to ambient measurement one can apply two different power levels, P_1 and P_2, and measure the junction temperature after temperature stabilization in each case. The two measurements yield:

$$T_{J1} = P_1\, R_{thJA} + T_A$$
$$T_{J2} = P_2\, R_{thJA} + T_A \tag{3}$$

so

$$(P_1 - P_2)\, R_{thJA} = T_{J1} - T_{J2} \tag{4}$$

$$R_{thJA} = (T_{J1} - T_{J2})/(P_1 - P_2) \tag{5}$$

This differential principle offers a lot of advantages. The temperature is measured at a single point of the system. As shown later, with this solution all offset problems at measurement and calibration cancel out.

In many cases, the X surface is an exposed cooling surface of a power device or module, the "case". In the simplest approach, the junction to case thermal resistance can be defined in a two-point measurement, measuring the "temperature of the case", T_C:

$$R_{thJC} = (T_J - T_C)/P \tag{6}$$

However, the measurement of a "case temperature" is far from being unambiguous, as presented in References [2–6] and in Sections 3 and 5 below.

Another way for finding R_{thJC} is, again, based solely on the change of the junction temperature. This method, called the Transient Dual Interface Measurement (TDIM), compares more complex but more repeatable thermal descriptors, such as the structure functions of a device-on-heat sink arrangement, and defines the junction to case thermal resistance as the point where the structure descriptors start do differ. This methodology is defined among others in the JEDEC JESD 51-14 standard [13].

It has to be emphasized that the TDIM method yields much more than just a single R_{thJC} value; it automatically generates a one-dimensional thermal compact model of the power device or module.

In real measurements, many factors influence the achievable accuracy of thermal data and of the thermal metrics calculated from them. In order to separate the measurement errors related to the composition of the assembly and the ones caused by the inaccuracies of the test equipment, we present below the results of a simulated experiment and of real tests.

3. Simulation Experiment on Static and Transient Metrics

The errors of thermal measurements have various sources. A bunch of the problems are associated to the transient behavior of the devices under test. Some other problems are related to the instrumentation and to the thermal tester equipment. These problems are investigated in Section 4.

Another set of inaccuracies is related to the test arrangement. These can be best investigated in a simulation experiment, where the device and instrument induced errors play no role.

For demonstrating the techniques used and the associated problems, we present the temperature changes of specific points in a typical assembly, an IGBT module mounted on a cold plate with various thermal interface material (TIM) layers under the base plate.

In this section, we focus on the measurement problem; for this reason, the actual dimensions, material parameters, and temperature monitor points are presented separately below in Appendix A (Tables A1 and A2, Figures A1 and A2). A simplified sketch of the arrangement is shown in Figure 3.

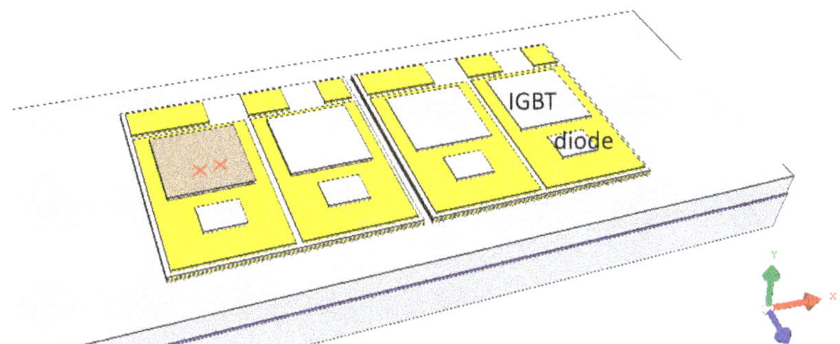

Figure 3. The IGBT module on a cold plate; the left IGBT is powered.

The IGBT chips were 11.2 mm × 11.2 mm in size, and this dimension is of interest for treating the displacement-related errors. The layers of the assembly were approximately the ones shown in Figure 1a. Under the silicon chips, a laminate of solder, copper, and ceramics layers was attached to an aluminum base plate. The cold plate was modelled with a constant heat transfer coefficient (HTC) of 3000 W/m^{2}K which is a realistic value for an aluminum surface with internal water cooling.

In order to examine the influence of the base plate to cold plate thermal interface, a 50 μm TIM layer was inserted between the module and the cold plate.

In this assembly, the transients were simulated in the FloTHERM tool [19] at a 100 W power step (heating), uniformly distributed on the die surface.

The monitoring points for the simulated transients were selected as follows:

- Ch0: center on the top of the powered semiconductor die, in the dissipating layer;
- Ch1: center on the top of the TIM, below the semiconductor die;
- Ch2: center on the bottom of the TIM, adjoining the cold plate;
- Ch3: as Ch1, but displaced from the center towards the right edge of the die, by 3 mm;
- Ch 4: as Ch2, but displaced from the center towards the right edge of the die, by 3 mm.

Obviously, Ch0 corresponds to the junction temperature.

The monitoring point Ch1 mimicked the ideal placement of the thermocouple for measuring the temperature of the "reference point" shown in Figure 1a. This was also the prescribed position for determining R_{thJC} in References [10,16].

The monitoring point Ch2 corresponded to the case when the probe does not (completely) penetrate the TIM layer. Both Ch3 and Ch4 represented small lateral displacement of the probe, now about half of the chip size, as it mostly happens at such measurements.

For illustrating different measurement methodologies, the TIM layer was represented by different thermal conductivities, such as *dry surface* (0.2 W/mK) and *different interface materials* (1 W/mK, 4 W/mK). The two latter conductivity values corresponded to different qualities of thermal grease materials.

Figure 4 shows the change of temperature at the monitoring points at the different TIM conductivities. Besides the obvious fact that the improved thermal interface reduces the temperature elevation from 50 K to 26 K, the figure also proves that a good TIM also makes it less essential whether the reference probe really touches the module baseplate or it is just "somewhere near" (Ch0–Ch1 versus Ch0–Ch2 distance).

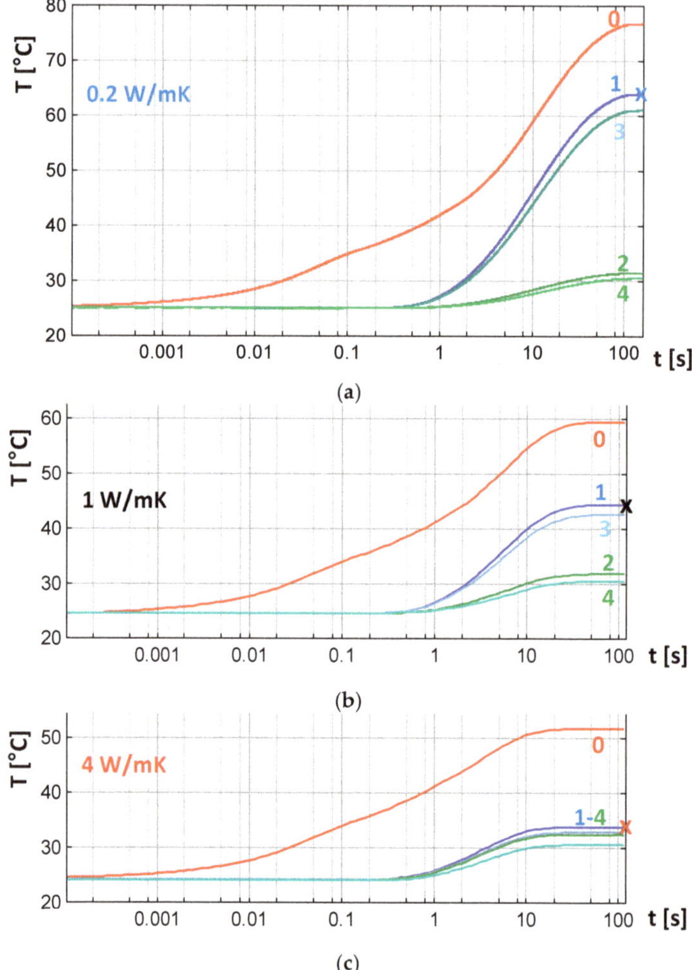

Figure 4. Simulated temperature change at 50 W, thermal conductivity of the TIM: (**a**) 0.2 W/mK, (**b**) 1 W/mK, (**c**) 4 W/mK. Ch0: junction, Ch1: case center, Ch2: cold plate top position. Ch3 and Ch4 represent small lateral displacement of the probe.

The figure also indicates that the external monitor points reacted on the power change with a 0.5 s delay; accordingly, also in a live system, a slow data acquisition of the reference temperatures with a few samples measured in a second was appropriate. One can observe that more intensive cooling resulted in earlier stabilization of the temperature, and steady state was approximately reached at 140 s, 50 s, and 30 s for the interface layers of 0.2 W/mK, 1 W/mK, and 4 W/mK thermal conductivity, respectively.

It would be hard to provide the full three-dimensional temperature distribution in the assembly as it develops in time; Figure 4 is restricted to a few characteristic points.

Another informative chart presents the typical bell-shaped temperature distribution of the case_bottom/TIM_top interface in steady state (Figure 5). The peak temperature under the chip center corresponds to the final transient value at Ch1, shown as a blue "x" marker for the "dry" assembly in Figure 4a and as black "x" and red "x" markers in Figure 4b,c, respectively, for different TIM qualities. Note the large temperature difference even within the chip area.

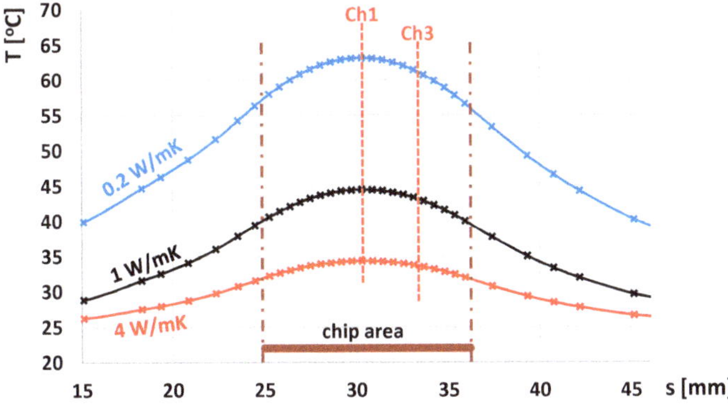

Figure 5. Temperature distribution on the case_bottom/TIM_top interface in stationary state. The peak temperature under the chip center corresponds to the final transient value at Ch1, shown as the blue, black, and red "x" in Figure 4a–c, respectively. The temperature at the displaced location Ch3 is also shown.

The temperature record in Figure 4 depicts only the outcome of one certain powering at three given boundaries. The results can be interpreted in a more general way calculating the Z_{th} *thermal impedance* curves which are derived normalizing the time-dependent temperature *change* by the applied power:

$$Z_{th}(t) = \Delta T_J(t)/P \tag{7}$$

The Z_{th} curves are popular thermal descriptors of a system. They can be used already for back of the envelope calculations; knowing an actual P_{act} heating power in the system, the temperature change in time will be approximately $T_J(t) = P_{act} Z_{th}(t) + T_{ref}$, where T_{ref} is the temperature of the whole assembly at low powering. Moreover, further thermal descriptors can be derived from Z_{th} as shown in References [1,5,13] and in further sections below.

In Figure 6, we can see the Z_{th} curves (normalized temperature change) of the arrangement with the three different TIM materials.

With a TIM layer of λ = 0.2 W/mK, first we can observe that the R_{thJA} total junction to ambient thermal resistance of the assembly is 0.52 K/W. This is the only true physical quantity in such a thermal measurement, based on the objective measured data without further assumptions on locations, divergence threshold, and other artificial elements introduced later on for other thermal metrics. The only approximation is assuming a uniform T_J junction temperature. Some considerations on the validity of this assumption are given in Reference [22].

Measuring separately the junction and an external probe yields R_{thJC} = 0.13 K/W junction to case thermal resistance if the probe penetrates the TIM, and R_{thJC} = 0.45 K/W if the probe just touches the lower surface of it ("**B = 0–1**" and "**A = 0–2**" in Figure 6a, respectively).

With a TIM layer of λ = 1 W/mK, separate measurements at the junction and at the external probe yield R_{thJC} = 0.16 K/W if the probe penetrates the TIM, and R_{thJC} = 0.28 K/W if the probe just touches the lower surface of it ("**B = 0–1**" and "**A = 0–2**" curves in Figure 6b, respectively). At this TIM quality for the whole assembly, R_{thJA} is 0.36 K/W.

With a TIM layer of λ = 4 W/mK, the two-point method yields R_{thJC} = 0.18 K/W junction to case thermal resistance if the probe penetrates the TIM, and R_{thJC} = 0.19 K/W if the probe just touches the lower surface of it ("**B**" and "**A**" curves in Figure 6c, respectively); R_{thJA} is now 0.28 K/W.

We can observe that, with better TIM and cold plate qualities, the measured junction to case thermal resistance grows as the heat flow is more attracted to the center of the die–die attach–insulator–base plate sandwich, and the base plate temperature is more uniform (Figure 5). With real thermocouples where the probe tip is coated with an insulator layer and the wires draw some of the heat from the sensor tip, the measured thermal resistance can be well 100% larger than the ideal value obtained in a simulation.

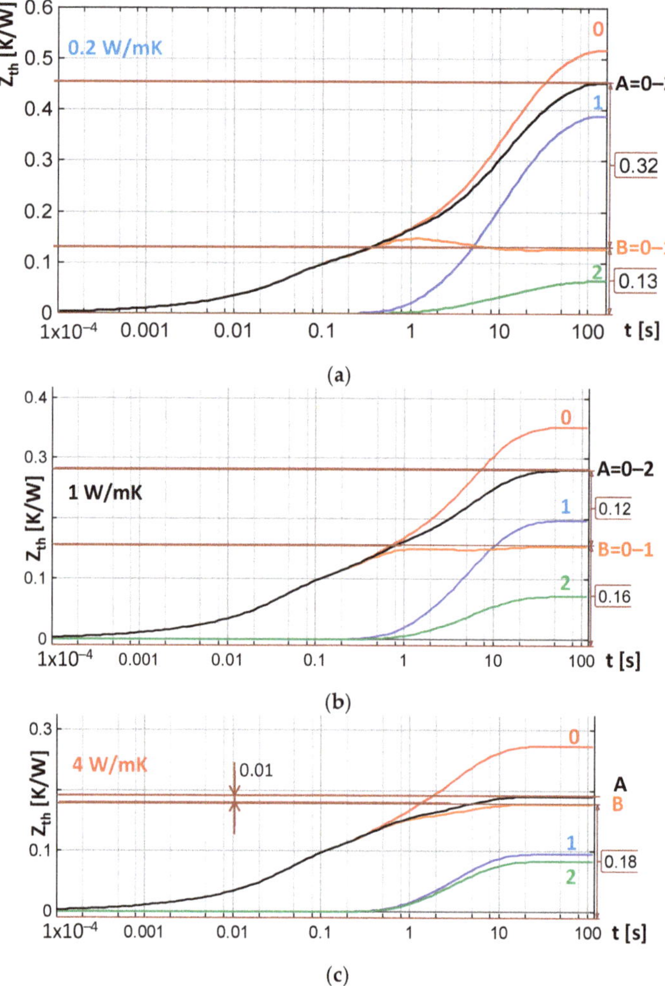

Figure 6. *Zth* curves, at junction and sensor locations, at TIM thermal conductivity: (**a**) 0.2 W/mK, (**b**) 1 W/mK, (**c**) 4 W/mK. Ch0: junction, Ch1: case center, Ch2: cold plate top position.

The TDIM methodology is a transient method which is based on measurement at a single point. This technique is based on the comparison of the change of the junction temperature at different boundaries.

Figure 7 compares the Z_{th} curves belonging to the junction at different TIM qualities. We can observe that the heat flow arrived at the base plate at 1.7 s, and the curves deviated a bit below 0.2 K/W.

This difference is much more expressed in the structure functions which can be derived from the Z_{th} plot of to the hottest point (junction).

Figure 8a shows the equivalent RC chain circuit of thermal resistances and capacitances which corresponds to the exponential decomposition of the Z_{th} curves (Foster network). This RC chain can always be converted into a ladder-type network shown in Figure 8b (Cauer network).

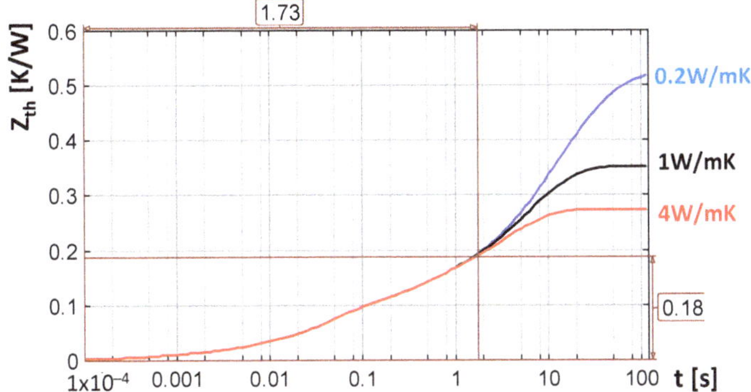

Figure 7. Z_{th} curves at thermal conductivities of the TIM at 0.2 W/mK, 1 W/mK, and 4 W/mK.

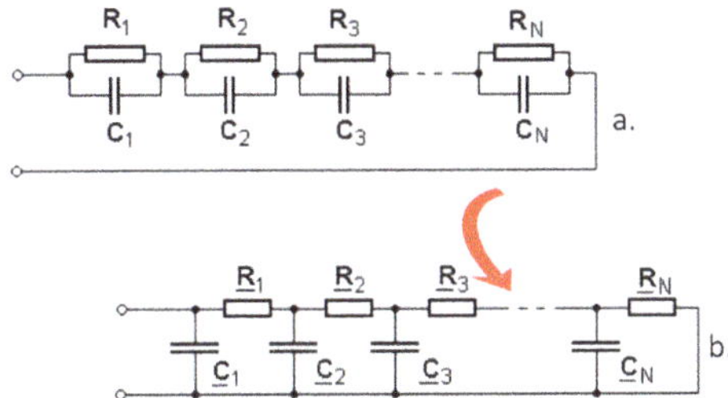

Figure 8. Foster- (**a**) and Cauer- (**b**) type representations of a 3D thermal RC net (based on Reference [4]).

The Foster–Cauer RC transformation is a systematic process of consecutive steps of division and subtraction, presented in detail in Reference [13]. The theoretical background of the technique is outlined in Reference [1], and many practical hints on its use are given in References [2–8]. Moreover, an interesting treatment of a modified method is presented in Reference [18].

The Cauer network can be visualized in a structure function (Figure 9). In this plot, we summed up the thermal resistances in the ladder, starting from the heat source (junction) along the *x*-axis and the thermal capacitances along the *y*-axis.

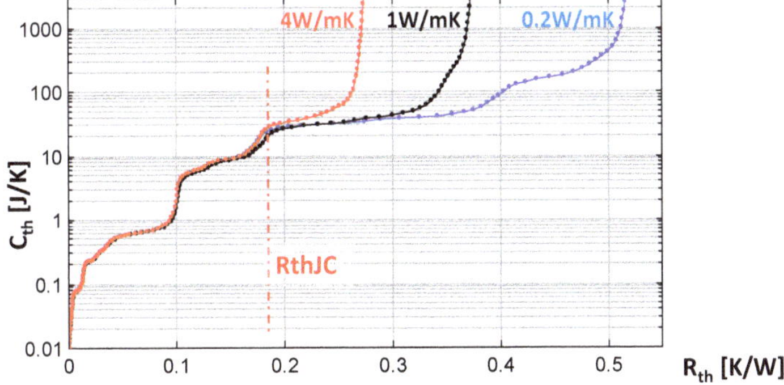

Figure 9. Structure functions with thermal conductivities of the TIM at 0.2 W/mK, 1 W/mK, and 4 W/mK. Junction to case thermal resistance is shown.

Thermal capacitance is proportional to the mass and volume of a material layer through its specific heat and density. *Low gradient sections* in the chart mean that a small amount of material having low capacitance causes large change in the thermal resistance. These regions have *low thermal conductivity* or a *small cross-sectional area*. *Steep sections* correspond to material regions of *high thermal conductivity* or a *large cross-sectional area*, as even a large bulk of material corresponding to high thermal capacitance is of low thermal resistance only. Sudden breaks of the slope belong to material or geometry changes. Thus, thermal resistance and capacitance values, geometrical dimensions, heat transfer coefficients, and material parameters can be directly read on structure functions.

In Figure 9, the structure functions generated from the Z_{th} curves of Figure 7 are compared. The curves belonging to different thermal conductivities started to diverge after 0.17 K/W. Until this point, we see the characteristic steps in the structure function corresponding to the sandwich-like internal structure of the module composed of materials of highly different thermal conductivities.

One can note that the figure also describes well, besides the device under test, also the test fixture and the external cooler. The separation between the device and the outer environment occurs around the $R_{thJC} = 0.17$ K/W, $C_{thJC} = 30$ J/K point. Further on, we see the change of the thermal conductivity and specific heat in the external domains of the test equipment.

The approximate thermal capacitances of the components of the material stack are listed in Table A2. We cropped Figure 9 above 3000 J/K, as all the lines turn vertical, and no further change in the thermal resistance can be observed. In the case of a real measurement on real cold plate as presented in Section 4 below, this capacitance would correspond to 700 liters of water, driven through the cold plate of the tester for more than 10 min at typical pump rates, we can rightfully assign this thermal capacitance to the "ambient".

At this high thermal capacitance, the structure functions end at the R_{thJA} junction to the ambient values (i.e., 0.52 K/W, 0.36 K/W, and 0.28 K/W) established previously.

In the case of real measurements, some noise-induced perturbation occurs on the curves; for this reason, the TDIM measurement, as outlined in the standard [13], requests an ε threshold to be defined in the thermal capacitance, after which the structure functions can be treated as different.

Figure 10 presents the difference of the structure functions in Figure 9. The figure demonstrates that selecting a threshold between 0.05 J/K and 2 J/K, being of a ratio of 40, we can state that the R_{thJC} junction to case thermal resistance is between 0.17 K/W and 0.19 K/W. In real cases with actual measured transients instead of simulated ones, this difference is less steep, as shown in Reference [23], but still gives a sharp detection of the R_{thJC} quantity.

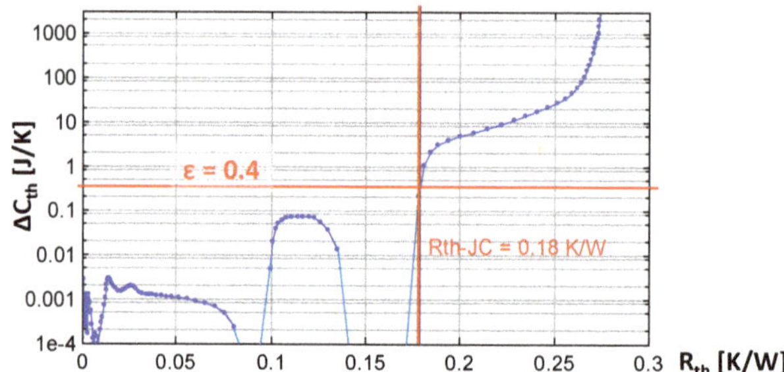

Figure 10. Difference of structure functions belonging to TIM thermal conductivities of 4 W/mK and 0.2 W/mK.

The JEDEC JESD51-14 standard defines the details of the TDIM methodology and identifies two alternative metrics by which the divergence point of the measured curves can be quantified. One such metric is the difference in the *derivative* of the Z_{th} curves, and the other is the difference of structure functions. It has to be noted that both metrics are related to "edge-enhancing" techniques of image processing which are famous also for their noise enhancing nature.

Sources of Uncertainty, According to the Simulation Experiment

As a result of the above presented simulations, we can conclude that when using two-point methodologies for determining the R_{thJC} thermal metrics, the obtained value depends on the TIM quality, lateral displacement of the probe measuring the "case" temperature, penetration of the probe through the TIM, the heat transfer coefficient of the cold plate, and other factors.

In the case of a single-point test, the assembly is totally destroyed and rebuilt between the two measurements. The differences in TIM quality belong to the essence of the technique. Still, although the structure functions are highly reproducible, a decision on the ε threshold used has to be made to define at which divergence point it is considered to be the R_{thJC} value.

In a rigorous simulation model, the temperature transient at Ch2 would be valid only if the probe does not protrude into the TIM layer. This assumption is true when elastomer foils, metal laminates or similar TIMs are used.

If the TIM used is some thermal paste, the probe tip is pressed into it by its elastic support. However, other effects (listed below in Section 4) would cause a systematically lower recorded temperature in the same way as shown for Ch2 in the simulation experiment.

4. Thermal Transient Tests

In the case of measurements, the consequences of the simulation experiment remain valid, but now inaccuracies of the device characteristics and of the test system have to be considered in addition.

Thermal transient measurements need one or more heater elements and one or several temperature sensors in a system. In most cases, the heat source is a piece of semiconductor, typically called a "chip" in the literature on system design and "die" in works on semiconductor technology and packaging.

Normally, the hottest point in the circuitry is the powered thin material layer of the semiconductors, traditionally called "junction". For many device categories (diodes, MOSFETs, IGBTs), both the heat source and the sensor are, in fact, pn junctions which are driven into forward operation (Figure 11). A sudden power change on the junction can be created by *switching down* from a high I_H heating current to a low I_M measurement current level.

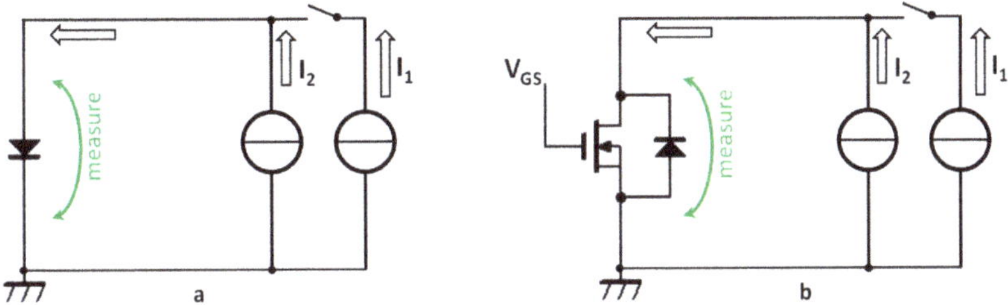

Figure 11. Powering scheme for the thermal transient measurement of a diode (**a**) and an IGBT in saturation mode (**b**).

In actual realizations of the thermal test instruments, I_M is realized as a steady source of programmable low I_2 current. A programmable high I_2 current can be swapped between the device under test and an external shunt; I_H is composed as $I_1 + I_2$.

First, we demonstrate the basics of the thermal transient testing in an actual test of a power IGBT module. The actual device type and measurement equipment are not the focus of the present study, the description of the test setup, the environment, and photographs are presented again in Appendix A.

With trial measurements, we found that a relevant test can be carried out at a 50 A heating and 100 mA measurement current.

The measurement current was used in two related steps of the transient testing. In a *calibration* process, the forward voltage (or other temperature-sensitive parameter) at I_M was recorded in a thermostat at different T_J junction temperatures; such a voltage to temperature mapping is provided. Figure 12 presents the $V_{CE}(T_J,I_M)$ calibration curve of the actual device.

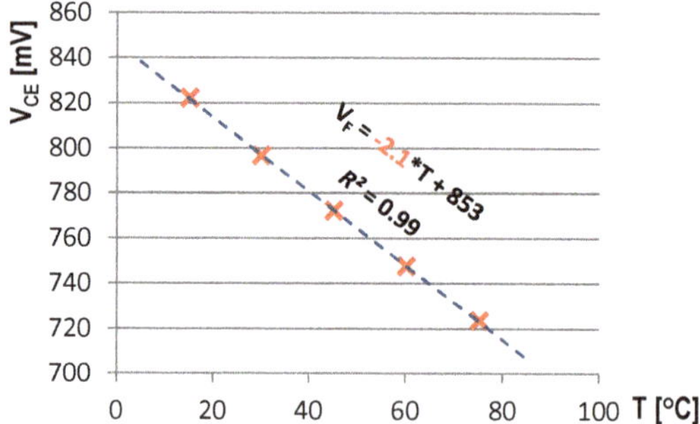

Figure 12. Calibration result: forward voltage of a power IGBT at a $I_M = 100$ mA measurement current.

The test started with a longer equalization period until the V_{CE} voltage at constant I_H stabilized. When steady state was reached, the $P_H = V_{CE}(I_H) \cdot I_H$ power on the device was stored, and after switching down to I_M, the change of $V_{CE}(T_J,I_M)$ was recorded. During the transient recording there was also a low $P_M = V_{CE}(I_M) \cdot I_M$ power on the device; the ΔP power step was calculated as the difference of P_H and P_M. We found that the power step on the actual device was around 55 W when switching down from 50 A to 100 mA. The power step slightly depends on the actual thermal boundary which obviously influences $V_{CE}(T_J,I_H)$ at the same I_H. Details of the switching process are presented in Reference [4].

Figure 13 presents the change in the saturation voltage of the power module at $\Delta P = 55$ W, attached to a dry cold plate and then to a cold plate wetted by grease as prescribed in the standard of Reference [13]. This voltage change can be mapped to the temperature change of Figure 14 using the calibration data in Figure 12.

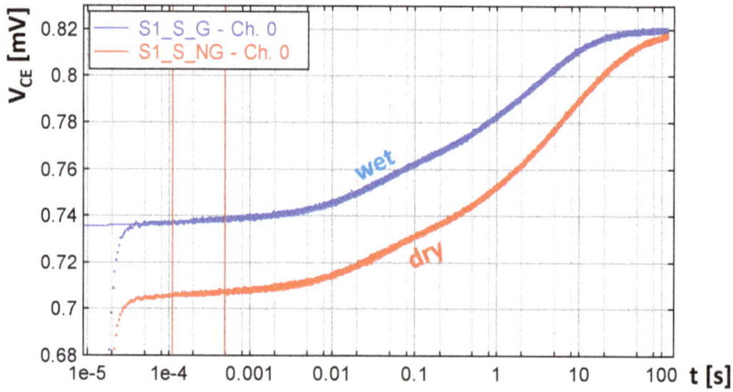

Figure 13. Measured transient of the V_{CE} saturation voltage of an IGBT on dry and wet cold plates.

In an ideal case, one can record P_H in a "hot device at high current" state in the last moment before switching down, and then the voltage/temperature change can be sampled from the first moment in a "hot device at low current" state. In Figure 13 we can observe that switching among different current levels causes a long electric transient in the device voltage which lasted for 50 µs in the actual case.

The temperature change in Figure 14 depicts only the outcome of one certain powering at two given boundaries. The results can be interpreted in a more general way calculating the Z_{th} curves which are derived dividing the temperature change by the applied power, $Z_{th}(t) = \Delta T_J(t)/\Delta P$.

The Z_{th} curves (Figure 15) can be converted to structure functions, as shown in Section 3, and all considerations treated there apply again.

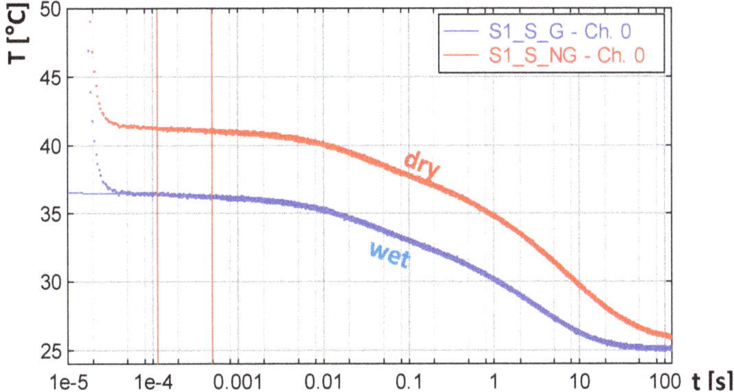

Figure 14. The recorded voltage transient converted to temperature change using the mapping of Figure 12.

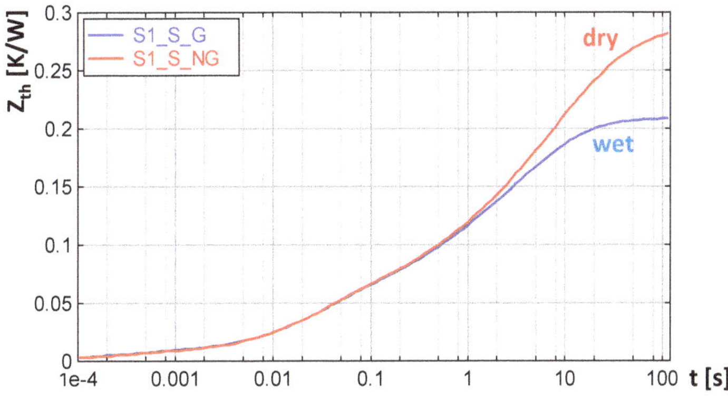

Figure 15. Z_{th} curves calculated from Figure 14.

Many details of the powering and temperature sensing principles are treated in Reference [7], and considerations on the appropriate transient test planning are given in References [5,6].

4.1. Sources of Uncertainty in the Case of Transient Thermal Testing

In the real tests, all sources of error which were discussed in Section 3 still apply. However, we had further sources of uncertainty.

4.1.1. Electric Transient

Power devices typically have a long electric transient when switching among different current levels. In Figures 13 and 14, we have no direct information on the temperature until 50 µs; we just see the collapse of V_{CE} due to the recombination of charge in the IGBT junction. There exist extrapolation techniques to restore the missing thermal signal based on the analytic solution of the homogeneous heat spreading in a block which is powered on its surface. The result is given in Reference [13] as a square root of time function:

$$\Delta T_J(t) = \Delta P / A \cdot k_{therm} \cdot \sqrt{t} \tag{8}$$

where $\Delta P/A$ is the power density on the heated surface, and k_{therm} cumulates several material parameters. However, the use of Equation (8) for IGBTs which are not surface heated is at least doubtful.

Generally, this equation can be only used if the heat flow from a 2D junction is one directional. If there are other highly conductive structures on top of the heated die (top metallization, clip, chip-on-chip, etc.), it cannot be used either.

4.1.2. Noise on the Recorded Signal

The signals are slightly noisy as proved in Figures 13 and 14, but this can be cured with high sampling rate and averaging.

4.1.3. Power Measurement Uncertainty on the Device

The measurement of the power on the device is based on voltage and current measurements, this way it is quite accurate for discrete devices.

At large power modules, the internal wiring is more intricate, and some compromises cannot be avoided. Applying a higher current on the device, the voltage on the internal pn junction grows logarithmically; theory says that current growth by a factor of 10 results in 60 mV voltage elevation at room temperature. Based on the series resistance of the semiconductor device and on the wiring, the voltage grows proportionally. As a result, we experience quadratic growth of the power dissipation in the wiring, while similar power growth on the internal chip is rather flat.

For this reason, we typically see a shrinking effect in the Z_{th} curves at higher currents and also in structure functions. During the cooling, we recorded the correct chip temperature. When composing the Z_{th} curves or structure functions, we divided the temperature by the power which is measured across the whole module including the portion dissipated in the internal wiring.

In Figure 16 the Z_{th} curves of a power module at several I_H heating currents between 10 A and 40 A can be seen. Supposing that we can neglect the power component on the wires at 10 A current, Figure 16 indicates that at 40 A already 13% of the heating occurs away from the chip.

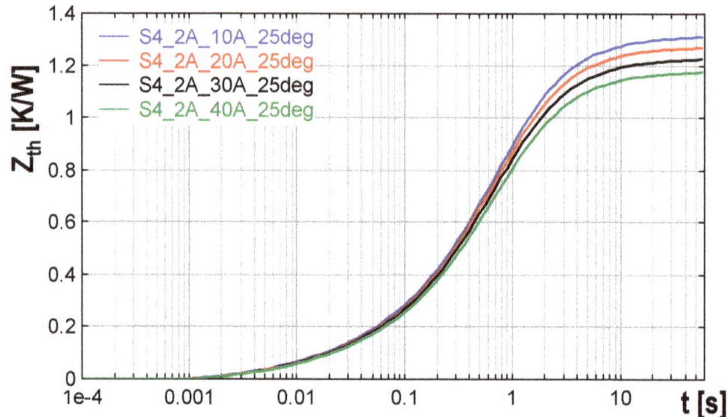

Figure 16. Z_{th} curves of a power module at I_H heating currents, 10 A to 40 A [4].

Another contribution to the decreasing R_{th} with increasing current is that the increasing surface temperature in the case of higher power levels enhances the heat loss through convection and radiation as well.

4.1.4. Offset and Gain Errors in the Data Acquisition

The data acquisition channels of the measurement instrument also have some errors; these can be classified typically as gain and offset errors. Theoretically, in the calibration process (Figure 12) all these cancel out; the errors in the mapping will be reversed during the measurement. However, while the gain of a data acquisition channel is largely constant, a tiny drift in the offset of the acquisition system is typical, and it cannot be guaranteed that the same acquisition channel is used in the calibration process and in the transient measurement.

The raw electric signal which can be acquired is typically tiny, 1–2 mV/K on pn junctions and 40–50 µV/K on thermocouples. We pointed out in Reference [4] that the major factor which undermines measurement accuracy is the offset of the data acquisition channel of the test equipment which is also in the few mV range representing a difference of a few degrees. In Section 3, we demonstrated that this source of inaccuracy can be eliminated by thermal transient tests at a single "hot" point; the differential measurement of the temperature automatically cancels out acquisition channel offsets. This can also be formulated in a way that the differential measurement principle introduced in Section 2 relieves measurements of high repeatability but poor accuracy (Figure 17) from their constant error.

Figure 17. Illustration of the concepts of accuracy and repeatability of measurements repeated within a short period of time. Reproducibility can be illustrated in the same way but over longer time periods and eventually at different laboratories using different instrumentation. Resolution can be best formulated in the case of measurements where results are transformed to digital values at some point; in this case, it may correspond to the thickness of the black and white rings of the target.

4.1.5. Reproducibility Issues of the Selected Sample

The selected samples have slightly different mechanical features such as die attach thickness, base plate roughness, and planarity. These cause random differences in the measured thermal metrics.

4.1.6. Reproducibility Issues of the Test Environment

Different laboratories have different materials and geometries of the cold plate used, other formations of the liquid flow, various surface roughness and planarity levels, types, and positions of external temperature sensors. Using the same equipment, the type and thickness of the applied thermal paste varies. Some hints on the proper construction of cold plates are given in Reference [13].

Some sources of inaccuracy related to the probe position for two-point measurements were already highlighted in the previous simulation experiment in Section 3. In a real measurement, further error sources can be identified such as:

- The thermal contact resistance between the case surface and probe tip can be quite large, especially since the contact area in the case of a spherical probe is just a point;
- The heat flow from the tip through the thermally conductive material of a thermocouple diminishes the probe tip temperature;
- There is a temperature drop inside the alloy joint of the thermocouple, since the thermocouple does not measure the temperature at its tip but at the point where the two wires of different alloys separate, etc.

5. Static Thermal Tests

In light of the former sections, the static tests seem to be simple. For example, for establishing the R_{thJC} junction to case thermal resistance, one has to determine the T_J junction temperature and the temperature reading of one of the sensors attached to the appropriate cooling surface as T_C, as presented in Figure 1. From Equation (6) it can be deduced that $R_{thJC} = (T_J - T_C)/P$, where P is the applied power.

Sources of Uncertainty in the Case of Static Thermal Testing

Regarding T_C, it is really just a simple reading of a sensor; but how can one determine T_J? In all cases it is an average value of the actual temperature distribution on the semiconductor surface. Moreover, there are only indirect ways to gain information on the chip temperature; for this reason, several standards call this quantity "virtual junction temperature" and denote it as T_{VJ}.

Taking a closer look at the measurement schemes in References [9,10,16,17], we find that T_{VJ} is determined by:

- Putting a low I_2 current on the device under test in a thermostat and composing a chart in the style of Figure 12;
- Adding a high I_1 current to the device bias and heating it up by $I_H = I_1 + I_2$ as proposed in Figure 11;
- Periodically switching off I_1 and measuring the voltage on the device at low $I_M = I_2$ at a "proper" time.

Proper time is not clearly defined in the standards; there is some hint that the measurement should take place after an eventual electric transient but before considerable cooling of the chip.

We can recognize that determining T_{VJ} is a transient test, at least a shortened one. In Figure 13, "proper time" would be somewhere between 100 μs and a few milliseconds. The transient measurement can be aborted after that time, but there is no statement in the standards for when it should be stopped, if at all. The voltage meter used typically has some integration time for suppressing noise; this way, actually, an average of the transient signal is recorded.

All standards prescribe an iterative process for the "virtual" junction temperature measurement but in a different way. The JEDEC JESD51 standards [12,13] aim at thermal characterization only; they tacitly assume that the cold plate in the measurement is kept at stable T_{cp} temperature, and a few trials are needed to find a proper I_H current which induces a "high enough" ΔT_J temperature elevation to keep low the influence of the limited accuracy of the test equipment (such as the offset errors mentioned previously).

The guidelines in the CIE Technical Report 225:2017 [17] comprise measurement of thermal and optical parameters of solid-state light sources. The light output of these devices strongly depends on the current and temperature, accordingly; the optical parameters have to be measured at a constant (T_J, I_F) pair. For this reason, the T_{cp} cold plate temperature is regulated at forced $I_H = I_1 + I_2$ driving current, until the pulsed voltage measurement at low $I_M = I_2$ corresponds to the target temperature determined in the calibration curve.

A comparative study on the T_J regulation defined in the JEDEC standards and CIE guidelines is presented in Reference [24].

The IEC 60747 standards [9,10] and the MIL-STD-750 standard [11] aim at measuring many various semiconductor parameters such as breakdown voltage, recovery time, etc. For all of these measurements the T_{VJ} value, at which the measurement is carried out, has to be specified. The measurement of the virtual T_{VJ} is carried out mainly in the same way as in the CIE guidelines [17]. Still, the depicted measurement sequence in IEC 60747 is a bit obscure; it is not clear whether the iterative regulation of the cold plate temperature targets a predefined T_{VJ} or if two different predefined T_{cp1} and T_{cp2} values at freely selected I_1 and I_2 currents.

Although the measurement of T_J does not conceptually differ in transient and static (that is truncated transient) measurements, the static approach needs simpler instrumentation, because the noise on the signal can be suppressed with integration along a short time period.

6. Brief Overview of Thermal Measurements Standards

We referred to several measurement standards in the previous sections, now we give a short but more systematic overview of them.

When the purpose of the measurements is building a properly accurate package model there are no specific prescriptions on the number and style of the measurements needed. However, there exist guidelines for successful combination of measurement and simulation at various boundary conditions which yield a two resistor model [14] or a compact thermal model consisting of a net of thermal resistances connecting simplified geometrical faces of a package [15].

On the other hand, when the purpose of the measurement is to produce *comparable* thermal data on packaged devices, a meticulous procedure has to be followed as listed in the appropriate standards.

Many relevant semiconductor test procedures, such as measurement of isolation voltages, parasitic inductances, capacitances, etc., are defined in the set of IEC 60747 (EN 60747) standards (e.g., [9,10]).

In Reference [10], several aspects of the thermal measurement of power modules are treated. The measurement of the virtual junction temperature and for static methods also the position of thermocouples is specified. The transient methods are restricted to a short mentioning of Z_{th} curves as "transient thermal impedance".

The set of IEC 60747 standards differentiates between type tests and routine tests. Type tests are carried out on selected samples of new products in order to determine the electrical and thermal ratings of a type and for establishing test limits for further tests. The type tests are repeated regularly on a given number of samples taken from manufacturing batches at the manufacturer or delivery batches at the end-user in order to confirm the quality of the product. Routine tests are carried out on each sample of the production or delivery.

Thermal tests as routine tests are carried out only in mission critical industries (e.g., military, space).

The MIL standards [11] give some hint on the powering of the device for reaching a required temperature elevation in thermal tests, but the actual selection of voltages and currents for different semiconductor device categories seem to be ad hoc and sometimes poorly defined. A detailed review on the powering options is given in Reference [7].

The most developed set for thermal testing is at present the JEDEC JESD51 family [12,13]. Especially, the JEDEC JESD51-14 standard [13] treats many aspects of the transient testing including the problem of removing eventual short-time electric perturbations from the thermal signal. Moreover, it introduces the concept of structure functions and the transient dual interface methodology (TDIM) as used before in Section 3.

The new European Center for Power Electronics (ECPE) AQG324 guidelines for the automotive industry, "Qualification of Power Modules for Use in Power Electronics Converter Units (PCUs) in Motor Vehicles", serve validation purposes for different parameters of automotive power modules. They restrict the thermal qualification to two-point methods, but, besides the junction to case thermal resistance of the module, junction to heatsink and junction to fluid thermal resistances are also defined for devices with an integrated cooling mount.

It has to be noted, however, that although thermal testing becomes more and more important in order to achieve reliable operation over a long lifetime, still, the construction of complete appliances often overlooks thermal testability aspects. Consequently, these tests often need a workaround for accessing devices that are relevant for their power consumption or can be used as sensing points.

7. Comparison of the Results Gained from Static and Transient Measurements

We previously listed a number of different standards and guidelines which aim at providing thermal descriptors bearing identical names in different standards but not necessarily covering the same content. Still, the similarity of the results gained in different ways is expected.

As exposed in Section 1, in the case of electric measurements, it is common to get highly uniform results for repeated measurements with different instrumentation, but for the thermal measurements this is not the case. Accordingly, we cannot save defining "similarity" in a more definite way.

The similarity of measurements can be interpreted in the terms of the following concepts:

- **Accuracy** is the degree of closeness of measurements of a quantity to that quantity's true value;
- **Precision** is the degree to which repeated measurements under unchanged conditions show the same results; precision can relate to:

  o   *Repeatability*—the variation of measurements with the same instrument and operator and repeated in a short time period;

  o   *Reproducibility*—the variation among different instruments and operators and over longer time periods.

- **Resolution** is the smallest change which can be detected in the quantity that is measured (especially when the output of the measurement is of a digital nature).

Below we compare the results of static and transient methods in general. If specific details are needed, we turn to AQG324 as the static guideline [16] and JEDEC JESD51-14 as the transient standard [13].

We referred formerly to the static method as a *two-point* method because the temperature of the junction and of an external point was involved in a measurement. We can define *multi-point* methods if more temperature sensors are attached to dedicated accessible points of the structure. This distinction is only needed because the JEDEC JESD51-1 standard [12] uses the term "static method" in a quite odd way for describing the transient method.

In order to quantify whether the results of two methods are "similar", first, we have to define the acceptable tolerance of the methods.

7.1. Tolerance Expectations in the ECPE Guideline AQG 324

The AQG 324 guideline [16], in its Section 4.7 "Standard tolerances", specifies the following acceptable tolerances (Table 1):

Table 1. Definitions of standard tolerances in Table 4.6 of [16].

Measured temperatures	±2 °C
Indirectly determined temperatures	±5 °C

We can state that the *two-point method* accepts data of limited accuracy, as we see a rather loose definition. For example, if the true temperature difference between two points is 50 °C and one measurement produces 57 °C and another 43 °C, both measurements will be accepted as valid (a 32% difference).

In practice, the actual difference is much lower if the measurement is carried out with the same instrumentation and by the same operator. Unfortunately, the difference can already be even higher if done by two different operators. We experience this range of differences when comparing numbers coming from different companies where the instrumentation is also dissimilar (round robin tests).

In reality, a well calibrated thermocouple can be accurate to within 0.1 °C. We can typically reproduce the virtual temperature *change* of a semiconductor junction within 3% over a 50 °C temperature span which makes a ±1.6 °C of uncertainty.

Still, the expectations of Table 1 are very realistic due to the following problems as exposed before:

- R_{thJC} is not a physical quantity like a voltage difference between two points;
- The obtained T_{VJ} virtual junction temperature is an average of the actual non-uniform temperature distribution on the chip. Simulations assuming homogeneous power distribution on the chip allege that a bell-shaped temperature distribution similar to Figure 5 develops on the surface. However, the series resistance of real semiconductor devices has a positive temperature coefficient at the high I_H current. This effect repels the current threads towards cooler portions of the semiconductor block and equalizes the temperature distribution to an extent. Infrared measurements still attest some inhomogeneity. On the "case" surface of the device, the typical bell-shaped temperature distribution of Figure 5 develops. The shape of this temperature curve depends on the roughness and planarity of the surface, interface material, liquid cooling quality in the cold plate, and other parameters. The actual location found by the external probe can differ in repeated measurements, and it is more likely diverse among different laboratories;
- The hole drilled for the probe distorts the shape of heat spreading. Figure 1 suggests that the thermal interface layer has to be penetrated by the probe tip; this can be more or less successful at different materials (grease, elastomer foil, etc.). The tip of the probe is typically coated by an electric insulation layer [25]. The material and thickness of this will be different in different laboratories, and the force with which the probe is pressed against the device case will also be different;
- The type of the probe influences the measured value [26].

An even weaker constraint is given in the actual IEC 60747 standards such as in References [9,10]. There, the accuracy to be reached is given with the following prescription: "The accuracy of the method is not specified. However, adequate precautions should be taken" ([9], Section 7.2.2.1, page 81).

7.2. Actual Performance of the TDIM Method as Specified in JEDEC JESD 51-14

In this methodology, the following quantities are measured directly:

- Two power levels based on voltage and current measurements. This can be done at 1% or better accuracy;
- The temperature change in time when the switching among power levels occur. In this procedure, all offset and gain errors cancel out automatically; only the repeatability of the calibration process influences the result. As stated above, here, 3% repeatability can be reached.

From the raw measurements, the transient thermal impedance, $Z_{th} = \Delta T(t)/\Delta P$ can be derived (as described in JEDEC JESD 51-14 and similarly in IEC 60747-15—Section 6.2.4.5 and IEC 60747-2—Section 7.2.2.3). This accuracy is inherited by the structure functions calculated from the Z_{th} curves, regarding their endpoint (R_{thJA} junction to ambient thermal resistance). Theoretical considerations [1] hint that the calculation process can add a further 5% uncertainty to the reading of the partial resistance (divergence point in Figure 9).

Consequently, the repeatability of the structure functions is much better than that of the temperature differences measured by probes in the previous section. The *reproducibility* is something that cannot be interpreted for the whole length of structure functions. The method is based on completely destroying the measurement arrangement between the dry and the wet step, lifting the sample, changing the surface quality or using another cold plate. The actual structure functions will be different *after* the separation point in each measurement, but the part belonging to the internal structures of the device is stable and highly reproducible.

As previously discussed, in the *two-point method*, the three-dimensional heat conducting path is distilled automatically into a single (rather uncertain) number. In the TDIM methodology, we get a highly repeatable 1D projection of the 3D structure.

The software distributed with the present standard prescribes actual thresholds only for small packages of discrete devices.

As the standard does not explicitly state the size of the *package*, this way it stays for characterizing larger modules for which the realistic ε threshold is a few tens of millijoule/kelvin.

The robustness of the TDIM methodology is verified by the large user community of the JEDEC JESD 51-14 standard. A round robin test with statistical distribution results is presented, among others, in Reference [27].

8. Case Study: Comparison of R_{thJC} Values Gained from Different Methodologies in Actual Tests

In a first case study, n-channel power MOSFET devices (HUF75639G3 from ON Semiconductor, [28]) were tested in several arrangements.

The device is available in different packages. It is designed for fast switching at high current and voltage, with the maximum ratings of 56 A and 100 V. For this reason, the chip is thin and the silicon nearly fills up the approximately 6 mm × 8 mm available space in the small TO263 package in which it is also offered.

The TO247 package version was selected for the measurements, because this was the largest available with a cooling area of 13 mm × 13 mm at its bottom. Presumably the lateral displacement of the probe will cause the smallest error in two-point measurements with this package.

The data sheet specified a 0.74 K/W maximum R_{thJC} value for the packaged device, and typical values were not provided.

The TDIM measurement result of a typical device is shown in Figure 18. The internal structures can be well observed in the fully coinciding structure functions until 0.3 K/W. An R_{thJC} junction to case thermal resistance of 0.31–0.38 K/W can be deduced from curves using different ε divergence criteria.

An alternative technique can be introduced in TDIM analysis for providing a highly reproducible single number thermal descriptor. As stated, in Figures 9 and 18b, the structure functions coincide until the divergence point. Choosing a C_{th} thermal capacitance value just below the divergence point, for example, C_{th} = 20 J/K in Figure 9 or 0.3 J/K in Figure 18b, we shall get a repeatable number for a partial thermal resistance, independently from the quality of the TIM and cold plate used in the measurement setup.

This quantity is still unnamed and could be denoted as $R_{th@Cth}$. Its use can be easily extended to a population of devices from the same type. In power devices, some structural layers are of high thermal capacitance and of precise geometrical dimensions such as silicon, ceramics, and copper plates. Some layers are thin but of varying thickness and have lower thermal conductivity but negligible thermal capacitance. Such layers are the die attach and other TIM. These features imply that reading out $R_{th@Cth}$ at fixed C_{th} yields a relevant measure on the scatter of the production quality in type tests.

Simple back of envelope calculations also support the validity of the thermal capacitance values read in Figure 18. The copper tab of the TO247 package is approximately of 15 mm × 12 mm × 2 mm size, and its volume is approximately 360 mm^3. This volume of copper yields 1.2 J/K thermal capacity for the copper block. However, the silicon chip on the top of the copper is significantly smaller; it is also encapsulated into small packages like DPAK. The heat propagates in a truncated pyramid from the top to the bottom of the copper block, and the pyramid has a volume of approximately one-third of the total block. This volume corresponds to C_{th} = 1.2/3 J/K = 0.4 J/K, fitting well the reading in Figure 18.

Measuring a number of the devices in commercially available test fixtures [25], one can get rather different results. One such fixture has a solid copper mounting plate of high heat transfer coefficient ensured by liquid cooling (type highHTC below). A former version of the fixture (type lowHTC below) has a lower heat transfer coefficient (air cooling). Both fixtures have a spring-loaded PTFE-covered thermocouple probe under the package.

Seven samples of the MOSFET were measured in both fixtures as available stock parts from the distributor with case planarity and roughness as produced. Another seven samples were flattened and polished on their case surface. The measured R_{thJC} values are listed in Table 2.

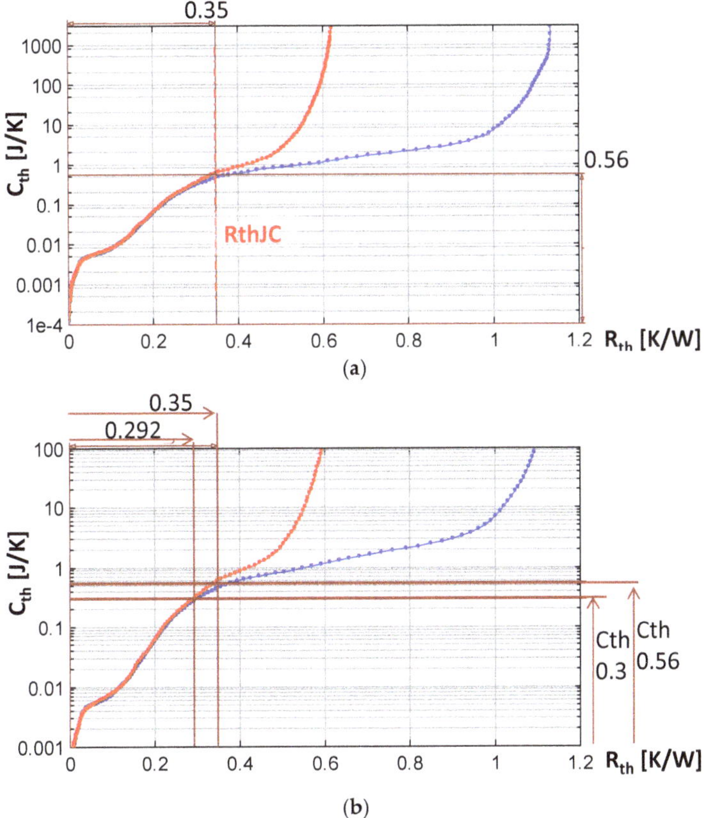

Figure 18. Dual interface measurement of HUF75639G3 structure functions. (**a**) R_{thJC} junction to case thermal resistance determined with $\varepsilon = 0.05$. (**b**)Enlarged detail of a $R_{th@Cth}$-style thermal parameter read-out at $C_{th} = 0.36$ J/K.

Table 2. Measured R_{thJC} values of HUF75639G3 samples, TO247 case, two-point method.

	Fixture Type	Mean of 7 Measured Samples (K/W)	SD
Part from stock	lowHTC	0.42	5.3%
	highHTC	0.57	8.7%
Case flattened and polished	lowHTC	0.29	3.9%
	highHTC	0.38	2.1%

We can observe that the R_{thJC} thermal metrics are not inherent constant values belonging to a packaged device, but are rather a function of external factors like the heat transfer coefficient of the measurement environment, probe construction, etc. The external conditions influence the shape of the heat spreading trajectories in the internal layers, too. A higher heat transfer coefficient at the device surface results in higher measured R_{thJC} (consequence of a flatter temperature distribution on the case in Figure 5). The TDIM method is less sensitive on the variation of conditions at the case surface.

It has to be noted that the datasheet of the part [28] also presents a Foster-style, one-dimensional compact model (Figure 8a) for the MOSFET, consisting of six RC stages; this was one of the reasons for the sample selection. We simulated the model in a realistic thermal boundary, and we found a poor match with Figure 18.

A deep analysis carried out at Infineon and presented in Reference [22] compares:

- Simulated R_{thJC} values with an ideal heat sink, considered as a fixed-temperature case surface corresponding to an infinite heat transfer coefficient;
- Simulated R_{thJC} values in a wide range of heat transfer coefficients;
- Measured values in two laboratories with two-point measurements;
- TDIM measurements.

A summary of the results is presented in Table 3 below and in Figure 19.

Table 3. Comparison of R_{thJC} of a MOSFET device obtained using different methods.

Method	R_{thJC} (K/W)	ΔR_{thJC1} (K/W)	ΔR_{thJC1} (%)	ΔR_{thJC2} (%)
FE simulation, floating case temperature BC	0.262	0	0	31%
FE simulation, constant case temperature BC	0.304	0.042	16%	−20%
1st Thermocouple measurement, apparatus I	0.35	0.088	34%	−8%
2nd Thermocouple measurement, apparatus I	0.38	0.118	45%	0%
1st Thermocouple measurement, apparatus II	0.42	0.158	60%	11%
2nd Thermocouple measurement, apparatus II	0.48	0.218	83%	26%
1st TDIM measurement	0.28	0.018	7%	−26%
2nd TDIM measurement	0.29	0.028	11%	−24%
3rd TDIM measurement	0.26	−0.002	−1%	−32%
4th TDIM measurement	0.29	0.028	11%	−24%
5th TDIM measurement	0.29	0.028	11%	−24%

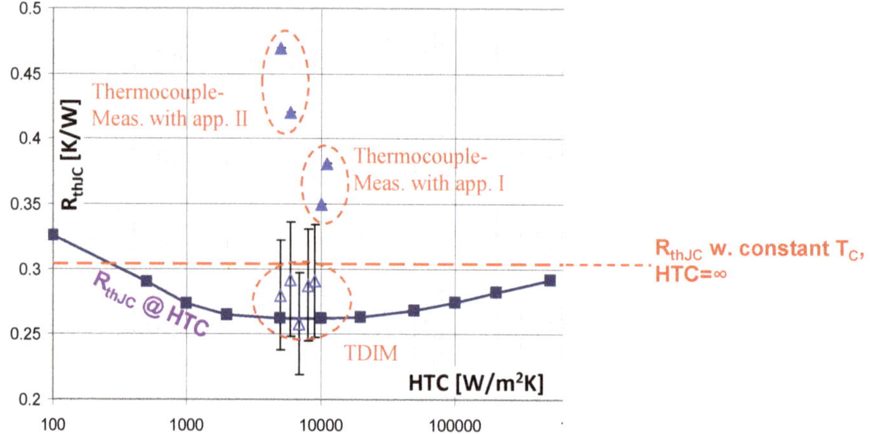

Figure 19. Comparison study at Infineon from Reference [22] and redrawn. The heat transfer coefficient (HTC) of the cold plate may vary within a certain range, and this is illustrated by fictive HTC values chosen for the abscissae of the measurement points. The ordinates are actual measured values (Table 3).

A realistic estimation of the heat transfer coefficient of a cold plate is approximately 3000–6000 W/m²K. For this reason, in a first comparison we took the result of the "floating case" FE simulation around these heat transfer coefficients as the basis for evaluating the values obtained with other methodologies. In the second column of Table 3, the measured R_{thJC} value is shown, and in the third and fourth columns, the ΔR_{thJC1} difference from the simulated reference value in absolute numbers and percentage, respectively. The reference values are highlighted in Table 3 with the bold border of the first row.

In an other approach we can compare the values from all methodologies to the two-point measurements (fifth column in the table, the fourth row highlighted with bold border as reference).

The R_{thJC} values from finite element simulation with both heat sink models were systematically lower than the values obtained by thermocouple measurements in two different thermal labs using different setups (apparatus I and II). Thermocouple measurements were 34%–83% higher than those predicted by simulation with realistic heat sink (floating case temperature boundary condition). The TDIM measurements provided only slightly higher values than the reference.

It was identified that the root cause of the large scatter in the measured values was that it was hard to accurately measure the case temperature with a thermocouple, as the operator cannot guarantee that the thermocouple actually measures the true T_C temperature of the package and not the temperature of the heat sink or some average value in between. Still, the repeatability of the measurements was surprisingly good at the same site, same equipment, and with operators having the same training.

On the other hand, the reproducibility of the values from thermocouple measurements at different laboratories was poor. Taking the higher value as reference from the site producing lower values systematically, we still experienced up to a 26% deviation as shown in the last column in Table 3.

The repeatability of the TDIM measurements was good, because the measurement of the case temperature was not involved.

The accuracy of the TDIM technique is limited by other factors, for example, by noise in the Z_{th} measurement, the influence of the thermal interface on the separation point [3], and the finite resolution of the structure function [2]. The assessment of the accuracy is always difficult, since there are no exact reference values for R_{thJC}. Based on the experience of several hundred measurements and on comparisons with simulations, it is estimated that the accuracy of the TDIM method is approximately 15% (see error bars in Figure 19). While this seems to be not overly accurate, it is still a lot better than the reproducibility of the two-point measurements shown in the table and chart above.

Laboratories having both kinds of instruments reported junction to case thermal resistances measured with the two-point method as 20% lower to 50% higher than the TDIM result [22].

An elaborated study on the repeatability of junction to case thermal resistance values for larger packages with complex internal structures (i.e., FCBGA, CABGA) is presented in Reference [26]. A sort of round robin testing was carried out with three operators using the two-point measurement concept. The series of tests was built up in a way that first all operators used the same piece of equipment and the same calibration data, then each operator recalibrated the devices under test, but they used the same equipment, then separate instruments of identical composition were used. The variation of measured data was below 8%. This variation quickly grew when the composition of the cold plate and the heat transfer coefficient of the measurement environment were changed. A similar study with associated simulation experiment is presented in Reference [29].

The impact of run-in effects in the fixture used for the transient measurement is highlighted in Reference [30].

A large round robin test involving several types of power LED devices was carried out in the European Delphi4LED project [27]. It included the measurement of the optical and thermal parameters of the same LED samples at five different European research and academic institutions. They used the same make of test equipment but carried out the calibration and the thermal transient tests independently. The reproducibility of the measured thermal resistance values was surprisingly good, within 1%–2% [27].

9. Discussion

Thermal testing has always been an integral part of the testing scheme of active components, but its importance has significantly grown with the advent of newer discrete devices and modules which are built of large and thin chips and package materials of high thermal conductivity.

Thermal tests are needed during all phases of development, and similar tests have to be carried out in the production again. Present trends extend thermal testing to the whole life cycle of an actual component including its live operation in the field. In the development phase, the performance of intermediate products can be revealed by thermal testing. At the end of the development data sheet values have to be provided for the ready product. However, single descriptive numbers like the R_{thJC} junction to case thermal resistance cannot be used for adequate selection of a part for an actual design, as their definition is based on supposing isothermal surfaces which almost never exist in practice. Moreover, they are often based on measurements of poor reproducibility, and for this reason the values in data sheets are published with an unknown safety margin.

More complex compact thermal models composed of a net of thermal resistances can be better used in thermal characterization to enable the reliable design of equipment. These models reflect the behavior of the components in a more precise way without revealing confidential structural details. Such models can be derived from a set of thermal measurements and simulations.

In production, a larger number of tests have to be carried out. Related standards distinguish between *type tests* and *routine tests*.

Type tests are carried out on samples of new products in order to determine the electrical and thermal ratings of a type and for establishing test limits for further tests. Such tests are often of destructive nature. The type tests are repeated regularly on a given number of samples taken from manufacturing batches at the manufacturer or delivery batches at the end-user in order to confirm the quality of the product. *Routine tests* are carried out on each sample of the production or delivery.

The type tests repeated at regular production intervals and the routine tests have to be relatively simple and should not be time consuming. For this reason, so far, it seemed to be satisfactory to provide only simple numbers describing component quality derived from temperature measurements at dedicated accessible points of the component.

Other related test categories can be reliability tests and failure tests on faulty components. Measurement of thermal parameters for health monitoring in live operational systems is also gaining importance; such tests can be quasi-continuous or can be repeated time by time.

In all cases, the minimum time needed for carrying out a thermal test is significantly longer than the comparable time needed for electrical tests. For a discrete device several seconds are needed, and for a module, at least tens of seconds are needed to reach thermal stability. The steady state is reached through a heating transient which is followed by an inherent cooling transient, and the two are needed for an accurate thermal transient measurement.

All test types aim at determining the most critical thermal parameter, the semiconductor chip temperature from a transient event.

The best way to gain information on the chip temperature is selecting a temperature-dependent electric parameter of the active device, such as the forward voltage of an internal pn junction or the threshold voltage of a MOSFET, and mapping the value of this parameter to the approximate temperature of the chip. This voltage to temperature calibration process occurs in a thermostat; the parameter value is recorded at several temperatures. In order to ensure that the chip temperature does not significantly differ from the external temperature in this process, low power has to be maintained on the device during the calibration. A typical way is applying a low "measurement current" on a pn junction and recording the corresponding voltage.

But the actual thermal parameters can be determined only in a high-powered state. The only way to gain the semiconductor temperature at high power is by switching to the low measurement current used at calibration and checking the actual value of the calibrated temperature-sensitive parameter.

Present day transient test schemes switch down to measurement current once and record the cooling at a high sampling rate until the cold steady state is reached. This way accurate temperature data are collected for the whole cooling process, except for the short-time interval around the switching when the electric perturbation distorts the temperature signal.

Static test schemes switch down repetitively and use a not too sharply defined "proper time" for measuring the calibrated parameter at low power. Proper time is where the electric distortion already decays but the temperature of the chip still does not significantly drop. Static techniques may abort the cooling transient record after this time, but this is not explicitly stated in the related standards.

In both schemes, extrapolation techniques can be used for estimating the starting temperature just after switching.

Static and transient thermal tests can both be carried out by measuring the temperature at a single point in an assembly or at multiple accessible points.

In the case of a static test, the only way to obtain the thermal characteristics of a specific device or module within a larger assembly is by making temperature measurements at multiple accessible points, otherwise the segment in the heat conducting path belonging to the very device cannot be distinguished from the other parts of the assembly. In transient tests of a layered structure, portions of different thermal conductivity and specific heat can be mapped, and such partial thermal resistances can be determined, and even the internal temperature distribution can be concluded.

In the standardized transient dual interface measurement methodology (TDIM), in each thermal measurement, the whole heat conducting path is characterized, from the heat source to the ambient. This way distinguishing between *component* and *test environment* is achieved by the intentional structural change at the geometrical interface separating the device from the test bench (such as a cold plate).

We used simulation experiments and actual tests to analyze the accuracy, repeatability, and reproducibility of thermal tests. For demonstrating the concept, we selected the simplest thermal descriptors, the junction to case thermal resistance of a device and the junction to ambient thermal resistance of an assembly.

We verified with simulation experiments that the R_{thJC} thermal metrics depend on the TIM quality used in the test bench, on the lateral displacement of the probe measuring the "case" temperature, on the penetration of the probe through the TIM, and on the heat transfer coefficient of the cold plate and other factors as well, resulting in a large uncertainty of the obtained value. In the case of a single-point transient test, the assembly is totally destroyed and rebuilt between the two measurements. Differences in TIM quality belong to the essence of the technique. Still, although the structure functions are highly reproducible, a decision on the threshold used has to be made in order to define at how large divergence it is considered to be the R_{thJC} value.

In actual thermal tests we found that the accuracy, repeatability, and reproducibility of static and transient tests depend on the following:

- Electrical transient at the switching process, as defined above;
- Power measurement uncertainty on the device causes real ambiguities only at large modules with complex internal wiring;
- Offset and gain errors in the data acquisition; these are the source of most reproducibility issues for multi-point measurements while indifferent in one-point transient measurements;
- Reproducibility of the selected samples, such as die attach thickness, base plate roughness, and planarity;
- Reproducibility of the test environment.

Regarding this last issue, different laboratories have different materials and geometries for the cold plate used in the measurements, other formations of the liquid flow, various surface roughness and planarity levels, and types and positions of external temperature sensors resulting in a large scatter of the obtained values.

We studied actual differences in static and transient measurements in several case studies. In the actual tests, we found that there was a systematic difference between the thermal data measured with the TDIM method and that measured with temperature probes, but this difference was smaller than the scatter in results measured at different laboratories with the latter method.

Author Contributions: G.F. and Z.S. carried out the transient tests and the simulation experiment presented in Sections 3 and 4. D.S. conducted the round robin tests presented in Section 7. G.F. formulated the bulk of the paper and designed the figures. M.R. provided the concept of the paper, elaborated the mathematical background and confirmed the validity of the results. All authors have read and agreed to the published version of the manuscript.

Funding: This research received no external funding.

Conflicts of Interest: The authors declare no conflict of interest.

Appendix A

Table A1. Thermal parameters of the materials used in the example of Section 3: λ—thermal conductivity, ρ—density, c—specific heat, c_V—volumetric specific heat.

	λ (W/mK)	ρ (kg/m³)	c (J/kgK)	$c_V = \rho \cdot c$ (kJ/m³K)
Silicon	450	2330	750	1750
Die attach *	100	8000	200	1600
Copper	385	8930	385	3440
AlN (Aluminum nitride) ceramics *	250	3500	740	2590
Solder compound *	100	8000	200	1600
Aluminum alloy	150	2710	910	2470
Thermal grease *	0.2, 1, 4	2000	1500	3000

Symbol * denotes estimated values based on literature.

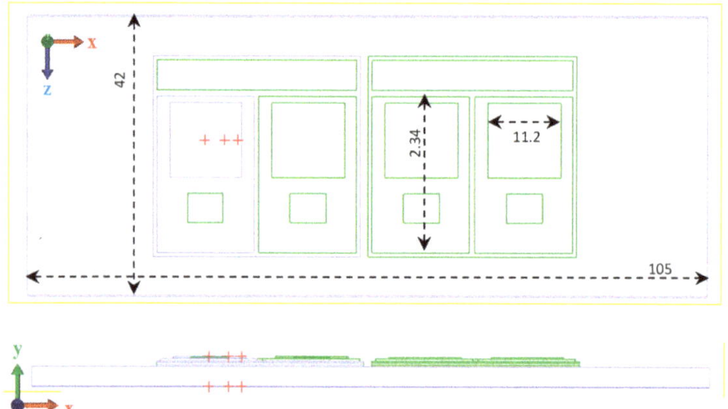

Figure A1. Sketch of the power module on the cold plate from the simulation study in Section 3. Stack composition and the size of elements are listed below in Table A2. Temperature monitor points are marked with "+".

Table A2. Stack composition and the size of elements in the example in Section 4: x, z—lateral size, y—thickness in the stack, V—volume of the element, c_V—volumetric specific heat, C_{th}—thermal capacitance of the element, ΣC_{th}—cumulative thermal capacitance from the chip top.

	x Size (mm)	z Size (mm)	y Size (mm)	V (mm³)	c_V (kJ/m³K)	C_{th} (J/K)	ΣC_{th} (J/K)
Silicon	11.2	11.2	0.3	37.6	1750	0.066	0.066
Die attach	11.2	11.2	0.1	12.5	1600 *	0.020	0.086
Copper	15	23.4	0.3	105	3440	0.36	0.45
AlN ceramics	32	30	0.3	288	2590 *	0.75	1.19
Copper	32	30	0.3	288	3440	0.99	2.18
Solder compound	32	30	0.3	288	1600 *	0.46	2.65
Aluminum	105	42	2.9	12,790	2160	27.6	30.3
Thermal grease	105	42	0.05	221	3000 *	0.00066	30.3
Aluminum	500	140	20	1,400,000	2160	3024	3054.3

Symbol * denotes estimated values based on literature.

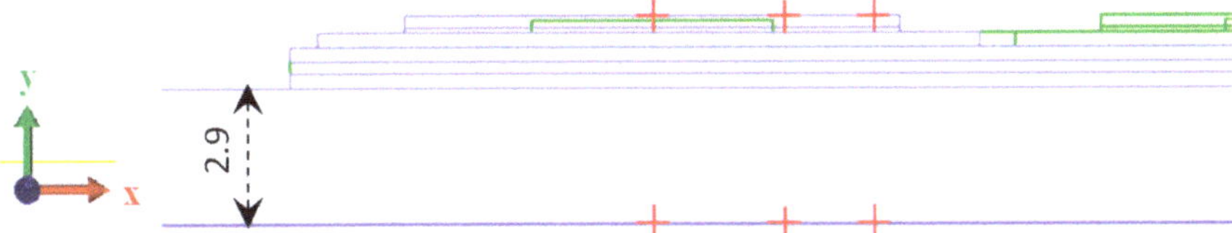

Figure A2. Power module on a cold plate from the simulation study in Section 3, excerpt from Figure A1. Temperature monitor points are marked as "+".

Figure A3 demonstrates a TDIM measurement of an IGBT module by a thermal transient tester. The measurement environment is a water-cooled cold plate, wetted by thermal grease in the figure. More images of the equipment can be found in References [20,21].

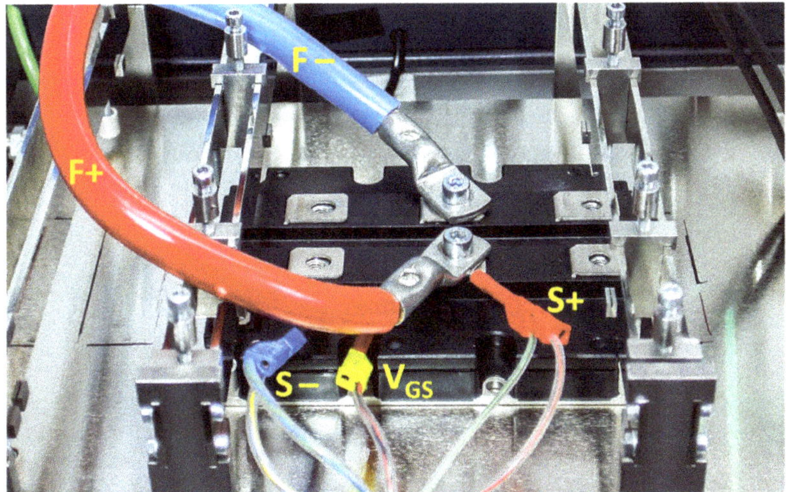

Figure A3. IGBT module prepared for TDIM measurement on cold plate. I_H and I_M applied on F+ and F− leads, measurement between S+ and S−. Eventual gate voltage applied to V_{GS}.

References

1. Szekely, V. Identification of RC networks by deconvolution: Chances and Limits. *IEEE Trans. Circuits Syst. Fundam. Number Theory Appl.* **1998**, *45*, 244–258. [CrossRef]
2. Schweitzer, D.; Pape, H.; Chen, L. Transient Measurement of the Junction-To-Case Thermal Resistance Using Structure Functions: Chances and Limits. In Proceedings of the 2008 Twenty-fourth Annual IEEE Semiconductor Thermal Measurement and Management Symposium, San Jose, CA, USA, 2 May 2008. [CrossRef]
3. Schweitzer, D. Transient Dual Interface Measurement of the Rth-JC of Power Packages. In Proceedings of the 14th Thermal Investigation of ICs and Systems, Rome, Italy, 24–26 September 2008. [CrossRef]
4. Farkas, G.; Sarkany, Z.; Rencz, M. Structural Analysis of Power Devices and Assemblies by Thermal Transient Measurements. *Energies* **2019**, *12*, 2696. [CrossRef]
5. Szabo, P.; Steffens, O.; Lenz, M.; Farkas, G. Transient junction-to-case thermal resistance measurement methodology of high accuracy and high repeatability. *IEEE Trans. Compon. Packag. Technol.* **2005**, *28*, 630–636. [CrossRef]
6. Steffens, O.; Szabo, P.; Lenz, M.; Farkas, G. Thermal transient characterization methodology for single-chip and stacked structures. In Proceedings of the Semiconductor Thermal Measurement and Management Symposium, San Jose, CA, USA, 15–17 March 2005. [CrossRef]

7. Farkas, G. Thermal transient characterization of semiconductor devices with programmed powering. In Proceedings of the Semiconductor Thermal Measurement and Management Symposium (SEMI-THERM), San Jose, CA, USA, 17–20 March 2013.

8. Rencz, M.; Szekely, V. Non-linearity issues in the dynamic compact model generation. In Proceedings of the Semiconductor Thermal Measurement and Management Symposium, San Jose, CA, USA, 11–13 March 2003.

9. IEC/EN 60747-2. Standard: "Semiconductor devices—Part 2: Discrete devices—Rectifier diodes". Available online: https://webstore.iec.ch/publication/24519 (accessed on 13 January 2020).

10. IEC/EN 60747-15. Standard: "Semiconductor Devices-Discrete Devices Part 15: Isolated Power Semiconductor Devices". Available online: https://webstore.iec.ch/publication/3255/ (accessed on 13 January 2020).

11. MIL-STD-750D. Test Methods for Semiconductor Devices. Available online: https://www.navsea.navy.mil/Portals/103/Documents/NSWC_Crane/SD-18/Test%20Methods/MILSTD750.pdf (accessed on 13 January 2020).

12. JEDEC Standard JESD51. Methodology for the Thermal Measurement of Component Packages (Single Semiconductor Devices). Available online: https://www.jedec.org/standards-documents/docs/jesd-51 (accessed on 13 January 2020).

13. JEDEC Standard JESD 51-14. Transient Dual Interface Test Method for the Measurement of the Thermal Resistance Junction-To-Case of Semiconductor Devices with Heat Flow Through a Single Path. 2010. Available online: www.jedec.org/sites/default/files/docs/JESD51-14_1.pdf (accessed on 13 January 2020).

14. JEDEC JESD15-3. Standard: Two-Resistor Compact Thermal Model Guideline. Available online: https://www.jedec.org/standards-documents/docs/jesd-15-3 (accessed on 13 January 2020).

15. JEDEC JESD15-4. Standard: Delphi Compact Thermal Model Guidelines. Available online: https://www.jedec.org/standards-documents/docs/jesd-15-4 (accessed on 13 January 2020).

16. ECPE Guideline AQG 324. Automotive Qualification Guideline. Available online: https://www.ecpe.org/research/working-groups/automotive-aqg-324/ (accessed on 13 January 2020).

17. CIE. *Optical Measurement of High-Power LEDs*; CIE Technical Report 225:2017; CIE: Vienna, Austria, 2017. [CrossRef]

18. Tang, Y. A Modified Single Pulse Method for Transient Thermal Impedance (TTI) Measurement of VDMOSFET Relates Gate Bias to the TTI Results. *J. Semicond. Technol. Sci.* **2018**, *18*. [CrossRef]

19. FloTHERM. Available online: https://www.mentor.com/products/mechanical/flotherm/flotherm/ (accessed on 13 January 2020).

20. T3Ster®. Available online: http://www.mentor.com/products/mechanical/products/t3ster (accessed on 13 January 2020).

21. Power Tester 1500A. Available online: https://www.mentor.com/products/mechanical/micred/power-tester-1500a/ (accessed on 13 January 2020).

22. Schweitzer, D. The junction-to-case thermal resistance: A boundary condition dependent thermal metric. In Proceedings of the Semiconductor Thermal Measurement and Management Symposium, Santa Clara, CA, USA, 21–25 Febuary 2010. [CrossRef]

23. Vass-Varnai, A. Issues in junction-to-case thermal characterization of power packages with large surface area. In Proceedings of the Semiconductor Thermal Measurement and Management Symposium, San Jose, CA, USA, 21–25 February 2010. [CrossRef]

24. Bein, M.C.; Hegedüs, J.; Hantos, G.; Gaál, L.; Farkas, G.; Rencz, M.; Poppe, A. Comparison of two alternative junction temperature setting methods aimed for thermal and optical testing of high power LEDs. In Proceedings of the 23rd International Workshop on Thermal Investigation of ICs and Systems (THERMINIC'17), Amsterdam, The Netherlands, 27–29 September 2017. [CrossRef]

25. Rjc Liquid Cooled Test Fixture. Available online: http://analysistech.com/semiconductor-thermal-tester/rjc-liquid-test-fixture/ (accessed on 8 October 2019).

26. Galloway, J.; de los Heros, E. Developing a ThetaJC standard for electronic packages. In Proceedings of the Semiconductor Thermal Measurement and Management Symposium, San Jose, CA, USA, 19–23 March 2018. [CrossRef]

27. D2.1-Report on Round-Robin Testing of LEDs. Available online: https://delphi4led.org/pydio/public/2f72dd (accessed on 5 October 2019).

28. HUF75639G3, HUF75639P3, HUF75639S3S, HUF75639S3. Available online: https://www.onsemi.com/pub/Collateral/HUF75639S3S-D.PDF (accessed on 8 October 2019).

29. Galloway, J.; Bhopte, S.; Nelson, C. Characterizing junction-to-case thermal resistance and its impact on end-use applications. In Proceedings of the Semiconductor Thermal Measurement and Management Symposium, San Diego, CA, USA, 30 May–1 June 2012. [CrossRef]
30. Deng, E.; Zhao, Z.; Zhang, P.; Li, J.; Huang, Y. Study on the Method to Measure the Junction-to-Case Thermal Resistance of Press-Pack IGBTs. *IEEE Trans. Power Electron.* **2018**, *33*, 4352–4361. [CrossRef]

Article

Thermal Modelling of a Prismatic Lithium-Ion Cell in a Battery Electric Vehicle Environment: Influences of the Experimental Validation Setup [†]

Jan Kleiner *, Lidiya Komsiyska, Gordon Elger and Christian Endisch

Technische Hochschule Ingolstadt, Institute of Innovative Mobility, Esplanade 10, 85049 Ingolstadt, Germany; Lidiya.Komsiyska@thi.de (L.K.); Gordon.Elger@thi.de (G.E.); Christian.Endisch@thi.de (C.E.)

* Correspondence: jan.kleiner@thi.de

† This paper is an extended version of our paper pressed in 2019 IEEE 25th International Workshop on Thermal Investigation of ICs and Systems (THERMINIC 2019), Lecco, Italy, 25–27 September 2019.

Received: 18 November 2019; Accepted: 18 December 2019; Published: 20 December 2019

Abstract: In electric vehicles with lithium-ion battery systems, the temperature of the battery cells has a great impact on performance, safety, and lifetime. Therefore, developing thermal models of lithium-ion batteries to predict and investigate the temperature development and its impact is crucial. Commonly, models are validated with experimental data to ensure correct model behaviour. However, influences of experimental setups or comprehensive validation concepts are often not considered, especially for the use case of prismatic cells in a battery electric vehicle. In this work, a 3D electro–thermal model is developed and experimentally validated to predict the cell's temperature behaviour for a single prismatic cell under battery electric vehicle (BEV) boundary conditions. One focus is on the development of a single cell's experimental setup and the investigation of the commonly neglected influences of an experimental setup on the cell's thermal behaviour. Furthermore, a detailed validation is performed for the laboratory BEV scenario for spatially resolved temperatures and heat generation. For validation, static and dynamic loads are considered as well as the detected experimental influences. The validated model is used to predict the temperature within the cell in the BEV application for constant current and Worldwide harmonized Light vehicles Test Procedure (WLTP) load profile.

Keywords: lithium-ion battery; thermal modelling; electro-thermal model; heat generation; experimental validation

---

1. Introduction

Realizing the vision of green mobility one major aspect is the reduction of worldwide car emissions by use of battery electric vehicles (BEVs). Automotive manufactures focus on lithium-ion based battery systems due to their high specific energy, low self-discharge and long cycle-life to meet the challenges of the continuously rising environmental regulations [1–3]. In BEVs different lithium-ion cell formats like pouch, cylindrical, and prismatic cells are used. Thereby, the thermal management is performed with various cooling concepts based on air, liquid and other approaches [3]. Cooling selection depends e.g., on the battery concepts using module architecture, or not. However, the temperature has a significant influence on the lithium-ion battery performance, ageing and safety [2]. Therefore, investigation of the heat generation and temperature development are important to examine thermal cell behaviour with regard to boundary conditions, especially the cooling conditions.

In order to answer the scientific issue of thermal battery behaviour, researchers develop different electro-(chemical)-thermal coupled simulation models [4–12]. Most of them are based on the general energy balance for lithium-ion based systems [13] in its detailed or simplified form.

In lithium-ion batteries, non-linear physico–chemical processes occur on different length scales from nanometre to millimetre. To face this challenge, model-based thermal investigations of battery cells are performed and reported in the literature in two fundamentally different ways: They use independent thermal models on the one hand with defined heat generation and on the other hand models with electrical or electrochemical-thermal coupling [6–8,10–12,14].

The heat generation of the former can be empirically determined [10], or e.g., modelled by a neural network [14]. The latter approach, electrically and thermally coupled behaviour, is described in the literature with different model approaches. Researchers use either the pseudo-two-dimensional modelling approach [7,8,12] based on the work of Newman et al. [15,16], equivalent circuit models (ECM) [9], and empirical models [11]. The thermal model can i.e., be executed as a point model [6] or can also be resolved three-dimensionally [7].

As in our previous research [17], in this work, an electrical ECM model with coupled 3D thermal model is used, because this approach represents a sufficient accuracy for thermal questions with at the same time acceptable computing time. The model uses the framework of the MSMD battery model architecture. This adopted general multi-scale multi-dimensional (MSMD) approach is described and used in [18,19].

In general, for validating and calibrating the model, the simulation results are compared to experimental data. Thereby, non-precise known model parameter is adjusted, i.e., optimized, to ensure correct model prediction over a defined range of model conditions. Often, important influence of the experimental setup is neither mentioned nor included in the boundary conditions within the model for validation. Especially convection conditions are often assumed to be constant [12] while validating the model's behaviour. This generates uncertainties in the predicted thermal behaviour of the battery cell after validation. Only few studies consider this issue [4,5,20] by special setup or boundary conditions in simulation. Erhard develops an experimental setup with a pouch cell in a wind tunnel to obtain knowledge of the convection coefficient for meaningful thermal validation [4]. Furthermore, several experimental influences are monitored during the experiment used as an input to the model for validation. Rieger et al. [20] use a boundary condition for a heat flux at the terminal connection of pouch cells in the experiment. Samad et al. [5] investigate forced and natural convection conditions with prismatic cells and find uneven convection conditions depending on the position of the cell in the temperature chamber.

In this work, we focus on the automotive use case of prismatic cells in module architecture on a cooling plate-like in [6,7]. Lundgren et al. [7] suggest a detailed 3D electrochemical-thermal model with experimental validation of a prismatic cell. However, they do not consider the influences of the experiment such as wiring to the battery cycling machine or the cell's environment. Damay et al. [6] introduce an electrical model coupled with a lumped thermal model for investigation of the thermal behaviour. Experimental influences such as insulation plates are taken into account but the constant convection condition based on literature results in uncertainty for model validation.

However, none of these studies for prismatic automotive cells take special care of experimental influences and high-resolution validation. Experimental investigations show equal temperature gradients of large-format cells [21,22] and, therefore, the necessity of spatially resolved temperature measurement and validation. Panchal et al. [21] show a temperature gradient on the casing of up to 4 K at the end of constant current discharge with a 20 Ah cell. Christen et al. [22] reveal a temperature gradient of 6.5 K at the casing of their 60 Ah prismatic cell after constant current cycling with cooling at the cell's bottom at a temperature of 25 °C. As a result of temperature gradients on the casing, neither a single temperature sensor in experiment nor an average temperature value for validation is conclusive enough. Temperature gradients occur due to cooling conditions and thermal resistances in- and outside the cell and need to be validated. Therefore, a validation concept is needed containing the differences of the experimental conditions and the use case of a prismatic cell in a BEV application.

The goal of this work is to investigate the commonly neglected influences of experimental validation setups in detail. Based on the gained knowledge a detailed validation for a 3D electro–thermal model of a prismatic lithium-ion cell is performed and the thermal behaviour in a BEV application is predicted. Therefore firstly, the focus is on the development of a single cell experimental setup for investigation of external influences. Secondly, the impact of boundary conditions on the behaviour of the used electro-thermal model is investigated. Based on this, a validation of spatially resolved temperatures and heat generation is performed for static as well as dynamic cases. Finally, the model is used to illustrate the thermal behaviour of the cell under real BEV application.

2. Modelling

2.1. General Approach

The studied cell in this work is a prismatic automotive cell. Geometric dimensions and electrical and thermal behaviour are determined experimentally and subsequently used for the parametrization of the model. Furthermore, physical properties are taken from [7,23]. Important for the thermal behaviour of the cell are the conditions of use, especially the cooling conditions. The general set up of a prismatic cell positioned in a real BEV battery module is shown in Figure 1a. In automotive application, prismatic cells are commonly stacked in modules and connected by busbars. This stack of cells is compressed for better performance. Cooling is realized by a fluid cooling system at the bottom of the cells [6]. The cooling system is commonly working with constant cooling capacity in various stages to keep the cells in the best case at constant temperature in the desired range. Simulating one cell in the middle of a module, the resulting boundary conditions of the outer walls are adiabatic beside the bottom wall where the cooling system is connected. The cell's floor is modelled as constant temperature boundary condition, neglecting the small temperature gradient in the fluid of the cooling system for one cell to avoid detailed computational fluid dynamics simulation (CFD) of the complex cooling system in a BEV.

It is obvious that the cell in any experimental setup is experiencing slightly different boundary conditions in comparison to the model setup for simulation. Validation of a single cell in a full battery module is often not practical. Therefore, for validation of single-cell models, a more realistic validation setup is necessary that takes account for the targeted stimulation conditions as well as the application influences. The schematic of the resulting laboratory setup on single-cell level is depicted in Figure 1b. The resulting boundary conditions for model validation are comparable with Figure 1a with regard to adiabatic conditions on the models outer surfaces and constant temperature cooling conditions at the cell floor. Additionally, the bracing components, as well as the heat removal by the wiring, are considered. The modelling approach and boundary condition details are discussed in the following Sections 2.2 and 2.3, while the experimental setup is explained in detail in Section 3.

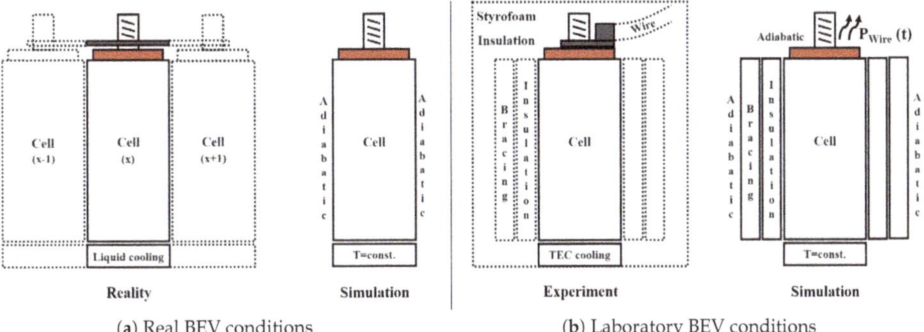

Figure 1. Schematic boundary conditions for a prismatic cell in battery electric vehicle (BEV) application for simulation and experiment: (**a**) real BEV scenario in a module of a battery system and (**b**) laboratory BEV scenario with single cell experimental setup using a thermoelectric cooler (TEC).

2.2. Electro-Thermal Co-Simulation

In this work, a multi-scale multi-dimensional (MSMD) approach was adopted and used, which is described in [18,19]. It is built in ANSYS CFD as MSMD battery model [24]. The model uses the framework of the MSMD Battery Model architecture and is parametrized for an electrical ECM and 3D thermal model.

The MSMD-approach makes it possible to resolve different phenomena on different length scales. It is not feasible to solve the cell on the smallest length scale in the nanometre range in FEM analysis without generating considerable computing times and networking problems. With MSMD approach individual phenomena with sub-models are solved without resolving the entire cell in detail. Nevertheless, it is possible to obtain information on cell level with a model less detailed. The MSMD approach does neither specify the type of sub-models nor their number but defines the information exchange between the model levels [18]. The main differential equations at cell level taken from [19,24] are

$$\frac{\partial \rho \cdot C_p \cdot T}{\partial t} - \nabla[k \cdot \nabla T] = \dot{q}_{ECh} + \dot{q}_{Ohm} \tag{1}$$

$$\nabla[\sigma_+ \cdot \nabla \phi_+] = -j_{ECh} \tag{2}$$

$$\nabla[\sigma_- \cdot \nabla \phi_-] = j_{ECh} \tag{3}$$

where (1) represents the spatial energy balance in the active material called jelly roll, the heart of a lithium-ion cell, where the electrochemic reactions and the heat generation take place. The internal energy is calculated by jelly roll's density ρ, specific heat capacity C_p, and Temperature T. The spatial heat conduction is expressed with heat transfer coefficient k. $\dot{q}_{Ohm}$ is the heat generation from ohmic losses of the current collector foils of copper/aluminium in the jelly roll [24]. $\dot{q}_{ECh}$ is the reaction heat from the electrochemical processes taking place such as charge transfer over-potentials at the interface, and mass transfer limitations [2] (see Equation (5) for details). Equations (2) and (3) are potential equations that describe the electrical behaviour. In these equations ϕ is the phase potential, j_{ECh} the volumetric current transfer rate, and σ the electrical conductivity of the materials [19,24].

Using an ECM for the electrical modelling of the cell's behaviour, the voltage is represented by

$$U(t) = U_{OCV}(SOC, T) - U_{serial}(SOC, T) - U_{RC}(SOC, T) \tag{4}$$

whereat the voltage U is calculated in the ECM sub-model by the open circuit voltage U_{OCV} and the voltage drop at the serial resistance U_{serial} and the RC-element U_{RC}. All used parameters depend on local temperature and state of charge (SOC) in the jelly roll.

The model structure of this work, implemented in the MSMD framework, is shown in Figure 2. In every time step, the model is iteratively solved for all components. The spatial temperature distribution is determined in the 3D thermal model. In every finite volume of the jelly roll, an electrical model is calculated containing the time- and local-depending temperature. Every single electrical model is implemented as an ECM. In the present model, the aim of the electro-thermal coupling is not to resolve the detailed chemical processes, but to map the resulting heat generation accurately for the thermal model. If a spatial resolution of the heat generation has aspired, ECM models represent a sufficient accuracy for thermal questions with at the same time acceptable computing time. The developed model structure is designed for first or second-order ECM. The necessary parameter sets for resistors and capacitors are determined from the current-voltage behaviour on discrete points in hybrid power pulse characterization (HPPC) test. During the simulation, these discrete data points are interpolated in the model. Additionally, the open-circuit voltage (OCV) of the used cell is determined experimentally.

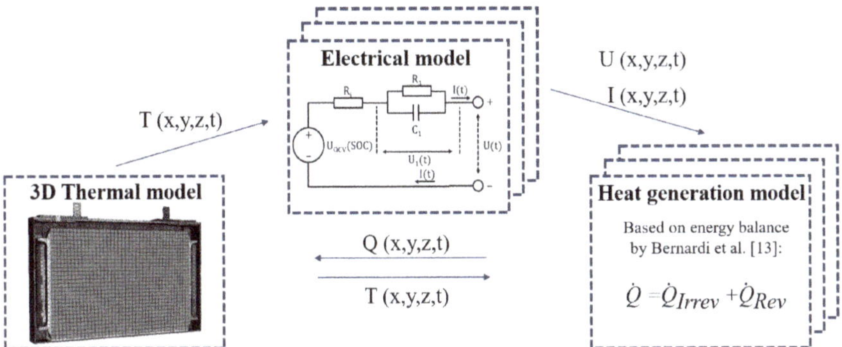

Figure 2. Model structure for electro–thermal co-simulation of 3D thermal model, multiple electrical equivalent circuit models (ECM) models and spatial heat generation models.

The present model uses the information for resistances and capacities as a function of temperature and SOC, but is designed for n-dimensional dependencies, i.e., if a relevant dependency of the parameters on other influencing variables, such as current or ageing, becomes apparent, corresponding parameter sets can be included in the model.

With available information of local voltage drop and current load, the power dissipation in the active material of the present model is determined locally by the approach suggested by Bernardi et al. with the following equation [19,24]:

$$\dot{q}_{ECh} = j \cdot [U - U_{OCV}(SOC, T)] + j \cdot T \cdot \frac{dU_{OCV}}{dT}(SOC). \tag{5}$$

The first term considers the power loss due to the voltage drop at the internal resistance of the cell. The voltage U below is calculated in the ECM sub-model, the open-circuit voltage U_{OCV} is part of the parameter set and the local current j results from the solution of the potential equations. This first part of the overall heat generation is irreversible. The second term considers the reversible heat generation based on an entropy change of the chemical system. The entropy change is shown e.g., in the temperature dependence of the OCV. In the literature, entropy change is determined experimentally for different combinations of anode/cathode pairs and recorded in the entropic heat coefficient dU/dT. This represents the reversible heat generation when multiplied with the local temperature T and the current j. The data set for the used NMC/Graphite chemistry is taken from [25].

Finally, the local heat generation is transferred in the 3D thermal model. The heat source changes temperatures, and therefore electrical resistances and heat generation, for the calculation of the next time step.

2.3. Thermal Modelling

Significant factors influencing the thermal model are the cell geometry, the associated material parameters, and the thermal boundary conditions. Geometrically, the thermal model of the prismatic cell is based on the information from the measured sectional models and a performed computer tomography and post-mortem analysis. This information is converted into a 3D CAD model, which is shown in Figure 3a. In the centre of the cell, the active material in the form of a jelly roll is modelled as one solid part with anisotropic properties of the thermal conductivity [7,23]. The copper and aluminium sheets characterize the high in-plane conductivity, while the separator in-between anode/cathode defines the lower through-plane conductivity. Furthermore, the real jelly roll is wounded. This is taken into account by a coordinate transformation regarding the conductivities in the outer rounding of the jelly roll geometry. The jelly roll's dimensions as a result of CT analysis are shown in Figure 3b. All components are modelled as solids without considering the fluids, e.g., the electrolyte at the cell's bottom or the air at the cell's top. The electrolyte is simplified and modelled as solid with heat conduction. Heat transfer by convection and radiation in the top part of the cell is negligible due to small temperature differences and small areas. Additionally, the following simplifications are made creating the geometry:

- Radii and chamfers of the cell components are neglected,
- Thin components are considered using thin wall modelling (e.g., insulation) [24],
- Jelly roll is modelled as a volume body with anisotropic thermal properties [7,23]
- Junction of jelly roll sheets and current collector is modelled as a volume body.

The materials and their associated physical parameters are listed in Table 1. The jelly roll with its individual components has already been combined into one material as mentioned above.

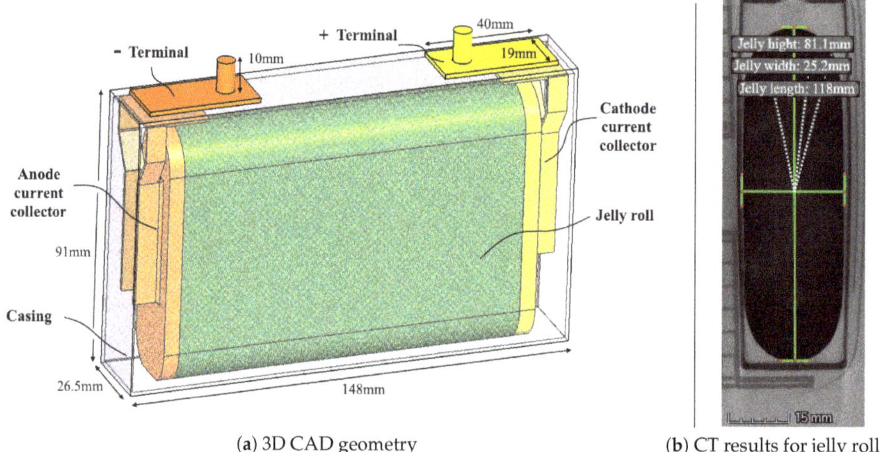

(a) 3D CAD geometry   (b) CT results for jelly roll

Figure 3. (a) The 3D CAD geometry and outer dimensions of prismatic 25 Ah cell with jelly roll, anode/cathode current collectors, +/− terminals, and casing. (b) Computer tomography (CT) results for jelly roll dimensions.

Table 1. Used materials and thermal properties of 3D thermal model of prismatic 25 Ah cell.

Material	Point of Use	ρ [kgm^{-3}]	C_p [Jkg^{-1}K^{-1}]	λ [Wm^{-2}K^{-1}]
Aluminium [a]	casing, collector, terminal	2700	900	238
Copper [a]	collector, terminal	8700	385	400
Insulation [b]	all insulations	1470	1190	0.18
Thermal pad [c]	connection cooling	2740	903	2.22
Electrolyte [d]	rest-electrolyte	1130	2055	0.6
Jelly roll [a]	jelly roll	2043	1371	in-plane 33 [e] trough-plane 0.7 [e]

[a] [23], [b] [7], [c] data-sheet, [d] [26], [e] cell manufacturer.

The general targeted thermal boundary conditions were previously shown in Figure 1a. These assumptions result in a model containing only the cell geometry including full adiabatic model boundary conditions on the outer surfaces, except for the constant temperature boundary condition at the bottom. Thereby, the heat exchange by the small side surfaces and the top of the cell is neglected due to small areas and temperature differences resulting in negligible impact. Nevertheless, for a correct model validation, the conditions of the experimental setup (see Figure 1b) must be mapped as thermal boundary conditions in the simulation model. The following thermal boundary conditions are implemented in the simulation model and investigated in the results section:

- Aluminium bracing plates and polyoxymethylene (POM) measurement plates are considered as solid bodies,
- Adiabatic conditions are modelled on all unattached surfaces of cell and bracing plates except terminals and cell bottom,
- Heat flow from/to the battery testers is considered at the terminal surfaces,
- Thermal pad is included in-between constant temperature cooling plate and cell floor.

The full adiabatic model boundary conditions on the outer surfaces result from the polystyrene insulation in the experiment. Only the heat sink is the constant temperature boundary condition at the bottom. The heat flow into the battery tester is determined in the experiment using two temperature sensors on the busbars. With knowledge of geometrical parameters and physical properties, the heat flux is calculated and considered as a time-dependent boundary condition in the simulation. The removed heat is monitored in every experiment as it depends on the temperature of the connected battery tester and can, therefore, have different impact for similar experiments.

In addition to the external boundary conditions, commonly neglected modelling parameters are considered within the model of the prismatic cell. Important for proper model behaviour are e.g., heat generation in the current collectors and electrolyte at the cell bottom. If the cell's heating is only located in the jelly roll, the local heat generation in the current collectors is missing. With increasing current collector length as in the case of the used prismatic cell, the simulation results show increasing temperature differences between model and experiment, especially at the terminals. The heat generation in the jelly roll takes place as source term in the elements of the jelly roll geometry and is calculated based on the electro-thermal coupling.

Computer tomographic analysis of the cell showed rest-electrolyte of 10–15 mL at the cell's bottom. In [27] electrolyte influence is neglected due to long distances for conduction and low thermal conductivity of $0.6\,\mathrm{Wm^{-1}K^{-1}}$ [26]. In the case of this cell, the distance between the jelly roll bottom and cell casing is small with a minimum spacing of 0.4 mm. Therefore, the rest-electrolyte creates an additional thermal path at the cell bottom from the jelly roll to the cooled cell bottom, modelled by a solid body with electrolyte properties (see Table 1).

The geometrical model of the cell with all associated boundary conditions was implemented in ANSYS Fluent 19.0. The mesh of the 3D cell model contained 138k elements. The maximal adaptive time step was set to 5 s. With changing current in the used constant current profile the time step was set back to 0.1 s starting the logarithmic rise up to 5 s. In the dynamic profile, a constant time step of 1 s was applied. The used computer was a Dell Workstation with 12x Intel(R) Xeon(R) Gold 6136 CPUs and 64 GB RAM.

3. Experimental

In this work, an automotive battery cell by SANYO PANASONIC is investigated. It is a lithium-ion cell with a nominal capacity of 25 Ah based on nickel–manganese–cobalt (NMC)/graphite chemistry. The cell has a nominal voltage of 3.7 V with upper and lower cut-off voltage of 4.1 V and 3.0 V, respectively. The goal of the experimental setup is to reproduce the mentioned real BEV conditions of a cell in Figure 1b as close as possible in a laboratory setup. Therefore, a single cell measurement setup is designed and implemented. In Figure 4, the sensor schematics (a), the setup concept (b), and the final test bench of the developed setup in the temperature chamber (c) are shown. A custom cooling plate by QuickCool with thermoelectric cooler regulation is used to guarantee constant temperature cooling condition. A proper thermal connection to the cell is achieved with a thermal pad (see properties in Table 1) and a vertical clamping by an external compression unit. Polystyrene foam is used as insulation material to receive mostly adiabatic conditions (see Figure 4c). The cell's optimal electrical performance is ensured by horizontal bracing plates made of aluminium with a thickness of 10 mm.

In-between the horizontal bracing and the cell, two measurement plates of 10 mm POM are added for thermal insulation and measurement.

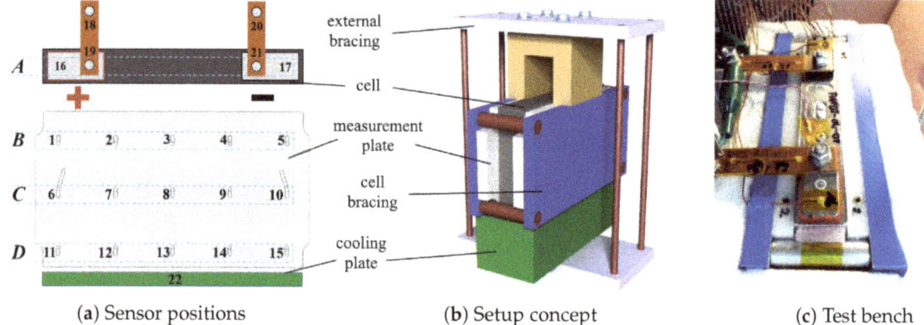

(**a**) Sensor positions   (**b**) Setup concept   (**c**) Test bench

Figure 4. Experimental setup with nomenclature and positions of temperature sensors (**a**), concept of experimental setup (**b**), and top view photograph of the cell on test bench without insulation at the top (**c**).

Overall 22 thermocouples type T class 1 by Omega with an accuracy of ±0.5 K are used with PicoLog TC-08 data loggers to measure the temperature in an experiment on several locations. We arranged 15 thermocouples three rows and five columns inside one measurement plate (see Figure 4a) to achieve spatially-dependent temperature distribution on the casing. Additionally, thermocouples are attached to each terminal and the cooling plate. For later investigations, average values per row are used named A-D. Two sensors on each terminal side are used to measure the temperature difference on the busbar to be able to calculate the related heat flux from the cell to the battery tester.

The entire setup is placed inside a Binder KB115 temperature chamber and the cell is connected to an Arbin battery tester (LBT 5 V/60 A). The temperature chamber was used to guarantee uniform start and operational conditions. Unless otherwise specified, all validation tests at laboratory BEV conditions are performed at a temperature of 30 °C including two hours pre-tempering prior to each experiment.

The used load profiles are either constant current cycling profiles or a commonly used dynamic profile based on the Worldwide harmonized Light vehicles Test Procedure (WLTP). The WLTP profile is used to validate the model for dynamic loads and is representative for cases as driving in the city. In Figure 5, exemplary a 75 A constant current cycling profile and the WLTP current profile are demonstrated. In case of constant current cycling, a fully charged cell is discharged to the cut-off voltage of 3 V. After a 10 s break the cell is charged with the same current until the upper cut-off voltage of 4.1 V is reached. After a second break of 10 s, one full cycle is finished and is subsequently repeated. For the dynamic load, the WLTP is transferred into a current profile per single cell. The resulting dynamic load is a transient input for the experiment and the simulation, as well.

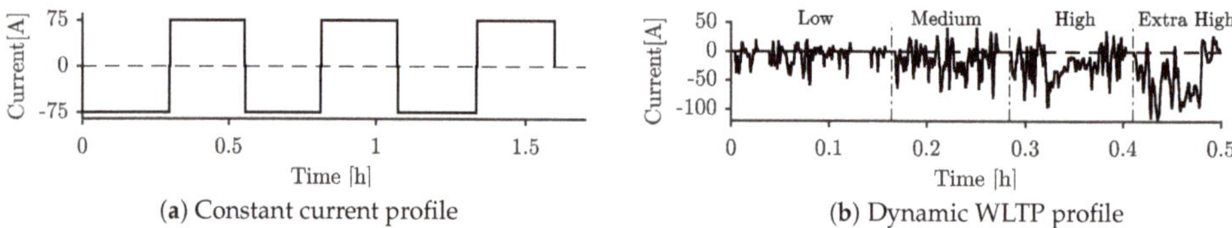

(**a**) Constant current profile   (**b**) Dynamic WLTP profile

Figure 5. Constant current profile (e.g., 75 A) and dynamic current WLTP profile used for experiment and simulation.

In addition to the developed experimental setup, experiments are performed under commonly used experimental conditions for thermal investigations of battery cells. The investigated setups, such as natural/forced convective cooling or adiabatic conditions, take place with the bracing and measurement plates. All corresponding experiments start with charging of the cell at 25 °C. After the charging process, the cell is tempered in the temperature chamber at 25 °C. In both cases of investigated convection conditions, the chamber's fan is either deactivated (Natural convection) or kept active (forced convection) for the duration of the cycling. A further investigated condition is the battery cell in a fully isolated setup targeting adiabatic conditions. The related experiment in this work is performed with the cell as well as measurement and bracing plates totally isolated in a box of 5 cm styrofoam plates.

4. Results and Discussions

To evaluate the impact of commonly used thermal boundary conditions, the prismatic 25 Ah cell has been subjected to multiple constant current charge/discharge cycles. In Figure 6a, the general temperature behaviour of the lithium-ion cell at 25 °C is shown using a cycling profile with 75 A.

In all measurements, the temperature increases with time due to heat generation during the cell's cycling. The chronological sequence of the used current profile (see Figure 5) results in local maxima at the end of every discharge. These maxima are obtained due to the high internal heat generation at low SOC. During the first discharge, the process of heating-up is similar for all setups. However, it is apparent that the following magnitude of the temperature rise depends strongly on the boundary conditions.

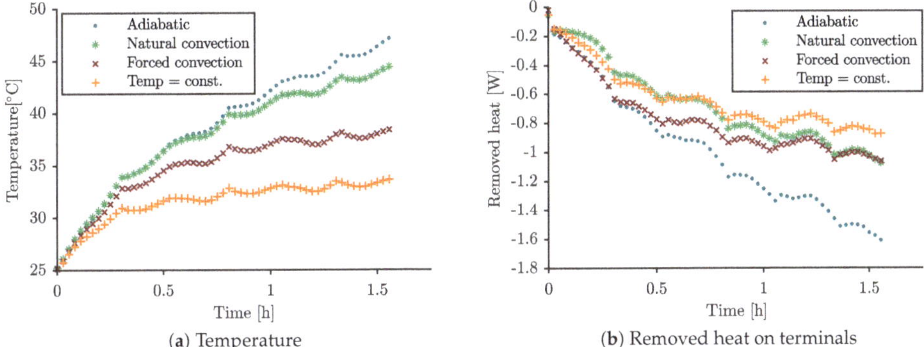

Figure 6. Measured transient behaviour of 25 Ah cell at 75 A cycling under varying thermal boundary conditions. Staring conditions as well as cooling plate and/or temperature chamber are set to 25 °C: (**a**) average cell temperature with time, and (**b**) related removed heat by the wiring on the terminals.

In the case of natural convection the cell reaches a temperature of 45 °C (20 K increase) at the end of the cycling. Whereas, under forced convection, the cell reaches a lower average temperature of 38 °C. Samad et al. [5] show comparable results investigating a prismatic cell in an active/inactive temperature chamber. Moreover, they show a dependency on the positioning of a cell in the chamber and assume a convection coefficient that is not constant on the cell's surface.

Thus, it is critical to use a setup like this for validation, because both, natural and forced convection, take place with unknown heat transfer conditions. Therefore, in the model validation, the associated parameters have a high uncertainty [5], are assumed as constant [12], or are commonly used as a fitting parameter. This reduces the significance of the validation and model behaviour. In [4] the issue of unknown convection coefficient is taken into account by a special experimental setup for pouch cells in a wind tunnel. In that case, the convection conditions are defined.

A solution for unknown convection is a fully isolated setup with adiabatic conditions. The fully thermally insulated cell (Adiabatic in Figure 6) shows the highest temperature increase of 22 K at the end of cycling. The main issue for validation with this setup is that targeted adiabatic conditions are not fulfilled in the experiment. In any experiment, the large wiring of the battery tester removes heat from the cell, which depends on the temperature increase in the experiment. The resulting heat removal is observable in Figure 6b. This (unwanted) heat sink removes up to 1.6 W of the heat generated in the cell during the experiment with fully insulated setup. That is up to 14% of the overall generated irreversible heat assuming an average cell resistance of 2 mΩ and 75 A cycling current. Heat removal by the wiring is an additional issue in all setups. In either case of convection compared to the full insulation setup, the overall temperature rise is lower than under adiabatic conditions. Therefore, less heat is removed by the battery tester connection. Both convection conditions lead to ~1 W heat removal at the end of the test, which is a significant influence. Investigating the different cell type of pouch cells in their work, Rieger et al. [20] and Erhard [4] considered heat removal on the terminals in simulation.

Avoiding uncertainties in convection conditions and monitoring heat removal by the battery tester, the laboratory single cell BEV setup in this work is developed (Temp = const. in Figure 6). With mostly adiabatic conditions and defined heat removal by conduction via the cooling plate, the uncertainties are significantly reduced in comparison to the investigated setups. The amount of (unwanted) heat removal is lower and the setup has defined cooling conditions with constant temperature cooling plate. Further benefit using the suggested setup for validation are the real use case of a battery electric vehicle. Using the laboratory BEV setup, the lowest temperatures increase of 9 K is observed. Simultaneously, 0.8 W are removed at the end of cycling, which still influences the thermal behaviour of the cell under investigation. Therefore, the heat flux at the terminals has to be monitored in every experiment because it affects the thermal behaviour significantly. It depends on the temperature of the connected battery tester and can, therefore, change for comparable experiments e.g., on different days. In the validation concept of this work, the measured heat removal is measured and used as time-dependent boundary conditions for the model validation.

In order to examine the impact of the experimental conditions of the laboratory BEV setup on the model's behaviour, several simulations are performed respectively not considering influences from the experimental setup.

In Figure 7 the experimentally measured average terminal temperature under laboratory BEV conditions is compared to the various simulations: The boundary condition on the terminals with heat flux from/to the battery tester is neglected (No heat flux), the cell is modelled without the bracing and the measurement plates (No bracing), and the constant temperature boundary condition is directly connected to the cell bottom without interstitial material (No thermal pad).

In all simulation curves, the spikes at the end of charge/discharge are due to the 10 s breaks without heat generation taking place. Due to the graphical representation of the experimental data, the spikes are not clearly visible in the black curve (Experiment], but they exist in both, experiments and simulations, for constant current cycling with 10 s break.

In the previous figure, the heat removal by the wiring in the experiment is discussed. Not considering heat removal as a transient thermal boundary conditions in the model, the temperature increases in the corresponding simulation results (No heat flux). Starting without visible differences during the initial cycles a significant temperature increase of 1–2 K mismatch is visible for the consecutive cycling. It is obvious that other experimental conditions for validation than laboratory BEV conditions would create a much higher deviation to the experiment (see Figure 6).

Neglecting the bracing components from the experimental setup (No bracing), the cell's average temperature shows a deviation of 2 K after the first discharge. The reason is the effect of the thermal masses in the setup on the transient process of heating up. Without the bracing components, the process of heating up takes place significantly faster, especially at the beginning. After the first cycle, the temperature differences are decreasing with consecutive cycling.

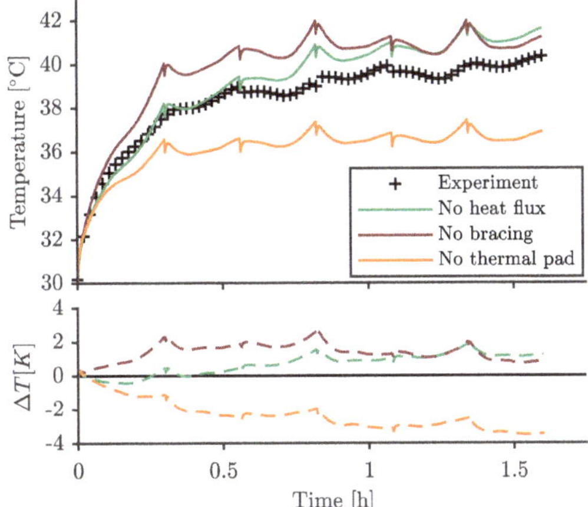

Figure 7. (top) Average terminal temperature for experiment and simulation at 75 A cycling of cell under laboratory BEV conditions with varying experimental influences. Staring conditions as well as cooling plate temperature are set to a temperature of 30 °C. **(bottom)** Corresponding temperature deviation to experimental results.

The thermal resistance between the cell and the cooling condition is realized in the experiment by a thermal pad. Assuming a direct thermal connection (No thermal pad), ideal thermal transfer leads to a maximum error of 3 K lower temperature in comparison to the experimental data. Moreover, the deviation increases with cycling time. The performed tests show, that in addition to modelling, the heat flux at the terminals and the thermal masses of bracing components, the thermal contact resistance needs to be taken into account, as well.

A thermal battery model is meaningfully validated if the heat generation and the resulting transient temperature development in the experiment and simulation show good agreement. In the used validation approach, the heat sources and the transient temperature distribution influences are validated together, considering the impact of the experimental setup.

At first, the preliminary investigated experimental and model influences are considered as boundary conditions. As a next step, the validation is performed for the heat generation and for the average and spatial dependent temperatures on different locations on the cell. Finally, an estimation of the temperature inside the jelly roll is possible.

The top of Figure 8a displays the transient temperature results of the experiments and the simulations for cycling profiles with three different currents of 25 A, 50 A, and 75 A. Both experiments and simulations, show in all cases an overall increase of the temperature and a typical temperature peak at the end of charge and discharge step. This curve shape is characteristic of the used cell. At the bottom of Figure 8, the transient absolute error is shown resulting in an average root-mean-square error (RMSE) of 0.1 K for 25 A, 0.2 K for 50A and 0.2 K for 75 A. The highest deviation for every profile exists at the end of the charge step. The error function is repetitive and does not increase with time. The errors are all below the accuracy of the used thermocouples of ±0.5 K. Furthermore, the magnitude of the errors are in the same range and the model's behaviour is load-independent. Thus, it can be concluded, that the model shows very good agreement with the experimental data.

To validate the heat source behaviour, the heat generation for all current profiles during the first cycle is calculated and presented in Figure 8b. The irreversible heat generation in the experiment is calculated as the measured voltage drop regarding the OCV of the cell. This calculation approach is very sensitive to the determined OCV in dependence of the SOC. Nevertheless, this approximation is sufficient to estimate the irreversible heat generation in the experiment. For the total heat generation in the experiment, the values for SOC-dependent reversible heat generation are cumulated. In the simulation, the irreversible and reversible heat generation are calculated all together by the previously mentioned co-simulation approach.

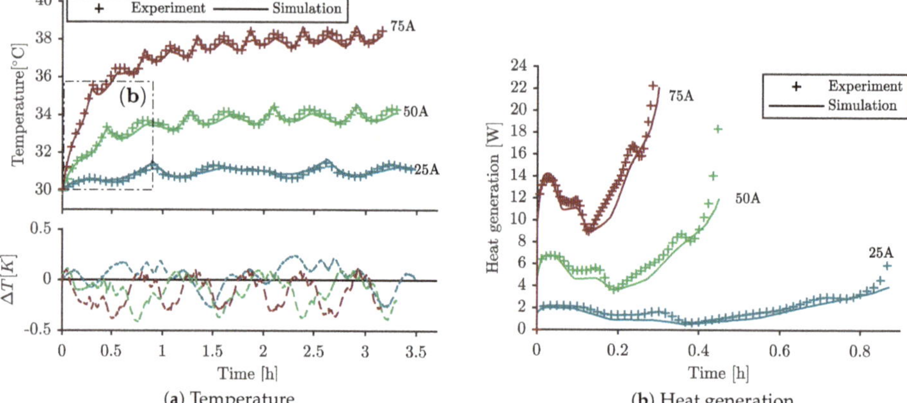

(a) Temperature

(b) Heat generation

Figure 8. Validation of experiment and simulation for different constant current cycling of cell under laboratory BEV conditions with: (**a**) overall temperature and corresponding temperature difference and (**b**) heat generation within the first cycle. Staring conditions as well as cooling plate temperature are set to a temperature of 30 °C.

The results in Figure 8b demonstrate that the calculated heat in the simulation has the same magnitude and transient curve form as the heat generations in the experiments. The relative shape depends on the magnitude of the current. With increasing currents, the irreversible heat generation dominates the reversible term. The maximum mean error is −1 W for 75 A with a maximum deviation lower than 2 W. Especially at the end of discharge, near 10% SOC, the measured OCV differs from the real behaviour and create some uncertainty in the calculation of experimental heat generation. The average error of the heat generation in the SOC-range of 100–10% is reduced by 30% resulting in an error of −0.7 W.

The main reason for the deviations due to the fitting error, when the RC-parameters of the ECM are fitted. Furthermore, RC-parameters exist only for discrete points with a resulting interpolation error. Nevertheless, with an acceptable heat generation error in the SOC-range of 100–10%, the established model describes the thermal cell behaviour very good under laboratory single cell BEV conditions.

As already mentioned, the goal is to investigate the temperature distribution within the cell during operation with high significance using the validated model. Therefore, the accuracy of local temperature prediction is validated in addition to the average temperature by the example of 75 A current profile. In Figure 9, the temperature distribution on the casing are shown for discrete points during the constant current profile. The geometries display the simulation results including the experimental results on discrete sensor positions (**x**) and the related deviation of experiment and simulation in percentages.

For discrete sensor values, the model's temperature prediction is in very good agreement. The highest measured deviation of 2.1% (±1.4% due to thermocouple accuracy) is localized at the bottom edges of the cell. As mentioned above, the thermal resistance between the cell bottom and the cooling has a great impact on simulative results. The underlying material parameters are affected by the outer clamping mechanism. This influences the results as it changes the thermal resistance of the used thermal pad. However, the overall deviation is rather small and the model reveals a good prediction of the local temperatures.

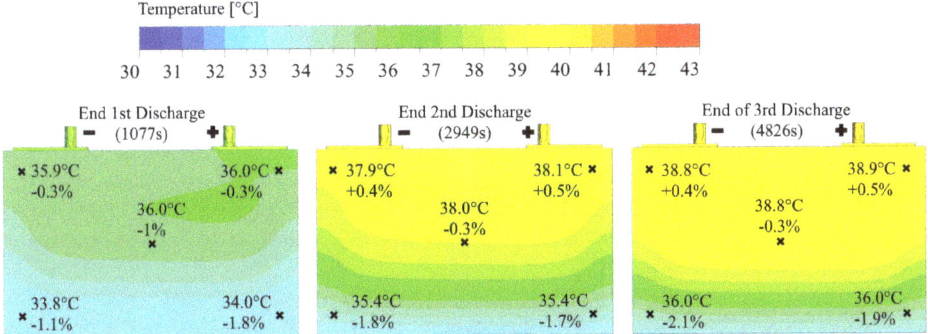

Figure 9. Temperature distribution on the casing of 25 Ah cell for discrete time steps (1077 s, 2949 s, and 4826 s) of constant current cycling with 75 A under laboratory BEV conditions. Staring conditions as well as cooling plate temperature are set to a temperature of 30 °C. On discrete sensor positions (**x**) experimental measurement results and deviation of simulation are included.

In order to investigate the thermal behaviour under dynamic operational conditions, the model is finally validated for the widely used vehicle driving profile WLTP. Starting at 95% SOC and 30 °C, the 1800 s WLTP profile is operated. The results for the spatial dependence of the temperature (see nomenclature in Figure 4) and the overall heat generation are presented in Figure 10.

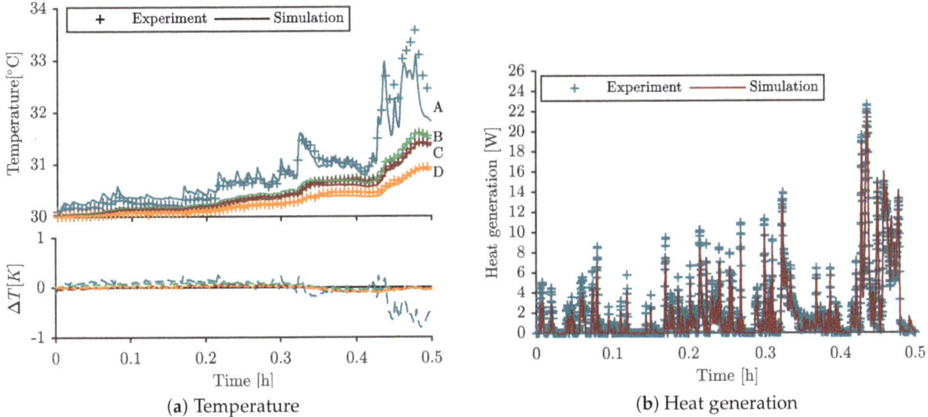

Figure 10. Validation of experiment and simulation for dynamic WLTP cycling of cell under laboratory BEV conditions with: (**a**) Overall spatially-resolved temperature (nomenclature A–D as in Figure 4) and corresponding spatially-resolved temperature difference and (**b**) heat generation during WLTP. Staring conditions as well as cooling plate temperature are set to a temperature of at 30 °C.

The simulative results in Figure 10a reproduce the temperature increase of the cell through the different sections of the WLTP (low–extra high). Very good agreement between simulation and experiment exists for the cell casing in B-D with an average RMSE values <0.1 K. The influence of liquid cooling at the bottom in BEV concepts is visible by the increasing temperature from the cell bottom (D) to the cell top (B).

The average terminal temperature (A) shows the most dynamic behaviour with a temperature increase >10% at the end of the final section in the WLTP profile. Furthermore, the largest deviation of 0.8 K is apparent at the end. Nevertheless, the observed deviation is small with an average RMSE of <0.3 K for section A. In general, the model describes the transient thermal cell behaviour very well under defined boundary conditions.

The comparison of related heat generation in Figure 10b shows increasing heat generation peaks with the charge throughput in the WLTP profile. At the end of the WLTP profile, the maximum peak heat generation is >20 W. With an average error of -0.3 W the model shows good prediction of the heat generation in WLTP use case. As earlier mentioned, the estimated experimental heat generation is strongly dependent on the measured OCV. A displacement of the OCV curve of 1% SOC results in increasing error of -0.2 W. However, for a validation of the magnitude and curve form of the heat generation that is certainly a sufficient accuracy.

In comparison to constant current cycling, the thermal strain is much lower in WLTP because of large periods with little or no current. Therefore, in the case of WLTP load, no additional knowledge is achieved compared to constant current cycling in Figure 9.

In order to investigate the thermal behaviour in a real BEV scenario after successful validation, the used boundary conditions from experiment and validation are transferred to the boundary conditions of the real BEV application (see schematics in Figure 1). To verify sufficient validation conditions, the varying boundary conditions are compared in Figure 11. Thereby, the transient temperature developments for the outer (a) and inner (b) cell temperatures are shown. For the laboratory single cell setup, the overall temperature on the terminals is lower and the process of heating up is different compared to the real BEV conditions (a). The reasons are the earlier mentioned influences by the experimental setup. Nevertheless, the maximum temperature increase on the casing and on the terminals is only 2 K and 3 K in comparison to real BEV conditions in the first cycle. Subsequently, the difference between the average temperatures decreases with cycling duration reaching a minimum of 1 K respectively 2 K at the end of the test.

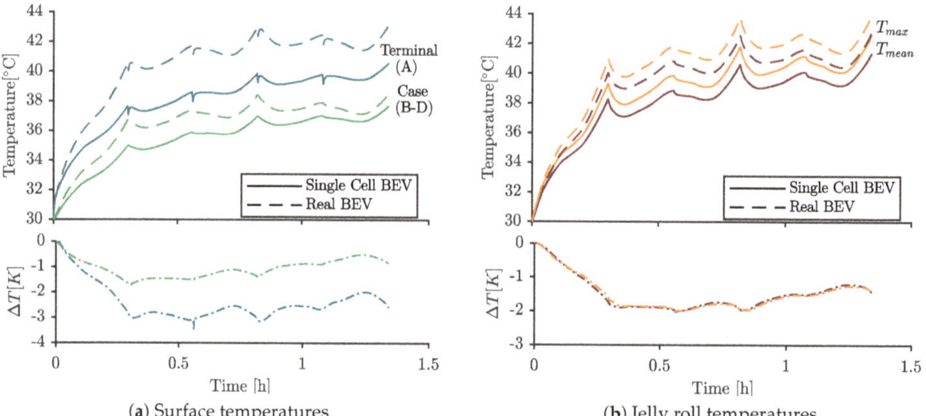

Figure 11. Simulation results for constant current cycling with 75 A of 25 Ah cell at 30 °C under varying BEV conditions: (**a**) overall temperature and corresponding temperature difference of outer temperatures (nomenclature A–D in Figure 4). (**b**) Overall temperature and corresponding temperature difference of jelly rolls max. and mean temperatures.

Most crucial temperature concerning safety, ageing, and cooling issues is the jelly roll temperature. In Figure 11b the estimated maximum and average jelly roll temperatures are given. There is a temperature increase for real BEV conditions of 1–2 K but the increase is similar for mean and maximum temperature. Overall, the used laboratory BEV setup reproduces the real BEV application very well and transfers the boundary conditions of a battery system to a practical single-cell setup.

Evaluating the effects of the boundary conditions in real BEV environment, the local temperature peaks at the end of discharge are more pronounced inside the jelly roll than on the surface. Moreover, the terminal temperature in case of real BEV conditions shows a similar transient behaviour compared to the mean jelly roll temperature. Therefore, the jelly roll temperature can be approximated by measuring the terminal temperature in real BEV application. This is important e.g., for proper thermal management systems.

The results for the temperature distribution on the central symmetry plane inside the cell are displayed in Figure 12. The stated value in each sub-figure ($\bigcirc$) is the maximum jelly roll temperature calculated in simulation under real BEV conditions. The heat generation of the cell with constant current cycling (Figure 12a) leads to a maximum temperature inside the jelly roll up to 44.1 °C at the end of the third discharge. The highest temperature is reached in the upper part of the jelly roll. The reason for the vertical location is the heat generation inside the jelly roll in combination with the cooling system located at the bottom. Additionally, the increase of the vertical temperature gradient with time is revealed on the casing. The gradient increases from 5 K after the first discharge up to 8 K after the third discharge. Whereas the vertical position is affected by the cooling conditions, the horizontal position is slightly located towards the cathode current collector due to different material properties on anode and cathode. This results in different temperatures in the current collectors. Similar temperatures compared to the jelly roll are detected in the positive current collector due to the ohmic heat generation at the location of high current density.

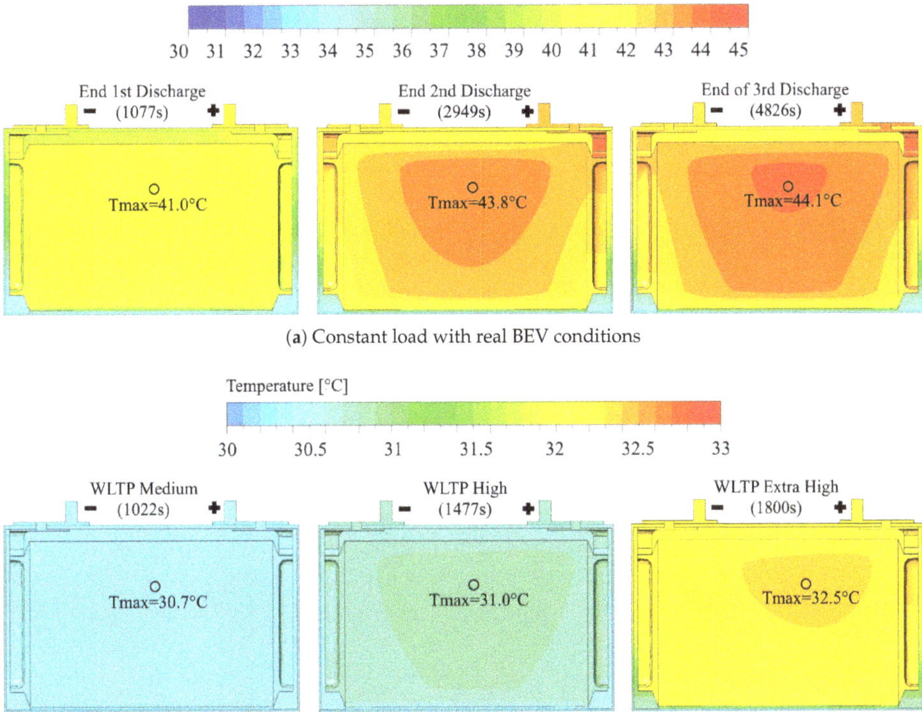

(a) Constant load with real BEV conditions

(b) WLTP profile with real BEV conditions

Figure 12. Temperature distribution on the central symmetry plane inside the 25 Ah cell under real BEV conditions at 30 °C for constant current cycling with 75 A (**a**) and dynamic WLTP profile (**b**). On discrete positions (**o**) the maximum jelly roll temperature in simulation is included.

Investigating the cell's temperature behaviour in the WLTP, the low thermal stress created by the load is revealed in Figure 12b. Especially in the first sections of the WLTP, the average heat load and the resulting temperature increase is small. Thus, a good regulation of the cell's temperature by the defined coolant temperature is possible. In the final section of the WLTP, the temperature inside the jelly roll increases up to 32.5 °C. In general, during the WLTP, the same temperature distribution results inside the jelly roll compared to constant current load but with much lower overall temperature. During periods of high loads towards the end of the profile at 1800 s, the terminal temperature increases much faster due to the direct connection to the heat generation within the current collectors.

The temperature behaviour with real BEV conditions is much more important to investigate under load conditions with high thermal impact than under conditions with low thermal impact. The temperature development in dynamic WLTP profile reveals that such cases are not challenging with the studied prismatic cells in BEV cooling conditions. High current loads in reality such as fast charging or continuous high velocity of vehicles are much more challenging for performance, safety and lifetime issues. With the developed and validated 3D electro–thermal model, further investigation in the field of cooling approaches and module influences are possible.

5. Conclusions

In this work, an experimental setup was developed that transfers the boundary conditions of a real BEV battery system with module architecture to a laboratory single cell setup. The setup is used for validation of a 3D thermal model of a prismatic 25 Ah lithium-ion battery cell. The influences of different commonly used experimental setups for thermal battery model validation are investigated and compared to the designed setup. The results show, that not only the general ambient conditions but also additional effects, such as the battery tester connection, have a strong impact on the measured cell temperature. In targeted adiabatic conditions over 14% of the produced heat are removed through the wiring. Therefore, the removed heat has to be monitored in the experiment and used as a transient boundary condition for validation. Further influences, such as thermal masses of the bracing component in the experiment are shown to be important for accurate model behaviour.

To guarantee the model's significance, a detailed validation approach was performed. The previously determined experimental and modelling influences on the thermal cell behaviour were considered as boundary conditions in the validation model. Subsequently, the model's thermal behaviour was validated for static loads as well as for dynamic profiles. The obtained simulation results are in very good agreement with the experimental data collected at various profiles. For proper validation, not only the average temperature but also local temperatures on discrete sensor positions are validated. The model reveals maximum local errors of ~2% at high current cycling. The maximum average RMSE value for dynamic as well as static profiles is <0.4 K and therefore under the accuracy of the used thermocouples of ±0.5 K. After successful validation, the model is used to predict the temperature in a real BEV application. First, the differences between laboratory setup and real application are evaluated by simulation. The evaluation indicates that the developed setup reproduces the real BEV conditions adequately. Considering real BEV conditions with fully validated model, the simulations reveal small thermal effects in the cell during the WLTP profile. During constant current profiles, exemplary for fast charging, the cell experiences stronger thermal effects with increasing temperature in the jelly roll up to 44.1 °C and gradients between anode/cathode, as well as in vertical direction towards the cooling system.

Further work with the validated model will focus on the effects of cooling conditions in BEVs and the module's specific conditions and their influence on the cell's thermal behaviour such as additional power loss by electronic components.

Author Contributions: conceptualization, C.E., G.E. and J.K.; methodology, J.K. and L.K.; software, J.K.; validation, J.K. and L.K.; investigation, J.K., L.K. and G.E.; writing—original draft preparation, J.K.; writing—review and editing, C.E., G.E., L.K. and J.K.; visualization, J.K.; supervision, G.E. and C.E.; project administration, C.E. All authors have read and agreed to the published version of the manuscript.

Funding: This work was funded by the AUDI AG within the scope of an ongoing research project.

Acknowledgments: The authors wish to acknowledge M. Keppeler (ZSW) for the CT-Analysis and D. Schneider (Technische Hochschule Ingolstadt), M. Hinterberger (Audi AG), B. Rieger (Audi AG) and R. Reinelt (ANSYS Germany) for the extensive discussions.

Conflicts of Interest: The authors declare no conflict of interest.

Abbreviations

The following abbreviations are used in this manuscript:

BEV	Battery electric vehicle
ECM	Equivalent circuit model
MSMD	Multi-scale multi-dimensional
NMC	Nickel–manganese–cobalt
OCV	Open circuit voltage
RSME	Root mean square error
SOC	State of charge
TEC	Thermoelectric cooler
WLTP	Worldwide harmonized Light vehicles Test Procedure

References

1. Rao, Z.; Wang, S. A review of power battery thermal energy management. *Renew. Sustain. Energy Rev.* **2011**, *15*, 4554–4571. doi:10.1016/j.rser.2011.07.096. [CrossRef]
2. Bandhauer, T.M.; Garimella, S.; Fuller, T.F. A Critical Review of Thermal Issues in Lithium-Ion Batteries. *J. Electrochem. Soc.* **2011**, *158*, R1. [CrossRef]
3. Wang, Q.; Jiang, B.; Li, B.; Yan, Y. A critical review of thermal management models and solutions of lithium-ion batteries for the development of pure electric vehicles. *Renew. Sustain. Energy Rev.* **2016**, *64*, 106–128. doi:10.1016/j.rser.2016.05.033. [CrossRef]
4. Erhard, S. Multi-Dimensional Electrochemical-Thermal Modeling of Lithium-Ion Batteries. Ph.D. Thesis, Institute EES, Technical University Munich, Munich, Germany, 2017. Available online: http://mediatum.ub.tum.de/doc/1338266/1338266.pdf (accessed on 5 November 2019).
5. Samad, N.A.; Siegel, J.B.; Stefanopoulou, A.G. Parameterization and Validation of a Distributed Coupled Electro-Thermal Model for Prismatic Cells. In *ASME 2014 Dynamic Systems and Control Conference*; American Society of Mechanical Engineers: New York, NY, USA, 2014; p. V002T23A006. doi:10.1115/DSCC2014-6321. [CrossRef]
6. Damay, N.; Forgez, C.; Bichat, M.P.; Friedrich, G. Thermal modeling of large prismatic LiFePO$_4$/graphite battery. Coupled thermal and heat generation models for characterization and simulation. *J. Power Sources* **2015**, *283*, 37–45. doi:10.1016/j.jpowsour.2015.02.091. [CrossRef]
7. Lundgren, H.; Svens, P.; Ekström, H.; Tengstedt, C.; Lindström, J.; Behm, M.; Lindbergh, G. Thermal Management of Large-Format Prismatic Lithium-Ion Battery in PHEV Application. *J. Electrochem. Soc.* **2016**, *163*, A309–A317. doi:10.1149/2.09411602jes. [CrossRef]
8. Tourani, A.; White, P.; Ivey, P. A multi scale multi-dimensional thermo electrochemical modelling of high capacity lithium-ion cells. *J. Power Sources* **2014**, *255*, 360–367. doi:10.1016/j.jpowsour.2014.01.030. [CrossRef]
9. Giegerich, M.; Koffel, S.; Filimon, R.; Grosch, J.L.; Fuhner, T.; Wenger, M.M.; Gepp, M.; Lorentz, V. Electrothermal modeling and characterization of high capacity lithium-ion battery systems for mobile and stationary applications. In Proceedings of the IECON 2013, Vienna, Austria, 10–13 November 2013; IEEE: Piscataway, NJ, USA, 2013; pp. 6721–6727. doi10.1109/IECON.2013.6700245.
10. Forgez, C.; Vinh Do, D.; Friedrich, G.; Morcrette, M.; Delacourt, C. Thermal modeling of a cylindrical LiFePO$_4$/graphite lithium-ion battery. *J. Power Sources* **2010**, *195*, 2961–2968. doi:10.1016/j.jpowsour.2009.10.105. [CrossRef]
11. Yi, J.; Kim, U.S.; Shin, C.B.; Han, T.; Park, S. Three-Dimensional Thermal Modeling of a Lithium-Ion Battery Considering the Combined Effects of the Electrical and Thermal Contact Resistances between Current Collecting Tab and Lead Wire. *J. Electrochem. Soc.* **2013**, *160*, A437–A443. doi:10.1149/2.039303jes. [CrossRef]
12. Ye, Y.; Shia, Y.; Cai, N.; Lee, J.; He, X. Electro-thermal modeling and experimental validation for lithium ion battery. *J. Power Sources* **2012**, *199*, 227–238. doi:10.1016/j.jpowsour.2011.10.027. [CrossRef]
13. Bernardi, D.; Pawlikowski, E.; Newman, J. A General Energy Balance for Battery Systems. *J. Electrochem. Soc.* **1985**, *132*, 5. doi:10.1149/1.2113792. [CrossRef]

14. Panchal, S.; Dincer, I.; Agelin-Chaab, M.; Fraser, R.; Fowler, M. Experimental and simulated temperature variations in a LiFePO$_4$-20 Ah battery during discharge process. *Appl. Energy* **2016**, *180*, 504–515. doi:10.1016/j.apenergy.2016.08.008. [CrossRef]

15. Doyle, M.; Fuller, T.; Newman, J. Modeling of Galvanostatic Charge and Discharge of the Lithium/Polymer/Insertion Cell. *J. Electrochem. Soc.* **1993**, *140*, 1526. doi:10.1149/1.2221597. [CrossRef]

16. Fuller, T.F.; Doyle, M.; Newman, J. Simulation and Optimization of the Dual Lithium Ion Insertion Cell. *J. Electrochem. Soc.* **1994**, *141*, 1. doi:10.1149/1.2054684. [CrossRef]

17. Kleiner, J.; Komsiyska, L.; Elger, G.; Endisch, C. Modelling of 3D Temperature Behavior of Prismatic Lithium-Ion Cell With Focus on Experimental Validation Under Battery Electric Vehicle Conditions. In Proceedings of the IEEE 25th International Workshop on Thermal Investigation of ICs and Systems (THERMINIC 2019), Lecco, Italy, 25–27 September 2019.

18. Kim, G.H.; Smith, K.; Lee, K.J.; Santhanagopalan, S.; Pesaran, A. Multi-Domain Modeling of Lithium-Ion Batteries Encompassing Multi-Physics in Varied Length Scales. *J. Electrochem. Soc.* **2011**, *158*, A955. doi:10.1149/1.3597614. [CrossRef]

19. Madani, S.S.; Swierczynski, M.J.; Kaer, S.K. The discharge behavior of lithium-ion batteries using the Dual-Potential Multi-Scale Multi-Dimensional (MSMD) Battery Model. In Proceedings of the 2017 Twelfth International Conference on Ecological Vehicles and Renewable Energies (EVER), Monte Carlo, Monaco, 11–13 April 2017; IEEE: Piscataway, NJ, USA, 2017; pp. 1–14. doi:10.1109/EVER.2017.7935915. [CrossRef]

20. Rieger, B.; Erhard, S.V.; Kosch, S.; Venator, M.; Rheinfeld, A.; Jossen, A. Multi-Dimensional Modeling of the Influence of Cell Design on Temperature, Displacement and Stress Inhomogeneity in Large-Format Lithium-Ion Cells. *J. Electrochem. Soc.* **2016**, *163*, A3099–A3110. doi:10.1149/2.1051614jes. [CrossRef]

21. Panchal, S.; Dincer, I.; Agelin-Chaab, M.; Fraser, R.; Fowler, M. Experimental temperature distributions in a prismatic lithium-ion battery at varying conditions. *Int. Commun. Heat Mass Transf.* **2016**, *71*, 35–43. doi:10.1016/j.icheatmasstransfer.2015.12.004. [CrossRef]

22. Christen, R.; Rizzo, G.; Gadola, A.; Stöck, M. Test Method for Thermal Characterization of Li-Ion Cells and Verification of Cooling Concepts. *Batteries* **2017**, *3*, 3. doi:10.3390/batteries3010003. [CrossRef]

23. Bohn, P.; Liebig, G.; Komsiyska, L.; Wittstock, G. Temperature propagation in prismatic lithium-ion-cells after short term thermal stress. *J. Electrochem. Soc.* **2016**, *313*, 30–36. doi:10.1016/j.jpowsour.2016.02.055. [CrossRef]

24. ANSYS INC. ANSYS Fluent MANUAL, no. 19.0. Available online: http://www.ansyshelp.ansys.com (accessed on 5 November 2019).

25. Schuster, E.; Ziebert, C.; Melcher, A.; Rohde, M.; Seifert, H.J. Thermal behavior and electrochemical heat generation in a commercial 40 Ah lithium ion pouch cell. *J. Power Sources* **2015**, *286*, 580–589. doi:10.1016/j.jpowsour.2015.03.170. [CrossRef]

26. Chen, S.C.; Wan, C.C.; Wang, Y.Y. Thermal analysis of lithium-ion batteries. *J. Power Sources* **2005**, *140*, 111–124. doi:10.1016/j.jpowsour.2004.05.064. [CrossRef]

27. Hopp, H. *Thermal Management of High-Performance Vehicle Traction Batteries using Coupled Simulation Models*; Research; Springer: Wiesbaden, Germany, 2016. doi:10.1007/978-3-658-14247-6. [CrossRef]

Article

A Novel Method for Thermal Modelling of Photovoltaic Modules/Cells under Varying Environmental Conditions

Ali Kareem Abdulrazzaq *, Balázs Plesz and György Bognár

Department of Electron Devices, Budapest University of Technology and Economics,
Hungarian Scientists Tour 2, H-1117 Budapest, Hungary; plesz@eet.bme.hu (B.P.); bognar@eet.bme.hu (G.B.)
* Correspondence: kareem@eet.bme.hu; Tel.: +36-1-463-3073

Received: 12 May 2020; Accepted: 20 June 2020; Published: 29 June 2020

Abstract: Temperature has a significant effect on the photovoltaic module output power and mechanical properties. Measuring the temperature for such a stacked layers structure is impractical to be carried out, especially when we talk about a high number of modules in power plants. This paper introduces a novel thermal model to estimate the temperature of the embedded electronic junction in modules/cells as well as their front and back surface temperatures. The novelty of this paper can be realized through different aspects. First, the model includes a novel coefficient, which we define as the forced convection adjustment coefficient to imitate the module tilt angle effect on the forced convection heat transfer mechanism. Second, the new combination of effective sub-models found in literature producing a unique and reliable method for estimating the temperature of the PV modules/cells by incorporating the new coefficient. In addition, the paper presents a comprehensive review of the existing PV thermal sub-models and the determination expressions of the related parameters, which all have been tested to find the best combination. The heat balance equation has been employed to construct the thermal model. The validation phase shows that the estimation of the module temperature has significantly improved by introducing the novel forced convection adjustment coefficient. Measurements of polycrystalline and amorphous modules have been used to verify the proposed model. Multiple error indication parameters have been used to validate the model and verify it by comparing the obtained results to those reported in recent and most accurate literature.

Keywords: module temperature; solar energy; thermal modelling; heat transfer mechanisms

---

1. Introduction

The increasing need for electricity and the risks of environmental pollution and global warming are the main problems increasing the interest in renewable and clean energy sources [1]. Solar energy sources using photovoltaic (PV) modules recently have the main focus among other renewable sources. This is due to several reasons such as the abundance of the solar irradiance, the photovoltaic (PV) phenomenon, by which a direct conversion is achieved from solar radiation to electricity, employable at both small and large scale, non-polluting, clean and reliable energy sources. The increase in the temperature of the silicon-based technology PV modules has direct effect on the current–voltage (I–V) characteristics of the device, that is, adversely affecting the power production and causes a significant drop in efficiency [2,3]. Therefore, it is insufficient to rely only on the rated efficiency to estimate the output power. One has to consider the operating temperature of the PV module as well as other environmental conditions and structural parameters [4]. The temperature of the PV module is affected by the module material compositions, mounting structure and the environmental conditions [5,6]. Multiple heat sources are physically contributing to the increment of the module temperature [5].

The first is the incoming short-wave solar irradiance, where only up to 20% will be converted to electrical energy, and the rest will be converted to thermal energy [4,7]. The second heat source is the long-wave infrared radiation. Accurate temperature prediction is not only needed for a precise prediction of the output power, but is also essential for estimating lifetime and quantifying the degradation of PV modules [7–10].

The heat generated in the PV module is conducted through the stacked layers of the PV module to the external surfaces (front and back surface). Radiation, forced convection and free convection heat transfer mechanisms are involved in dissipating the generated thermal energy from the surfaces to the surrounding environment. Therefore, a robust PV thermal modelling is required to estimate the operating temperature of the PV module under the given environmental, physical and structural conditions. These conditions are represented by physical parameters, which act as an input for the model.

The main objective of this work is to propose a novel thermal model to estimate the PV module temperatures at three different planes: The semiconductor p-n junction (electronic junction temperature), the front and the back surface of the PV module. The proposed model is constructed by new combination of effective sub-models found in the literature and including a novel solution for considering the effect of the module tilt angle on the forced convection heat transfer mechanism.

2. Thermal Modelling General Considerations

This section discusses the main environmental, physical and structural parameters that determine the thermal behaviour of a PV module.

2.1. Physical Structures of Pv Modules

An accurate description of the PV module is fundamental to achieve precise estimation for the operating temperature as well as its profile through different layers. Although the photovoltaic technologies are advancing rapidly with higher efficiency and lower cost, the basic solar module physical structure has not changed much over the years [11]. Figure 1 shows the basic structure of a typical PV module.

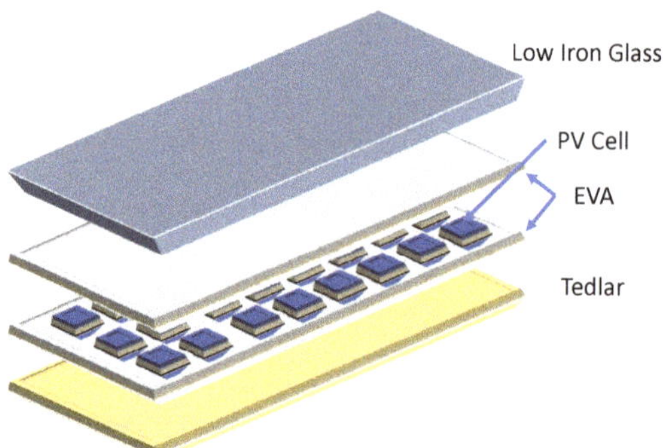

Figure 1. Schematic structure of a basic photovoltaic (PV) module.

The active semiconductor layer is consisting of several photovoltaic cells interconnected in series and parallel depending on the required output current and voltage levels.The active layer is encapsulated between two layers of, as a most often used material, ethylene-vinyl acetate (EVA) to bind the PV cells to the top and bottom layers and provide moisture resistance and electrical insulation [6,12]. Fundamentally, the glass layer is tempered (to increase the mechanical strength of the module), highly transparent, low iron content and has a textured upper-surface (to reduce the solar irradiance reflection and absorption losses). The back layer is usually made of tedlar polymer that is functioning as irradiance blocker and also providing moisture resistance [13]. Anti-reflection coating (ARC) layer is typically added to the PV layer for efficient light trapping (not shown in Figure 1) [14]. The active semiconductor layer may consist of materials like mono-crystalline, polycrystalline or amorphous silicon.

2.2. Parameters That Affect the Pv Module Thermal Behaviour

A well-known fact is that the temperature has a direct effect on the output power of the PV module. The maximum output power is decreased by 0.3 to 0.5% per Kelvin of temperature increase [15,16]. This is because the open-circuit voltage decreases significantly with increasing temperature while the short-circuit current increases only slightly [17]. However, several parameters affect the PV module temperature. These parameters have a different impact on the temperature value; therefore, some of them are essential to be considered when constructing the thermal model. Following is a list of these parameters [4,5,13,18–23].

- The amount of solar irradiance captured by the module and its spectral distribution.
- Ambient temperature.
- Wind speed, direction and air flow pattern.
- Relative humidity.
- The PV module electrical conversion efficiency.
- The PV module materials optical and thermal parameters such as irradiance absorptivity, thermal conductivity, etc.
- Mounting structure of the PV module.
- Homogeneity of the irradiance over the module surface.
- The connected electrical load.
- PV technology.

Some of these parameters are strongly influencing the module temperature; however, other parameters effect the thermal properties of the module only to a smaller extent [24]. For example, the module temperature is highly sensitive to the wind speed and much less to the wind direction [25,26]. Some of these parameters are not easy to be included in a general approach for estimating the module temperature since the module thermal behaviour is changing for different technologies [17].

3. Thermal Modelling Concepts

This section will review different thermal modelling techniques and focuses mainly on the energy balance and heat transfer mechanisms.

3.1. Classification of Pv Modules Thermal Modelling Concepts

The wide range of parameters that affect the PV module temperature (material and environmental parameters) as well as different heat transfer mechanisms that take place through the module or on its surfaces give rise to the need of complex models for estimating the junction temperature. However, for commercial products, the manufacturers do not provide all of the required information. Generally speaking, the module temperature is a dynamic, nonlinear and implicit function incorporating the controlling parameters [9]. Factors like the required level of the accuracy, details of the temperature changing profile and the model complexity produce different types of modelling approaches. Many researchers treated temperature variation as a static function; hence, it is abruptly changing to reach a steady state. That is, neglecting the material thermal capacity effect and discarding the lag in temperature variation with respect to one or more of the affecting parameters [6]. Based on this concept, the main classification of the PV temperature modelling is whether it is a static (steady-state) [12,15–17,22,25,27,28] or dynamic model [2,5–7,18,29–33]. Although the static model requires lower computational cost, its accuracy level could be affected in case of rapid changing of the controlling parameters. Temperature requires time between 4 and 10 minutes to reach the steady-state from its initial value, depending on the difference between the initial and the final temperatures and the PV module technology [5,34]. When the model input parameters are available with a frequency below this range, the dynamic model will be applicable for an accurate evaluation of the temperature [12].

The various existing thermal models in the literature, which are different in accuracy and complexity, can be grouped depending on their nature to be represented by the following.

- Direct physical equations based on theoretical expressions to incorporate different environmental, physical and structural parameters [12,15–17]. To create such an explicit relation, physical assumptions and mathematical approximations have to be made.
- Empirical expressions which are mainly based on observations and experimental measurements [25,27,28,35–37]. However, these models are optimised to represent the behaviour of the system under observation and difficult to be generalised to describe other systems, which are based on different technologies. Although these types of models require a low number of input parameters, their output accuracy is questionable [15,38]. Empirical approaches are also used to evaluate the heat transfer mechanisms to be substituted in models that have been constructed using different approaches [5,8,18,31].
- Dimensional analysis of the PV module [4,6,23,34,39]. This type of modelling will provide the capability to investigate the temperature profile and its changing rate through the PV module structure, including different thermal loss mechanisms as boundary conditions. However, it requires relatively high computational cost.
- Evaluating the heat balance equation for each structural layer of the PV module [18,40]. Different layers temperatures are estimated by substituting the effect of various heat transfer mechanisms, including thermal conduction between these layers.
- Treating the PV module as a single block of material and employ a single heat balance equation, including different heat loss mechanisms [2,3,5,7,8,13,19,22,29–33,41,42]. The thermal resistivity and thermal capacity (in case of a dynamic model) will be summed to find the component of the heat generated inside the module. Therefore, the model results will provide the module temperatures, but typically, without details about the temperature profile. The heat balance equation will be used as the core of the model, incorporating different heat loss mechanisms from the module surfaces.

The latter approach, recently, attracts the researchers focus and interest because of its applicability and high level of accuracy for estimating the module temperature. However, different researchers consider different methodologies when building up their thermal models. Based on this approach, this paper aims to propose a new thermal model. Sections 3.2 and 4 will discuss its physical translation and review the existing methods in the literature, respectively.

3.2. Energy Balance and Heat Transfer Mechanisms

It is a well-known fact that electronic junction temperature is not accessible from outside and cannot be directly measured using normal methods. Instead, models are used to estimate its value. One of the widely used methods considers the PV module as a single block of material and employ a single thermal heat balance equation (HBE), in which the absorbed energy ($q_{absorbed}$) should equal to the sum of the converted ($q_{converted}$) and the lost (q_{lost}) energies.

$$q_{absorbed} = q_{converted} + q_{lost}. \tag{1}$$

The absorbed energy results from collecting the irradiance by the front surface of the PV module represent the overall input energy for the PV system. The converted energy includes the output produced electrical energy and the heat energy generated within the PV module. The last term in

Equation (1) is related to the heat losses to the surrounding environment by different heat transfer mechanisms. The heat losses can be classified in two groups. The first group is mainly driven by the temperature difference between the module and its surrounding environment. The second group involves different effects such as joule heat in the wire contacts, diode losses and dirt accumulation. Considering all types of heat loss mechanisms gives rise to a complex modelling design that requires various parameters, which are related to the material properties and also the surrounding environment. Such a detailed model is impractical to be used for commercial products. Therefore, some of these losses are neglected due to their minor effects [9]. These minor losses include the energy initiated due to partial shading, low irradiance, dirt accumulation, joule heat through the wire contacts and diodes losses. Conduction heat transfer between the PV module and the holding structure is also neglected because of the small contact area and relatively small temperature difference [3,13,41].

Each of the heat balance equation terms (including their different components) need to be modelled to estimate their values individually because their effects cannot be directly measured [7,29]. Such an analysis should be based on the instantaneous temperature of the PV module. Therefore, an iterative process is needed. Conduction (between the PV module layers), convection and radiation (from front and back surfaces) are the three main heat transfer mechanisms that have to be evaluated to calculate the amount of losses.

Even in case it is not explicitly mentioned, the majority of the existing models share common assumptions, which are listed as follows [3,4,6–9,18,22,31,34,41].

- The temperature has a homogeneous distribution over the surface of the PV module.
- The ground temperature is equal to the ambient temperature.
- The thermal losses from the side edges of the PV module are negligible due to its small area compared to the front and back surfaces.
- The effect of the ARC layer is neglected due to its small thickness compared to other physical layers.
- The effect of the metallic frame that surrounds the PV module structural layer is neglected.
- Each of the PV module physical layers is treated as isothermal, that is, neglecting the boundary effects.
- The optical and physical parameters of the PV module different layers materials are homogeneous, isotropic and not changing with temperature nor the irradiance wavelength.
- The ambient temperature is homogeneous all around the PV module.
- The solar irradiance is reaching the front surface of the PV module equally.
- Neglecting the conduction heat transfer between the PV module and the holding structure.
- Neglecting energy losses due to partial shading, low irradiance, dirt accumulation, joule heat through the wire contacts and diodes losses.

Some researchers are going further in simplifying their models by eliminating other effects or parameters as a result of dealing with specific environmental conditions, materials properties or mounting structures. Table 1 shows some of these simplifications and assumptions.

Table 1. Special case assumptions found in the literature.

#	Introduced Simplifications	Ref.
1	Neglecting the radiation heat losses from the back surface based on the assumption that the back surface of the module is at same temperature of the building fabric it faces.	[29]
2	Back surface emissivity assumed to be equal to the front surface emissivity.	[8]
3	The heat transfer by free convection is assumed to be the same for both top and bottom surfaces. This assumption is applicable for a near vertical angles but introduces error for PV modules mounted flat.	[8]
4	The cell temperature is assumed to be the same as the front surface temperature and it is linearly related to the back surface temperature.	[5]
5	Neglecting the radiation heat losses of both surfaces.	[5,15,33]

Table 1. *Cont.*

#	Introduced Simplifications	Ref.
6	The total value of convective heat losses is the total of forced convection loss from only the PV module front surface and free convection loss from only the back surface.	[33]
7	The temperature is assumed uniform throughout the PV module five layers.	[13,15]
8	To avoid disturbing influences of fast irradiance changes at sunrise and sunset, the authors only analysed data from 10 am to 3 pm.	[17]
9	Neglecting the forced convection from the back surface.	[39]

4. Reviewing the Existing Sub-Models

As previously mentioned, each of the HBE terms (including their different components) need to be modelled and individually estimated. This section is dedicated to briefly discussing each term of the HBE with scanning the literature to review the typical methods adopted to estimate their values.

4.1. Absorbed Energy

The absorbed energy represents the amount of energy received by the PV module due to the total captured short wave irradiance. Different parameters are affecting the amount of absorbed energy, such as [2,4,6,7] the following.

- The intensity of the direct and the diffused irradiances.
- Optical parameters including the absorptivity, the reflectivity, the scattering and the transmittance of the front layers.
- Material defects and physical limitations.
- Mounting structure of the PV module.

A widely used equation to determine the short wave absorbed energy given as [2,3,29–33]

$$q_{absorbed} = \alpha \cdot \Phi \cdot A, \tag{2}$$

where α is the absorptivity of the front surface of the PV module, Φ is the total received irradiance and A is the surface area.

4.2. Converted Energy

The energy is converted into two forms: electrical energy (q_{elec}) and thermal energy (q_{therm}).

$$q_{converted} = q_{therm} + q_{elec}. \tag{3}$$

To determine the amount of produced electrical energy, the values of the current and voltage at the maximum power point (I_m, V_m) are required. Therefore, the fill factor (FF) and the efficiency of the PV module (η) play a major role as shown below [3,13],

$$q_{elec} = I_m V_m = (FF) I_{sc} V_{oc} = \eta \tau q_{absorbed}, \tag{4}$$

where τ is the front layer transmittance, and I_{sc} and V_{oc} are the short-circuit current and open-circuit voltage, respectively. Many authors, for the sake of a higher level of accuracy, tend to consider the environmental effects on the electrical performance of the PV module. Therefore, they include a dedicated electrical model for estimating the instantaneous value of the generated electric power [19,40]. These details are out of the scope of this paper.

The portion of the absorbed energy not converted to electrical energy is converted to heat, causing higher PV module temperature. With time, the thermal energy will be lost to the surrounding environment, mainly due to the temperature difference. However, this process requires some time before reaching a steady-state depending on the thermal properties of the PV module, represented by its thermal capacity and resistivity. In case of temperature evaluation is required within small periods, as described in Section 3.1, then the dynamic analysis is required to include different layers' thermal capacity. The module heat capacity (C_{module}) is determined as the sum of each layer's capacity [2,13,29–31] from the following formula,

$$C_{module} = \sum_{i=1}^{n} A \cdot d_i \cdot \rho_i \cdot c_i, \tag{5}$$

where n is the number of PV module physical layers and i is the layer index. Moreover, in Equation (5) we see, for each layer, A is the area, d is the layer thickness, ρ is the material density and c is the specific heat.

Another modelling approach adopts the concept of assuming that the temperature is abruptly following the changes in the absorbed energy. These methods are applicable in case the module temperature and its output power is required to be estimated with time resolution large enough to reach a steady thermal state, higher than its thermal time constant.

4.3. Heat Transfer Mechanisms

As indicated previously, different heat transfer mechanisms are involved in this context, including conduction (within the PV module), radiation and convection. The rest of this section presents a brief description of each one of these mechanisms.

4.3.1. Conduction Heat Transfer Mechanism

Typically, conduction is only considered between the structural layers of the PV module. Conduction to the holding structure is neglected due to the small contact area between the module and the holding structure and the low-temperature difference. The conduction heat transfer between the different layers is analysed based on the thermal resistivity and the thermal capacity of each layer of the PV module [9]. In such models, the HBE is derived for each layer [18].

4.3.2. Convection Heat Transfer Mechanism

Convection is a heat transfer mechanism between the surfaces of the PV module and the surrounding air based on Newton's law of cooling [43]. It is modelled by the corresponding heat transfer coefficients (h_c). The amount of heat convection per unit area (q_{conv}) is evaluated using the following equation:

$$q_{conv} = -h_c \cdot A \cdot (T_{module} - T_{ambient}), \tag{6}$$

where T_{module} and $T_{ambient}$ are module and ambient temperatures, respectively. The convection heat transfer involves two mechanisms—the forced convection mechanism and the free convection mechanism—which are characterised by forced convection coefficient ($h_{c,forced}$) and the free convection coefficient ($h_{c,free}$), respectively. Different researchers deal with their overall effect to be substituted in Equation (6) in different ways, as shown in Table 2.

Table 2. Determining the overall convection coefficient.

#	Used Expression	Eq. #	Ref.
1	$h_c = h_{c,forced} + h_{c,free}$	T 1.1	[3,29,32]
2	$h_c^3 = h_{c,forced}^3 + h_{c,free}^3$	T 1.2	[5,13,19,22,30]

The significance of both types of convection is differs under different environmental conditions [5,30]. The authors of [33] considered only forced convection for the front surface of the PV module and only free convection for the back surface. Other authors consider only free convection for both surfaces [42]. However, most of the recent literature work that aims for high accuracy is incorporating both free and forced mechanisms [2,6,31]. Modelling and estimating the value of each of the heat transfer coefficients is performed using various techniques [2,6]. Tables 3 and 4 summarise the well-known equations for estimating the free and forced convection, respectively. The following points are common between the different expressions listed in both tables. If any sub-model in the mentioned table use a different expression or parameter definition it will be explicitly mentioned.

- The module characteristics length (L_c) is taken as the longest dimension.
- ΔT is the temperature difference between the PV module surface and the ambient temperatures.
- β is the air thermal expansion coefficient, which is determined as $\beta = 1/T_f$. T_f is the average between the surface and the ambient temperatures, it is also known as the film temperature.
- Considering that the front and back temperatures are different, the film temperature and the thermal expansion coefficient are different for the two surfaces.
- The Grashof number (Gr) is determined as $Gr = \frac{g \cdot \rho_{air}^2 \cdot cos(\theta) \cdot \beta \cdot \Delta T \cdot L_c^3}{\mu_{air}^2}$, where g is the acceleration due to Earth's gravity, ρ_{air} is the air density, μ_{air} is the dynamic viscosity of air and θ is the angle of the module to the vertical direction.
- Gr_c is the critical Grashof number at which the Nusselt number starts deviating from laminar behaviour [5].
- Pr is the Prandtl number calculated as $Pr = \frac{cp_{air}\mu_{air}}{k_{air}}$, where cp_{air} is the specific heat at constant pressure of air and k_{air} is the air thermal conductivity.
- Ra is the Rayleigh number calculated as $Ra = Gr \cdot Pr$.
- Nu_{free} and Nu_{forced} are the Nusselt number of the free and forced convections, respectively. They are determined explicitly in each model.
- $h_{c,free} = \frac{Nu_{free} \cdot k_{air}}{L_c}$, $h_{c,forced} = \frac{Nu_{forced} \cdot k_{air}}{L_c}$.
- Re is the Reynolds number, defined as $Re = \frac{\rho_{air} L_c}{\mu_{air}} v_w$, where v_w is the wind velocity.

Table 3. Free convection equations.

#	Used Expression	Eq. #	Ref.
1	$h_{c,free} = 1.31 \cdot (T_{module} - T_{ambient})^{\frac{1}{3}}$	T 2.1	[29,32]
2	$Nu_{free} = M \cdot Ra^n$, where M and n are constants depend on the geometry of the surface.	T2.2	[42]
3	$Nu_{free} = 0.68 + 0.67 \cdot (Ra_L \cdot R)^{0.25}$, where R is a function tabulated as $R = \left[1 + \left(\frac{0.495}{Pr}\right)^{\frac{9}{16}}\right]^{\frac{-16}{9}}$. The characteristics length for this model is calculated as $L_c = \frac{A}{2 \cdot (H+W)}$, where A, H and W are the module area, length and width, respectively.	T 2.3	[7,8,30,31]
4	$Nu_{free\text{-}f} = 0.13 \cdot (GrPr)^{1/3} - (Gr_cPr)^{1/3} + 0.56 \cdot (Gr_cPr \cdot cos\theta)^{1/4}$, for $\theta < 60°$ $Nu_{free\text{-}f} = 0.13 \cdot Ra^{1/3}$, for $\theta \geq 60°$ $Nu_{free\text{-}b} = 0.56 \cdot (Ra \cdot cos\theta)^{1/4}$, for $\theta < 88°$ $Nu_{free\text{-}b} = 0.58 \cdot Ra^{1/5}$, for $88°\theta \leq 90°$ where $Nu_{free\text{-}f}$ and $Nu_{free\text{-}b}$ are the free convection Nusselt numbers for the front and the back surfaces, respectively. In this model the characteristics length is considered as the module dimension in the direction of the natural air flow. In case wind direction is irrelevant, the authors use the following form $L_c = 4 \cdot A/S$, where S is the perimeter.	T 2.4a T 2.4b T 2.4c T 2.4d	[22]
5	$Nu_{free\text{-}f} = [0.825 + \frac{0.387Ra^{1/6}}{[1+(0.492/Pr)^{9/16}]^{8/27}}]^2$ $Nu_{free\text{-}b} = 0.14[(GrPr)^{1/3} - (Gr_cPr)^{1/3}] + 0.56(Gr_cPrcos\theta)^{1/4}$ In this model the characteristics length is considered as the module dimension in the direction of the natural air flow.	T 2.5a T 2.5b	[2,5]
6	$Nu_{free} = 0.825 + \frac{0.387Ra^{1/6}}{[1+(0.492/Pr)^{9/16}]^{8/27}}$, for $Ra > 10^9$ $Nu_{free} = 0.68 + \frac{0.67(cos\theta)Ra^{1/4}}{[1+(0.492/Pr)^{9/16}]^{4/9}}$, for $Ra \leq 10^9$	T 2.6a T 2.6b	[3]

Table 4. Forced convection equations.

#	Used Expression	Eq. #	Ref.
1	The authors chose $h_{c,forced}$ to be $2\,\mathrm{Wm^{-2}K^{-1}}$ as a constant value.	T 3.1	[29]
2	$cp_{air} \cdot \rho_{air} = 1300.37 - 456{,}864 \cdot |T_f| + 0.0116391 \cdot T_f^2$ $h_{c,forced} = \dfrac{cp_{air} \cdot \rho_{air} \cdot 0.931 \cdot \left(\frac{v_w \cdot w_v}{L_c}\right)^{0.5}}{Pr^{\frac{2}{3}}}$ where $cp_{air} \cdot \rho_{air}$ is the specific heat and density product that found by curve fitting. The heat transfer coefficient is assumed to be the same from both surfaces; therefore, the overall coefficient will equal to the value determined in the above equation multiplied by 2.	T 3.2a T 3.2b T 3.2c	[8]
3	$h_{c,forced} = 3.83 \cdot v_w^{0.5} \cdot L^{-0.5}$, for $L_c/L \geq 0.95$ $h_{c,forced} = 5.74 \cdot v_w^{0.8} \cdot L^{-0.2}$, for $L_c \ll L$ $h_{c,forced} = 5.74 \cdot v_w^{0.8} \cdot L^{-0.2} - 16.46 \cdot L^{-0.1}$, for $L_c/L < 0.95$ where L is the normal length of the PV module and the characteristics length in this model is determined as $L_c = Rec \cdot v/v_w$.	T 3.3a T 3.3b T 3.3c	[22,44]
4	$h_{c,forced} = \dfrac{0.931 \rho_{air} v C_p Re^{1/2}}{L_c Pr^{2/3}}$.	T 3.4	[2,30,31]
5	$h_{c,forced} = 5.6212 + 3.9252 v_w$, for $v_w < 4.88$ m/s. $h_{c,forced} = (3.290 v_w)^{0.78}$, for $4.88 \leq v_w < 30.48$ m/s.	T 3.5a T 3.5b	[32]
6	$h_{c,forced} = 8.55 + 2.56 v_w$.	T 3.6	[33]
7	$h_{c,forced} = 2.8 + 3.0 v_w$.	T 3.7	[40]
8	$h_{c,forced} = \frac{k_{air}}{L_c}(2 + 0.41 Re^{0.55})$.	T 3.8	[13]
9	$h_{c,forced} = 2\frac{k_{air}}{L_c} \dfrac{0.3387 Pr^{1/3} Re^{1/2}}{(1+(0.0468/Pr)^{2/3})(1/4)}$, for $Re \leq 5 \cdot 10^5$. $h_{c,forced} = 2\frac{k_{air}}{L_c} Pr^{1/3}(0.037 Re^{4/5} - 871)$, for $Re > 5 \cdot 10^5$.	T 3.9a T 3.9b	[3]

4.3.3. Radiation Heat Transfer Mechanism

The heat exchange by radiation heat transfer mechanism involves the long-wave irradiance [9]. The amount of radiative energy per unit time per unit area (q_{rad}) is determined based on the Stefan–Boltzmann law as follows

$$q_{rad} = \epsilon \cdot F \cdot \sigma \cdot (T_{ob}^4 - T_{sur}^4), \tag{7}$$

where σ is the Stefan–Boltzmann constant, T_{ob} is the radiating object temperature, T_{sur} is the surrounding temperature, ϵ is the emissivity of a surface and F is the view factor. Table 5 summarises various existing methods from the literature for estimating the amount of thermal radiation. The following notes are common between the expression listed in Table 5 unless explicitly defined again: If any sub-model in Table 5 uses a different expression or parameter definition it will be explicitly mentioned.

1. The subscript *ground* refers to ground, earth or roof in the reference.
2. The subscript *sky* refers to sky.
3. The subscripts *fs* and *bs* refer to the PV module front surface and back surface, respectively.
4. The subscript *rad-front* refers to the radiation from the front surface of the PV module.
5. The subscript *rad-back* refers to the radiation from the back surface of the PV module.
6. The subscript *mfsky* refers to module front to sky.
7. The subscript *mfgr* refers to module front to ground.
8. The subscript *mbsky* refers to module back to sky.
9. The subscript *mbgr* refers to module back to ground.
10. $F_{mfsky} = \dfrac{(1+cos(\beta_{surface}))}{2}$, $F_{mfgr} = \dfrac{(1-cos(\beta_{surface}))}{2}$, $F_{mbsky} = \dfrac{(1+cos(\pi-\beta_{surface}))}{2}$, $F_{mbgr} = \dfrac{(1-cos(\pi-\beta_{surface}))}{2}$, where $B_{surface}$ is the tilt angle between the module and the ground.

11. The ground temperature (T_{ground}) is assumed to be equal to the ambient temperature ($T_{ambient}$).

12. Some authors define the radiative heat transfer coefficient (h_{rad}) as: $h_{rad} = \sigma \cdot F_{xy} \cdot \epsilon_x \cdot (T_x^2 + T_y^2)(T_x + T_y)$; therefore, the heat energy per unit time per unit area is $q_{rad} = h_{rad}(T_x - T_y)$

Table 5. Radiation thermal energy losses equations.

#	Used Expression	Eq. #	Ref.
1	$q_{rad} = \sigma(F_{mfsky} \cdot \epsilon_{sky} \cdot T_{sky}^4 + F_{mfgr} \cdot \epsilon_{ground} \cdot T_{ground}^4 - \epsilon_{module} \cdot T_{module}^4)$ In this model, the temperature of the module back surface is assumed to be very close to the building roof where the module is installed. Thus, the heat radiation exchange between the module back surface and both sky and ground are neglected. $T_{sky} = (T_{ambien} - \delta T)$ for clear sky condition where $\delta T = 20K$, $T_{sky} = T_{ambient}$ for overcast condition. $\epsilon_{sky} = 0.95$ for clear conditions; 1.0 for overcast condition, $\epsilon_{ground} = 0.95$, $\epsilon_{module} = 0.9$.	T 4.1	[29]
2	$q_{rad\text{-}front} = \sigma \cdot F_{mfsky} \cdot \epsilon_{front} \cdot (T_{fs}^4 - T_{sky}^4) + \sigma \cdot F_{mfgr} \cdot \epsilon_{front} \cdot (T_{fs}^4 - T_{ground}^4)$ $q_{rad\text{-}back} = \sigma \cdot F_{mbsky} \cdot \epsilon_{back} \cdot (T_{bs}^4 - T_{sky}^4) + \sigma \cdot F_{mbgr} \cdot \epsilon_{back} \cdot (T_{bs}^4 - T_{ground}^4)$ $T_{sky} = (\epsilon_{sky} \cdot T_{ambien}^4)^{0.25}$ $\epsilon_{sky} = 0.727 + 0.0060 \cdot T_{dew\text{-}c}$ during daytime; $0.741 + 0.0062 \cdot T_{dew\text{-}c}$ during nighttime, where $T_{dew\text{-}c}$ is the dew point temperature measured in degree Celsius. The emissivity of front side (ϵ_{front}) is between 0.9 and 1. The emissivity of the back surface (ϵ_{back}) is assumed to be equal to the front glass emissivity.	T 4.2a T 4.2b	[8]
3	$h_{rad\text{-}front} = \sigma\epsilon_{front}[F_{mfsky} \cdot (T_{fs}^2 + T_{sky}^2) \cdot (T_{fs} + T_{sky}) + F_{mfgr} \cdot (T_{fs}^2 + T_{ground}^2) \cdot (T_{fs} + T_{ground})]$ $h_{rad\text{-}back} = \sigma\epsilon_{back}[F_{mbsky} \cdot (T_{bs}^2 + T_{sky}^2) \cdot (T_{bs} + T_{sky}) + F_{mbgr} \cdot (T_{bs}^2 + T_{ground}^2) \cdot (T_{bs} + T_{ground})]$ $T_{sky} = 0.0552 \cdot T_{ambien}^{1.5}$ $\epsilon_{front} = 0.85$, $\epsilon_{back} = 0.91$.	T 4.3a T 4.3b	[22]
4	$q_{rad\text{-}front} = \sigma \cdot F_{mfsky} \cdot \epsilon_{front} \cdot (T_{fs}^4 - T_{sky}^4) + \sigma \cdot F_{mfgr} \cdot \epsilon_{front} \cdot (T_{fs}^4 - T_{roof}^4)$ $q_{rad\text{-}back} = \sigma \cdot F_{mbgr} \cdot \epsilon_{back} \cdot (T_{bs}^4 - T_{rack}^4)$ $T_{sky} = (T_{ambien} - \delta T)$ for clear sky condition in which $\delta T = 20K$, $T_{sky} = T_{ambient}$ for overcast condition. ϵ_{front} and ϵ_{back} is between 0.9 and 1. $F_{mbgr} = 1$, $F_{mbsky} = 0$. The rack temperature T_{rack} is approximated to be equal to the ambient temperature. The roof temperature T_{roof} is calculated as $T_{roof} = T_{ambient} + \alpha_r \Phi_h$, where α_r is the roof absorptivity coefficient and Φ_h is the incoming total solar irradiance on the horizontal plane.	T 4.4a T 4.4b	[30]
5	Same Equations T 4.2a and T 4.2b $\epsilon_{front} = 0.91$, $\epsilon_{back} = 0.85$. $T_{sky} = 0.037536 \cdot T_{ambien}^{1.5} + 0.32 \cdot T_{ambien}$		[3]

5. Detailed Construction of Thermal Model

Constructing the thermal model in this research work is based on the approach of treating the PV module as a single block of material and employing the HBE, including different heat transfer mechanisms. Figure 2 shows the thermal behaviour of a PV module that described by the HBE.

The model will provide the PV module junction temperature as well as the temperature difference to both surfaces. Therefore, both front and back surface temperatures will be estimated. This result will be useful in the validation phase because then we can compare the back surface estimated temperature to the measured one by a thermometer attached to the backside of the PV module. The model was constructed based on different, already existing models from the literature (see Section 4). However, the new model was constructed by incorporating sub-models of different existing models in a new and unique way to yield a new model with improved accuracy. For this, different sub-models of the above described models were combined and tested, and the combination with the best accuracy was chosen as the mode for this paper. The total absorbed energy is consisting of two components and given as

$$q_{absorbed} = q_1 + q_2, \tag{8}$$

$$q_1 = \alpha_{fg} \cdot A \cdot \Phi, \tag{9}$$

$$q_2 = \tau_{fg} \cdot \alpha_{PV} \cdot A \cdot \Phi \cdot (1 - \eta), \tag{10}$$

where q_1 is the rate of thermal energy absorbed by the tempered glass layer, q_2 is the energy absorbed by the semiconductor layer, τ_{fg} is the transmittance of the glass layer, α_{fg} and α_{PV} is the absorptivity of the front glass and semiconductor layers, respectively.

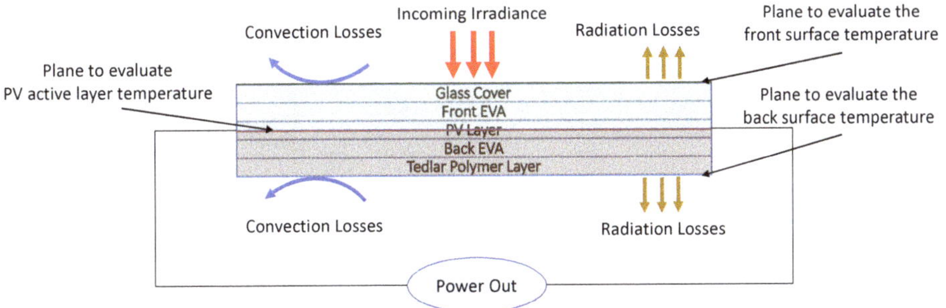

Figure 2. Thermal condition of the PV module.

In this paper, we consider a static model, that is, we assume that the module output power predicted by the proposed model is required only with a time resolution that enables the the temperature to reach a steady state. Therefore, the HBE component which is related to the material thermal capacity is neglected and the converted energy is limited only to the electrical produced component, which is given as

$$q_{converted} = \tau_{fg} \cdot \alpha_{PV} \cdot A \cdot \Phi \cdot \eta, \tag{11}$$

The radiation heat losses trough both front and back surfaces are calculated using the following expressions, respectively [8].

$$q_{rad-front} = \sigma \cdot F_{mfsky} \cdot \epsilon_{front} \cdot A \cdot (T_{fs}^4 - T_{sky}^4) + \sigma \cdot F_{mfgr} \cdot \epsilon_{front} \cdot A \cdot (T_{fs}^4 - T_{ground}^4). \tag{12}$$

$$q_{rad-back} = \sigma \cdot F_{mbsky} \cdot \epsilon_{back} \cdot A \cdot (T_{bs}^4 - T_{sky}^4) + \sigma \cdot F_{mfgr} \cdot \epsilon_{back} \cdot A \cdot (T_{bs}^4 - T_{ground}^4). \tag{13}$$

The view factors are calculated using the expressions given in Section 4.3.3 (point number 10). The sky temperature is, $T_{sky} = (T_{ambien} - \delta T)$ for clear sky condition where $\delta T = 20K$, $T_{sky} = T_{ambient}$ for overcast condition [29]. The ground temperature is assumed to be equal to the ambient temperature.

Both free and forced convection mechanisms are considered in creating this thermal model. Their overall effect is calculated by combining their effect using Equation T 1.2 from Table 1. For both mechanisms, we treat the front and back surface individually because the properties of the film layer at the boundary of each one are different. The free convection heat loss is determined using Equations (14) to (18), in which the subscript x refers to the front (f) or back (b) surface; therefore, during implementation, the equation has to be rewritten for each surface.

$$Gr_x = \frac{g \cdot \rho_{air,x}^2 \cdot cos(\theta) \cdot \beta_x \cdot \Delta T \cdot L_c^3}{\mu_{air,x}^2}, \tag{14}$$

$$Ra_x = Gr_x \cdot Pr_x, \tag{15}$$

$$Nu_{free,f} = 0.27 \cdot Ra_f^{0.25}, \tag{16}$$

$$Nu_{free,b} = 0.54 \cdot Ra_b^{0.25}, \tag{17}$$

$$h_{free,x} = Nu_{free,x} \frac{k_{air_x}}{L_c}, \tag{18}$$

For estimating the forced convection coefficients (for both surfaces), we modify the expressions used by Kayhan [13], given as

$$Re_x = \frac{V_w \cdot L_c \cdot \rho_{air_x}}{\mu_x}, \tag{19}$$

$$h_{forced,x} = \frac{k_{air_x}}{L_c} \cdot (2 + 0.41 \cdot Re_x) \cdot H_x. \tag{20}$$

The introduced modification can be seen in Equation (20) where we added a novel coefficient (H), which is defined as the forced convection adjustment coefficient for both front and back surfaces. This coefficient modulates the relationship between the tilt angle and the wind effect on the amount of heat loss by forced convection. This coefficient is calculated as

$$H_f = (1 + cos(\beta_{surface}))/m, \tag{21}$$

$$H_b = (1 - cos(\beta_{surface}))/m, \tag{22}$$

where m is an empirical factor estimated with the help of measurement data. The following points explain the fundamental concept behind the coefficient H by considering PV module mounted with different tilt angles and assuming that the value of m is equal to 2.

- 0° tilt angle:

    - The front surface will undergo a maximum effect of the wind that will sweep the hot air away. This fact is ensured by Equation (21), which will be evaluated to 1. That is, the expression used for calculating the heat loss by forced convection will not be disturbed by the tilt angle.
    - For a typical PV system, there are two facts: First, the system is consisting of many PV modules with a defined density. Second, PV modules are mounted close to the ground in case of flat and small tilt angles. Therefore, the wind will have no considerable effect on the back surface of the PV module. Equation (22) will be evaluated to zero for a flat surface; that is, the forced convection heat loss from the back surface will be neglected in this case.

- 60° tilt angle:

    - This implies that the wind will face resistance from the front surface of the PV module compared to the case of flat mounting. Therefore, reducing the ability to sweep out the hot air away from the surface. Equation (21) will be evaluated to 0.75. That is, the tilt angle will be a reason for reducing the amount of heat loss by forced convection.
    - The lower surface will be facing the wind, which was not the case for a flat-mounted module. Equation (22) will be evaluated to 0.25. Thus, heat loss by forced convection is much higher compared to flat or small tilt angles. However, it is still lower compared to the front surface.

- 90° tilt angle: Both front and back surfaces will be directly facing the air flow. Therefore, neglecting the wind direction for its minor effect compared to its speed [25,26], the wind will equally act on both surfaces. Both Equations (21) and (22) will be evaluated to 0.5.

Therefore, we consider that the PV module tilt angle will control the amount of heat losses from both surfaced. For tilt angles between 0° and 90°, the front surface heat loss by forced convection is higher compared to the back surface. Increasing the tilt angle (within this range) produces lower forced convection heat loss from the front surface and higher from the back surface.

We claim that m is a factor that affects the relationship between the tilt angle and the heat loss by forced convection by involving other installation parameters. These parameters include the PV modules installation density, the elevation from the ground and the thickness at the module edges at which the wind speed drops to zero. From experience, we found that this empirical factor has a value in the range between 1.5 and 2. Therefore, in this paper, we consider scanning this range with a specific resolution and running the model for each value. By increasing the resolution more, the value of m can be determined more accurately. Based on experience, We consider 0.1 as a resolution value considering a trade-off between the computational cost and accuracy. Therefore, we consider running the thermal model six times after which we decide what is the best value for m (by monitoring the error indication parameters) to be fixed for the module under investigation.

Once we have the value of the empirical factor m, we substitute it in Equations (21) and (22) to determine the forced convection adjustment coefficient for the front and back surface, respectively. For each surface, the overall convection coefficient and the corresponding rate of convection thermal energy losses can be calculated using the Equations T 1.2 from Tables 1 and 6.

Table 6. PV modules technical specifications, physical and installation parameters.

Parameter	Polycrystalline	Amorphous
Module dimensions	$1645 \times 990 \times 50$ mm	$350 \times 300 \times 25$ mm
Front side	Tempered glass	Tempered glass
PV layer	Polycrystalline Silicon	Amorphous Silicon
Encapsulating material	EVA	EVA
Back side	tedlar	tedlar
efficiency	12.64%	11.5%
Tilt angle	20	47
α_{fg}	0.04	0.04
α_{PV}	0.93	0.93
τ_{PV}	0.94	0.94
ϵ_{front}	0.91	0.91
ϵ_{back}	0.85	0.85

The proposed model also considers the following points.

- The characteristics length L_c is considered as the longest dimension of the PV module.
- The model operates to determine the PV electronic junction temperature. This temperature is correlated to the front and back surfaces employing temperature differences. Each temperature difference is defined as the total heat losses from the corresponding surface multiplied by the thermal resistivity of half of the PV structure (the volume between the half of the semiconductor layer plane and the corresponding surface plane), as shown in Figure 2.
- The Newton–Raphson iterative method is employed to solve the model and calculate the output PV layer, front surface and back surface temperatures.
- Therefore, with each iteration, the following two equations are evaluated to calculate the front and back surface temperatures, respectively,

$$\Delta T_f = q_{front\text{-}total} \cdot R_f. \tag{23}$$

$$\Delta T_b = q_{back\text{-}total} \cdot R_b. \tag{24}$$

where $q_{front\text{-}total}$ and $q_{back\text{-}total}$ are the total thermal losses from the front and back side of the PV module, respectively. R_f, and R_b are the thermal resistivity of the PV module, between the front and back surfaces and the active layer, respectively.

6. Results and Discussion

This model has been validated using a polycrystalline and an amorphous PV module. The validation data of the polycrystalline module has been taken from a reference [9]. Our measurement system has been used to collect the amorphous module validation data. This measurement system provides data such as PV module back surface temperature, full I–V curve, global solar irradiance, ambient temperature and wind speed. The global irradiance was measured by Delta Ohm LP RAD 03 piranometer that detect solar irradiance ranging from 0 to 2000 W/m^2. The temperature of the PV module is recorded using a circuit board attached to the back side of the module with a thermal conductive adhesive. The circuit includes a temperature sensor IC (MAX6603ATB+T). The accuracy of the circuit is $\pm$0.8 °C at +25 °C . The ambient temperature is measured using PT100 resistance thermometers. Wind speed data is taken from a wind turbine FD2.5-300 which is capable of measuring range of 0 to 60 m/s with an accuracy of $\pm$0.3 m/s.

Table 6 shows the technical specifications and physical parameters of both modules, which are required for running the model.

To verify the model, we use measurement data including irradiances, ambient temperatures and wind speed that have been recorded for two different full days for each module. For the amorphous module, the two days were the 5th and the 11th of October. For the polycrystalline module, the two days were the 3rd and the 26th of July. The main difference between the two days of each module is the wind speed. The average wind speed is 6.14 m/s on the 26th and 2.16 m/s on the 3rd of July, while it is 2.05 m/s on the 5th and 0.77 m/s on the 11th of October. As mentioned in Table 6, each module has a different tilt angle. Based on our experience, we claim that the module tilt angle has a significant effect on the value of the thermal energy losses by forced convection. As described in the model introduced in Section 5, we introduced a forced convection adjustment coefficient (H) and its empirical factor (m). In this regard, we report that the value of m typically takes a value between 1.5 and 2. We calculate this factor by scanning its range and running the model with a step of 0.1.

For evaluating the proposed model and validating the results we use two error indication parameters: one is the root mean square error ($RMSE$) and the other is the correlation coefficient (r). These parameters are used, as shown in Table 7, to compare the module's back surface temperature for each day (entire day measurement) of the two modules with the estimated values using the proposed model for different values of m.

Table 7. Error quantifying parameters corresponding different m values.

| | Polycrystalline | | | | Amorphous | | | |
| | 3rd July | | 26th July | | 5th July | | 11th July | |
m	$RMSE$ [°C]	r	$RMSE$ [°C]	r	$RMSE$ [°C]	r	$RMSE$ [°C]	r
1.3	1.724	0.997	3.099	0.964	1.455	0.965	2.089	0.966
1.4	1.329	0.996	1.740	0.989	1.312	0.971	1.906	0.969
1.5	1.051	0.997	1.090	0.994	1.208	0.976	1.832	0.970
1.6	0.927	0.997	0.937	0.997	1.140	0.978	1.707	0.972
1.7	0.964	0.996	1.228	0.997	1.106	0.980	1.646	0.973
1.8	1.126	0.996	1.558	0.997	1.101	0.980	1.577	0.975
1.9	1.315	0.996	1.891	0.996	1.119	0.979	1.585	0.975
2.0	1.534	0.996	2.216	0.996	1.310	0.971	1.606	0.974

Based on the results shown in Table 7, we chose a value of $m = 1.6$ for the polycrystalline module and $m = 1.8$ for the amorphous module to be used in this study, as these values give the best results. Figure 3 shows both the measured and the estimated PV module back surface temperature for the polycrystalline module, for the two investigated days: 3rd and 26th of July. Each curve includes 120 points as a result of recording the temperatures every 5 min between 9 am and 7 pm.

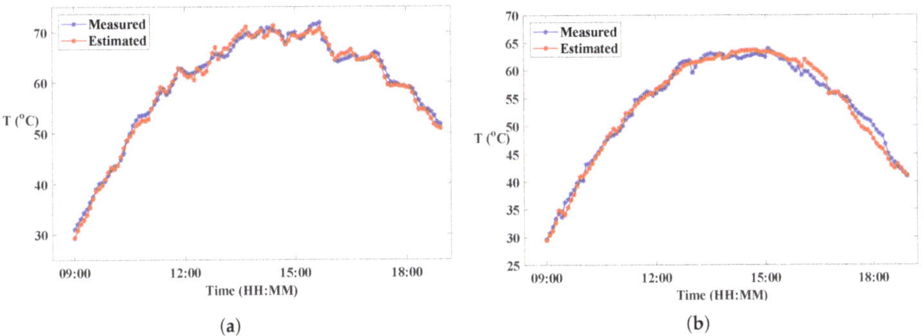

Figure 3. Measured and estimated polycrystalline module backside temperature. (**a**) 3rd July. (**b**) 26th July.

Figure 4 shows both the measured and the estimated PV module back surface temperature for the amorphous module, for the two investigated days, 5th of October (49 points between 9 am and 1 pm) and 11th of October (71 points between 9 am and 3 pm).

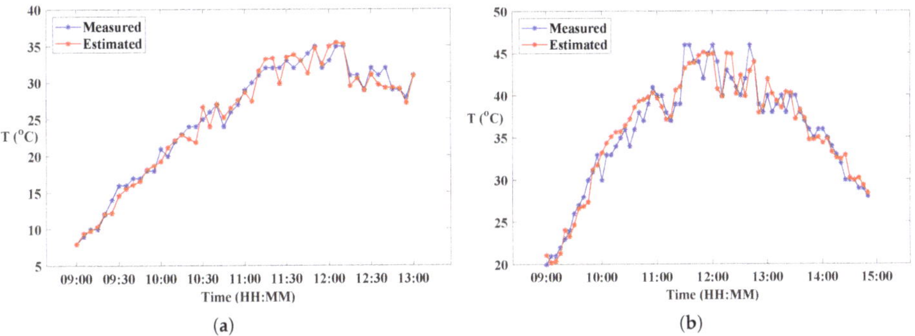

Figure 4. Measured and estimated amorphous module backside temperature. (**a**) 5th October. (**b**) 11th October.

Figure 5 shows the absolute value of the temperature difference between the measured and estimated values of the back surface temperature using the proposed model for both modules.

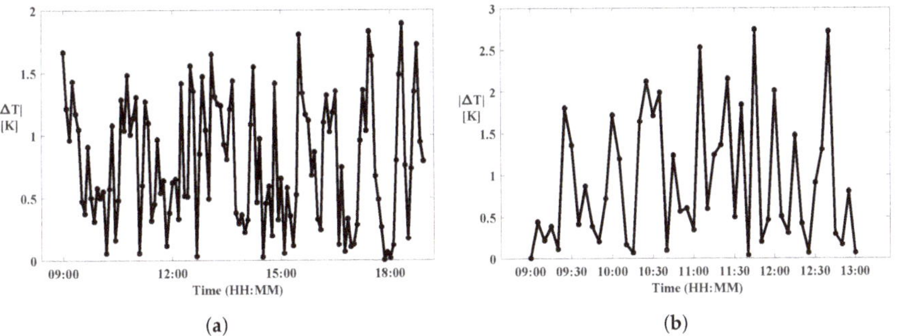

Figure 5. Absolute values of the temperature difference between the measured and the estimated values. (**a**) Polycrystalline 3rd October. (**b**) Amorphous 5th October.

Figure 6 shows the relationship between the junction temperature and both surfaces temperatures. It illustrates how these temperature differences are changing with the time of day. Therefore, it will provide a clear picture of the temperature profile across the PV module. The temperature difference to the front surface (ΔT_f) is ranging from 0.7 to 2.5, with an average value of 1.86 °C. The temperature difference to the back surface is between 1.67 and 0.22 with 0.96 °C as an average value.

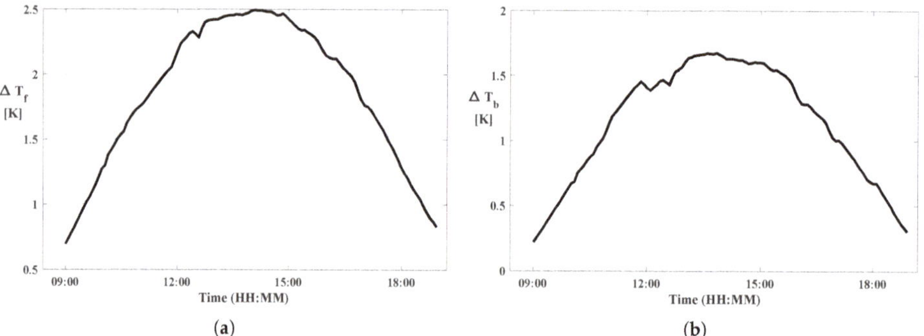

Figure 6. Temperature difference between the electronic junction and the polycrystalline module (measurements used for 3rd July) front and back surfaces. (**a**) Difference to the front surface. (**b**) Difference to the back surface.

Figure 7 highlight the importance of the novel coefficient and the consideration of the tilt angle in the forced convection, introduced in this paper. The same figure also shows the effect of neglecting the wind in the thermal model. Figure 7a compares the measured back surface temperature (blue colour) to the estimated temperature with the coefficient H (red colour), without the coefficient H (black colour), and without wind effect (green colour), for the polycrystalline module (measurements used for 3rd July). Figure 7b shows the absolute error to compare different situations.

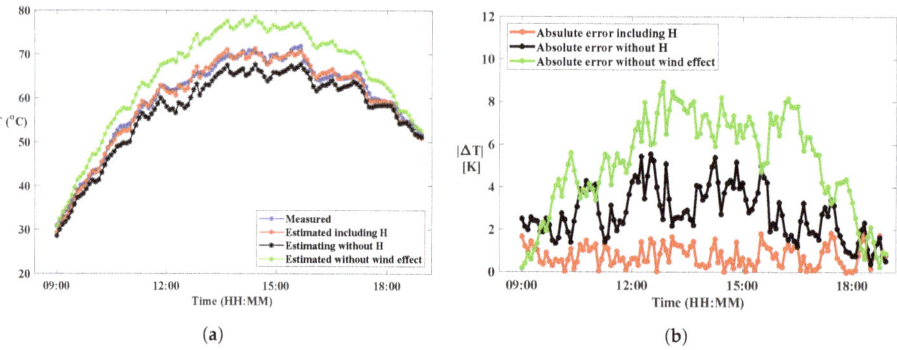

Figure 7. Evaluating the effect of including H in the PV thermal model (the model applied for the polycrystalline module, measurements used for 3rd July). (**a**) Compares back surface temperatures. (**b**) Compares the absolute error.

According to the results shown above, we summarise the discussion with the following points.

- One of the focus points of the proposed model is the special dependence of forced convection mechanism on the module tilt angle.
- Two modules made with different technologies and mounted with different tilt angles were used to validate the proposed model. A forced convection adjustment coefficient (H) has been

considered for this purpose, which includes an empirical factor (m). We calculate this factor by scanning its range as discussed above.

- For each module, the model has been validated using measurements of two days with different average wind speeds.

- Table 7 shows the proposed model's ability to estimate the temperature with high accuracy characterised by the two error quantifying parameters, namely, $RMSE$ and r. It is worth mentioning that the model provides low error rate for all values of m. However, the highest accuracy is realised at an optimum value of the factor m.

- From Figures 3 and 4, we see that the model results represented by the backside estimated temperature followed the experimentally measured values for PV modules of different technologies, different tilt angles and different wind speeds by measurements collected on two different days for each module.

- Figure 5 shows the absolute difference between the measured and the estimated values using the proposed model for both modules. Figure 5a shows that this value is always below 2 °C, with an average value of 0.78 °C for the polycrystalline module (measurements used for 3rd July). Figure 5b shows similar results for the amorphous module.

- Figure 7b shows the substitutional effect of the coefficient H on the thermal model, enabling accurate temperature estimation for the PV modules. The same figure also shows that large error is produced in case of neglecting the wind effect (neglecting the forced convection heat transfer mechanism).

- Figure 6 shows that the temperature difference between the PV electronic junction plane and both surfaces reach their maximum values around the midday time at which both ambient and electronic junction temperature are at their maximum values. This happens because the ambient temperature is higher at midday; thus, the temperature difference between the module surface and the ambient is smaller that will reduce the rate of the heat loss from the module to the surrounding.

- Table 8 shows the achieved accuracy of the proposed model using the two modules under different environmental conditions, represented by two introduced error quantifying parameters $RMSE$ and r.

Table 8. Thermal model accuracy achieved for the two modules.

PV Module	Date	$RMSE$ [°C]	r
Polycrystalline-module	3rd July	0.927	0.997
	26th July	0.937	0.997
Amorphous-module	5th October	1.101	0.980
	11th October	1.577	0.975

- For the same module, typically, $\Delta T_f > \Delta T_b$. Therefore, the top surface temperature is slightly lower than the backside temperature due to, relatively, more effective heat transfer mechanisms.

The rest of this section is dedicated to highlighting the scientific improvement that has been introduced in this work. We made a comparison between the proposed model and the results reported by different thermal models from the recent and most accurate literature using various error quantifying parameters. In this comparison, we will refer to the best results reported by the references and compare it to our model using the measurement recorded on the 3rd of July for the polycrystalline module.

- Several thermal models found in the literature use the root mean square error ($RMSE$) as an error quantifying parameter to validate the results. The models presented in [16,18,23,28,31,32,38,42,45,46] have reported $RMSE$ values ranging between 4.9 and 1.1 °C. However, in our presented model we report a value of 0.927 °C.

- The correlation coefficient (r) is another parameter used in the literature. Thermal model presented in [16,23] reported $r = 0.98$, and 0.95, respectively. In our thermal model, we calculate a correlation coefficient of 0.997.

- The authors of [40] used the relative error to validate their proposed thermal model by compression with other models. They reported an average deviation of 2.7%. Calculating the same error indication parameter using our proposed model gives 1.26%.

7. Conclusions

In this paper, we introduced a novel thermal model to predict the PV electronic junction, front surface and back surface temperatures. The model has been verified using on-site measurement for two modules made with two different technology and mounted with different tilt angles. The measurements have been recorded for each module for two different days in which the average wind speed is the main difference. A novel concept has been introduced to consider the module tilt angle effect on the amount of heat loss by forced convection. The result presented in Table 8 shows that the model is able to estimate the PV module temperature with high accuracy represented by $RMSE = 0.927\,^{\circ}\text{C}$ and $r = 0.997$ as the best results for both modules under the considered environmental conditions. From the same table, calculating the average of these parameters give $RMSE = 1.1\,^{\circ}\text{C}$ and $r = 0.987$, which are comparable, but slightly better compared to the best results review from the literature. During the model validation phase, we found that obtaining a high level of accuracy is only possible by including a novel forced convection adjustment coefficient (H). Using the same proposed model without this coefficient gives $RMSE = 3.02\,^{\circ}\text{C}$ when applying the model to estimate the junction temperature of the polycrystalline module (3rd July). The back surface temperature absolute differences between the measured and the estimated values have been calculated for both modules, which give an average value below $1\,^{\circ}\text{C}$ for both studied modules, considering the two days measurements for both. The following points summarise this work's conclusion.

- Based on the introduced and discussed results, the proposed model shows the ability to estimate the PV module temperature of different technologies, mounting tilt angles and environmental conditions.
- Based on the proceeding discussion and the information delivered in Figure 7, it is evident that the coefficient H introduces a significant improvement to the result accuracy of PV thermal modelling.
- We have also concluded that wind is an essential parameter to be considered in PV thermal modelling. Running our model with neglecting the wind effect raises the $RMSE$ from $0.927\,^{\circ}\text{C}$ to $5.62\,^{\circ}\text{C}$.
- The presented work proves that considering the static approach in this model provides excellent accuracy level of PV module temperature estimation even with a resolution of 5 min for the polycrystalline module.
- The electronic junction temperature as well as both front and back surface temperatures delivered by the model could be used in studying the PV module temperature profile, mechanical properties and lifetime.

It worth highlighting at this point that the novelty of the proposed paper is realized by introducing the new forced convection adjustment coefficient, and by reviewing the most often used existing expressions for calculating the different forms of the PV module heat losses and the related parameters and finding the proper combination of these expressions to be employed in the presented model.

Author Contributions: Individual author contributions are as follows: Conceptualization, A.K.A and G.B.; software, A.K.A. and G.B.; validation, A.K.A., G.B. and B.P.; writing—original draft preparation, A.K.A.; writing—review and editing, A.K.A., G.B. and B.P. All authors have read and agreed to the published version of the manuscript.

Funding: The research reported in this paper was supported by the BME Nanotechnology and Materials Science TKP2020 IE grant of NKFIH Hungary (BME IE-NAT TKP2020), by the Stipendium Hungaricum Scholarship

Programme, the grant EFOP-3.6.1-16-2016-00014 and by the Science Excellence Program at BME under the grant agreement NKFIH-849-8/2019 of the Hungarian National Research, Development and Innovation Office.

Conflicts of Interest: The authors declare no conflicts of interest.

List of Symbols and Abbreviations

HBE	Heat balance equation
FF	Fill factor (-)
I_{sc}	Short circuit current (A)
V_{oc}	Open circuit voltage (V)
I_m	Current at the maximum power point (A)
V_m	Voltage at the maximum power point (V)
T	Temperature (K)
C	Thermal capacitance ($J\,K^{-1}$)
C_{module}	Total module thermal capacitance ($J\,K^{-1}$)
d	Layer thickness (m)
c	Specific heat ($J\,kg^{-1}\,K^{-1}$)
cp_{air}	Specific heat at constant pressure of air ($J\,kg^{-1}\,K^{-1}$)
h_c	Convection heat transfer coefficients ($W\,m^{-2}\,K^{-1}$)
q	Heat flux ($W\,m^{-2}$)
q_1	Heat flux absorbed by the tempered glass layer ($W\,m^{-2}$)
q_2	Heat flux absorbed by the semiconductor layer ($W\,m^{-2}$)
q_{conv}	Convection heat flux ($W\,m^{-2}$)
q_{rad}	Radiation heat flux ($W\,m^{-2}$)
$h_{c,forced}$	Forced convection coefficient ($W\,m^{-2}\,K^{-1}$)
$h_{c,free}$	Free convection coefficient ($W\,m^{-2}\,K^{-1}$)
L_c	PV module characteristics length (m)
ΔT	Temperature difference between the PV module surface and the ambient temperatures (K)
ΔT_f	Temperature difference between the PV module junction and its front side surface (K)
ΔT_b	Temperature difference between the PV module junction and its back side surface (K)
δT	Constant value (K)
T_f	The average between the surface and ambient temperatures (K)
Gr	Grashof number (-)
Nu	Nusselt number (-)
Nu_{forced}	Nusselt number of the forced convection (-)
Nu_{free}	Nusselt number of the free convection (-)
Ra	Rayleigh number (-)
Pr	Prandtl number (-)
Re	Reynolds number (-)
k_{air}	Thermal conductivity of air ($W\,m^{-1}\,K^{-1}$)
W	Module width (m)
S	Module perimeter (m)
A	Module area (m^2)
F	View factor (-)
H	Forced convection adjustment coefficient (-) or module length (m)
m	Empirical factor of the forced convection to the tilt angle and wind relationship (-)
$RMSE$	Root mean square error ($^\circ$C)
r	Correlation coefficient (-)

Greek letters

α	Absorptivity (-)
α_{fg}	Absorptivity of the glass layer (-)
α_{PV}	Absorptivity of the semiconductor layer (-)
Φ	Total received irradiance (W/m^2)
η	Efficiency (-)
τ	Transmittance (-)
τ_{fg}	Transmittance of the glass layer (-)
ρ	Density (kg m^{-3})
β	Module tilt angle (°) or air thermal expansion coefficient (K^{-1})
θ	Angle of the module to the vertical axis (°)
μ_{air}	Dynamic viscosity of air (kg m^{-1} s^{-1})
ν_w	Wind speed (m s^{-1})
σ	Stefan–Boltzmann constant (W m^{-2} K^{-4})
ϵ	Emissivity (-)

References

1. Motiei, P.; Yaghoubi, M.; GoshtashbiRad, E.; Vadiee, A. Two-dimensional unsteady state performance analysis of a hybrid photovoltaic-thermoelectric generator. *Renew. Energy* **2018**, *119*, 551–565. [CrossRef]
2. Chopde, A.; Magare, D.; Patil, M.; Gupta, R.; Sastry, O.S. Parameter extraction for dynamic PV thermal model using particle swarm optimisation. *Appl. Therm. Eng.* **2016**, *100*, 508–517. [CrossRef]
3. Kant, K.; Shukla, A.; Sharma, A.; Biwole, P.H. Thermal response of poly-crystalline silicon photovoltaic panels: Numerical simulation and experimental study. *Sol. Energy* **2016**, *134*, 147–155. [CrossRef]
4. Aly, S.P.; Ahzi, S.; Barth, N.; Figgis, B.W. Two-dimensional finite difference-based model for coupled irradiation and heat transfer in photovoltaic modules. *Sol. Energy Mater. Sol. Cells* **2018**, *180*, 289–302. [CrossRef]
5. Armstrong, S.; Hurley, W.G. A thermal model for photovoltaic panels under varying atmospheric conditions. *Appl. Therm. Eng.* **2010**, *30*, 1488–1495. [CrossRef]
6. Aly, S.P.; Ahzi, S.; Barth, N.; Abdallah, A. Using energy balance method to study the thermal behavior of PV panels under time-varying field conditions. *Energy Convers. Manag.* **2018**, *175*, 246–262. [CrossRef]
7. Zhao, B.; Chen, W.; Hu, J.; Qiu, Z.; Qu, Y.; Ge, B. A thermal model for amorphous silicon photovoltaic integrated in ETFE cushion roofs. *Energy Convers. Manag.* **2015**, *100*, 440–448. [CrossRef]
8. Balog, R.S.; Kuai, Y.; Uhrhan, G. A photovoltaic module thermal model using observed insolation and meteorological data to support a long life, highly reliable module-integrated inverter design by predicting expected operating temperature. In Proceedings of the 2009 IEEE Energy Conversion Congress and Exposition (ECCE 2009), San Jose, CA, USA, 20–24 September 2009; pp. 3343–3349. [CrossRef]
9. Santiago, I.; Trillo-Montero, D.; Moreno-Garcia, I.; Pallarés-López, V.; Luna-Rodríguez, J. Modeling of photovoltaic cell temperature losses: A review and a practice case in South Spain. *Renew. Sustain. Energy Rev.* **2018**, *90*, 70–89. [CrossRef]
10. Verma, V.; Kane, A.; Singh, B. Complementary performance enhancement of PV energy system through thermoelectric generation. *Renew. Sustain. Energy Rev.* **2016**, *58*, 1017–1026. [CrossRef]
11. Kosyachenko, L.A. *Solar Cells—New Approaches and Reviews*; IntechOpen: Rijeka, Croatia, 2015.[CrossRef]
12. Jacques, S.; Caldeira, A.; Ren, Z.; Schellmanns, A.; Batut, N. Impact of the cell temperature on the energy efficiency of a single glass PV module: Thermal modeling in steady-state and validation by experimental data. In Proceedings of the International Conference on Renewable Energies and Power Quality (ICREPQ'13), Bilbao, Spain, 20–22 March 2013.
13. Kayhan, Ö. A thermal model to investigate the power output of solar array for stratospheric balloons in real environment. *Appl. Therm. Eng.* **2018**, *139*, 113–120. [CrossRef]
14. Solanki, C.S.; Singh, H.K. *Anti-Reflection and Light Trapping in c-Si Solar Cells*; Green Energy and Technology; Springer: Singapore, 2017; pp. 17–41. [CrossRef]
15. Mattei, M.; Notton, G.; Cristofari, C.; Muselli, M.; Poggi, P. Calculation of the polycrystalline PV module temperature using a simple method of energy balance. *Renew. Energy* **2006**, *31*, 553–567. [CrossRef]

16. Akhsassi, M.; El Fathi, A.; Erraissi, N.; Aarich, N.; Bennouna, A.; Raoufi, M.; Outzourhit, A. Experimental investigation and modeling of the thermal behavior of a solar PV module. *Sol. Energy Mater. Sol. Cells* **2018**, *180*, 271–279. [CrossRef]

17. Schwingshackl, C.; Petitta, M.; Wagner, J.E.; Belluardo, G.; Moser, D.; Castelli, M.; Zebisch, M.; Tetzlaff, A. Wind Effect on PV Module Temperature: Analysis of Different Techniques for an Accurate Estimation. *Energy Procedia* **2013**, *40*, 77–86. [CrossRef]

18. Notton, G.; Cristofari, C.; Mattei, M.; Poggi, P. Modelling of a double-glass photovoltaic module using finite differences. *Appl. Therm. Eng.* **2005**, *25*, 2854–2877. [CrossRef]

19. Guerriero, P.; Codecasa, L.; D'Alessandro, V.; Daliento, S. Dynamic electro-thermal modeling of solar cells and modules. *Sol. Energy* **2019**, *179*, 326–334. [CrossRef]

20. Sahli, M.; Correia, J.P.M.; Ahzi, S.; Touchal, S. Multi-physics modeling and simulation of heat and electrical yield generation in photovoltaics. *Sol. Energy Mater. Sol. Cells* **2018**, *180*, 358–372. [CrossRef]

21. Weiss, L.; Amara, M.; Ménézo, C. Impact of radiative-heat transfer on photovoltaic module temperature. *Prog. Photovolt. Res. Appl.* **2016**, *24*, 12–27. [CrossRef]

22. Kaplani, E.; Kaplanis, S. Thermal modelling and experimental assessment of the dependence of PV module temperature on wind velocity and direction, module orientation and inclination. *Sol. Energy* **2014**, *107*, 443–460. [CrossRef]

23. Usama Siddiqui, M.; Arif, A.F.M.; Kelley, L.; Dubowsky, S. Three-dimensional thermal modeling of a photovoltaic module under varying conditions. *Sol. Energy* **2012**, *86*, 2620–2631. [CrossRef]

24. Vogt, M.R.; Holst, H.; Winter, M.; Brendel, R.; Altermatt, P.P. Numerical Modeling of c-Si PV Modules by Coupling the Semiconductor with the Thermal Conduction, Convection and Radiation Equations. *Energy Procedia* **2015**, *77*, 215–224. [CrossRef]

25. Kaldellis, J.K.; Kapsali, M.; Kavadias, K.A. Temperature and wind speed impact on the efficiency of PV installations. Experience obtained from outdoor measurements in Greece. *Renew. Energy* **2014**, *66*, 612–624. [CrossRef]

26. Gu, X.; Yu, X.; Guo, K.; Chen, L.; Wang, D.; Yang, D. Seed-assisted cast quasi-single crystalline silicon for photovoltaic application: Towards high efficiency and low cost silicon solar cells. *Sol. Energy Mater. Sol. Cells* **2012**, *101*, 95–101. [CrossRef]

27. Skoplaki, E.; Palyvos, J.A. Operating temperature of photovoltaic modules: A survey of pertinent correlations. *Renew. Energy* **2009**, *34*, 23–29. [CrossRef]

28. Muzathik, A.M. Photovoltaic Modules Operating Temperature Estimation Using a Simple Correlation. *Int. J. Energy Eng.* **2014**, *4*, 151–158.

29. Jones, A.; Underwood, C. A thermal model for photovoltaic systems. *Sol. Energy* **2001**, *70*, 349–359. [CrossRef]

30. Torres-Lobera, D.; Valkealahti, S. Inclusive dynamic thermal and electric simulation model of solar PV systems under varying atmospheric conditions. *Sol. Energy* **2014**, *105*, 632–647. [CrossRef]

31. Torres Lobera, D.; Valkealahti, S. Dynamic thermal model of solar PV systems under varying climatic conditions. *Sol. Energy* **2013**, *93*, 183–194. [CrossRef]

32. Tsai, H.F.; Tsai, H.L. Implementation and verification of integrated thermal and electrical models for commercial PV modules. *Sol. Energy* **2012**, *86*, 654–665. [CrossRef]

33. Kurz, D.; Nawrowski, R. Thermal time constant of PV roof tiles working under different conditions. *Appl. Sci.* **2019**, *9*. [CrossRef]

34. Sánchez Barroso, J.C.; Barth, N.; Correia, J.P.M.; Ahzi, S.; Khaleel, M.A. A computational analysis of coupled thermal and electrical behavior of PV panels. *Sol. Energy Mater. Sol. Cells* **2016**, *148*, 73–86. [CrossRef]

35. Faiman, D. Assessing the outdoor operating temperature of photovoltaic modules. *Prog. Photovolt. Res. Appl.* **2008**, *16*, 307–315. [CrossRef]

36. Franghiadakis, Y.; Tzanetakis, P. Explicit empirical relation for the monthly average cell-temperature performance ratio of photovoltaic arrays. *Prog. Photovolt. Res. Appl.* **2006**, *14*, 541–551. [CrossRef]

37. Ventura, C.; Tina, G.M. Utility scale photovoltaic plant indices and models for on-line monitoring and fault detection purposes. *Electr. Power Syst. Res.* **2016**, *136*, 43–56. [CrossRef]

38. Veldhuis, A.J.; Nobre, A.M.; Peters, I.M.; Reindl, T.; Ruther, R.; Reinders, A.H.M.E. An Empirical Model for Rack-Mounted PV Module Temperatures for Southeast Asian Locations Evaluated for Minute Time Scales. *IEEE J. Photovolt.* **2015**, *5*, 774–782. [CrossRef]

39. Barth, N.; Al Otaibi, Z.S.; Ahzi, S. Irradiance, thermal and electrical coupled modeling of photovoltaic panels with long-term simulation periods under service in harsh desert conditions. *J. Comput. Sci.* **2018**, *27*, 118–129. [CrossRef]

40. Gu, W.; Ma, T.; Shen, L.; Li, M.; Zhang, Y.; Zhang, W. Coupled electrical-thermal modelling of photovoltaic modules under dynamic conditions. *Energy* **2019**, *188*, 116043. [CrossRef]

41. Migliorini, L.; Molinaroli, L.; Simonetti, R.; Manzolini, G. Development and experimental validation of a comprehensive thermoelectric dynamic model of photovoltaic modules. *Sol. Energy* **2017**, *144*, 489–501. [CrossRef]

42. Palacio Vega, M.A.; González López, O.M.; Martínez Guarín, A.R.; Gómez Vásquez, R.D.; Bula, A.; Mendoza Fandiño, J.M. Estimation of the Surface Temperature of a Photovoltaic Panel Through a Radiation-Natural Convection Heat Transfer Model in Matlab Simulink. In Proceedings of the ASME 2016 International Mechanical Engineering Congress and Exposition, Volume 8: Heat Transfer and Thermal Engineering, Phoenix, AZ, USA, 11–17 November 2016; American Society of Mechanical Engineers: Phoenix, AZ, USA, 2017; doi:10.1115/IMECE2016-66769. [CrossRef]

43. Cengel, Y. *Heat and Mass Transfer: Fundamentals and Applications*; McGraw-Hill Higher Education: New York, NY, USA, 2014.

44. Sartori, E. Convection coefficient equations for forced air flow over flat surfaces. *Sol. Energy* **2006**, *80*, 1063–1071. [CrossRef]

45. King, D.L.; Kratochvil, J.A.; Boyson, W.E. *Photovoltaic Array Performance Model*; United States Department of Energy: Livermore, CA, USA, 2004.

46. Barykina, E.; Hammer, A. Modeling of photovoltaic module temperature using Faiman model: Sensitivity analysis for different climates. *Sol. Energy* **2017**, *146*, 401–416. [CrossRef]

Article

Thermo-Fluidic Characterizations of Multi-Port Compact Thermal Model of Ball-Grid-Array Electronic Package

Valentin Bissuel [1,*], Frédéric Joly [2], Eric Monier-Vinard [1], Alain Neveu [2] and Olivier Daniel [1]

[1]   Thales Corporate Engineering, 19-21 Avenue Morane Saulnier, 78140 Vélizy-Villacoublay, France; eric.monier-vinard@thalesgroup.com (E.M.-V.); olivier.daniel@thalesgroup.com (O.D.)

[2]   LMEE, Université d'Evry, Université Paris-Saclay, 91020 Evry, France; F.Joly@iut.univ-evry.fr (F.J.); a.neveu@iut.univ-evry.fr (A.N.)

*   Correspondence: valentin.bissuel@thalesgroup.com

Received: 31 March 2020; Accepted: 3 June 2020; Published: 9 June 2020

Abstract: The concept of a single-input/multi-output thermal network was proposed by the Development of Libraries of Physical models for an Integrated design environment (DELPHI) consortium more than twenty years ago. The present work highlights the recent improvements made to efficiently derive a low-computing-effort model from a fully detailed numerical model and to characterize its performances. The temperature predictions of a deduced ball-grid-array (BGA) dynamic compact thermal model are compared to those of a realistic three-dimensional representation, including the large set of internal copper traces, as well as its board structure, which has been validated by experiment. The current study discloses a method for creating an amalgam reduced-order modal model (AROMM) for that electronic component family that allows the preservation of the geometry integrity and shortening scenarios computation. Typically, the AROMM method reduces by a factor of 600 the computation time needed to obtain the solution while keeping the error on the maximum temperature below 2%. Then, a meta-heuristic optimization is run to derive a more practical low-order resistor capacitor model that enables a thermo-fluidic analysis at the board level. Based on the calibrated numerical model, a novel AROMM method was investigated in order to address the chip behavior submitted to multiple heat sources. The first results highlight the capability to enforce a non-uniform power distribution on the upper surface of the silicon chip. Thus, the chip design layout can be analyzed and optimized to prevent thermal and reliability issues.

Keywords: BCI-DCTM; ROM; modal approach; BGA; experimental validation

---

1. Introduction

The thermal behavior of electronic components can be finely predicted by thermal and fluidic numerical simulations [1]. However, as the geometrical and functional description of the component grows in complexity, the time needed for simulations becomes unbearable for parametric studies if fully detailed numerical representations are used.

To overcome the inherent time-consuming computation of such a detailed model, new methods for creating surrogate models able to properly reproduce its steady-state response, as well as transient ones, in a shorter time, were developed. This statement leads research, at first, toward dynamic compact thermal models (DCTMs) [2], which aim to predict the key thermal characteristics of a component. The Development of Libraries of Physical models for an Integrated design environment (DELPHI) approach promotes the use of a matrix of thermal resistances that link the sub-divided exterior surfaces of a component to its junction, which is the highest temperature of the component. The construction of a DCTM required training data, obtained by numerical simulations or experiment results.

Modal models are an alternative to DTCMs. They can be seen as an extension of the classical Fourier decomposition. The temperature is searched as a sum of known elementary spatial functions, called the modes, weighted by unknown coefficients. Different methods are based on this principle: The most popular is the proper orthogonal decomposition method (POD), which requires knowledge of thermal fields from experimental or numerical data [3,4]. However, the challenge is to find a modal base independent of boundary conditions. In that perspective, Codecasa et al. used a multi-point moment-matching method algorithm [5]. Joly et al. [6] used 'branch modes' to solve problems associated with time-dependent boundary conditions. The originality of the branch modes is that 'the branch eigenvalue problem' uses Steklov boundary conditions. This method has been extended to the amalgam reduced-order modal model (AROMM) method: A modal base is calculated by solving an eigenvalue problem. The reduced model is obtained by reducing the initial base by the amalgam method [7]. These developments gave birth to a new hybrid methodology to build DELPHI-inspired DCTMs by replacing the full detailed models by a reduced-order model based on the modal approach [8,9].

However, more complex configurations demand more sophisticated methods, in which simple reduced models are built and then connected to each other. Following that idea, Grosjean et al. developed a substructuring modal method that allows the reduction of an electronic board with several active components [10,11]. Codecasa et al. also coupled boundary-independent reduced models of components [12]. However, in those studies, the elementary components were simple (Quad Flat No-leads (QFN) package 16 or 32 with a single heat source, or Insulated Gate Bipolar Transistor (IGBT) with heat sources activated together), as the challenge was to couple those independent reduced models efficiently.

A single component with independent multiple heat sources has also been recently considered using the MPMM method [9] or AROMM [13]. Those studies have been limited to a component with a couple of independent sources. The objective of this paper is to present a reduced modal model of a ball grid array package with nine independent heat sources located on the top of the chip.

The paper is organized as follows. The position of the problem, as well as the studied material, is presented first. Then, the experimental setup and measurements are introduced. In the second step, the numerical model is built and experimentally validated for different environments when a uniform power distribution is localized on top of the chip. The creation of the reduced-order model by the modal method is then presented, and two different use cases are outlined. The first use case highlights how reduced-order modal models can be used to replace the detailed model for the creation of the boundary condition-independent dynamic compact thermal model. The second one presents a purely numerical study in which the power distribution is not uniformly applied on the top surface of the chip and demonstrates the relevance of this approach applied to this complex configuration.

2. Position of the Problem

The ball grid array package with 208 solder balls is a very popular package for integrated circuits (IC) with a large interconnection number. That kind of IC package usually consists of an active centered semiconductor chip that is glued on a two-copper-layer laminate, as described in Figure 1.

The semiconductor chip, as well as its gold wire bonds, is over-molded with a plastic resin. The diameter of the gold wires is 25.4 µm, which highlights the aspect ratio constraints.

The laminate structure routes functional signals from the input/output of the chip to the solder balls and also acts as a radiator for heat spreading. It is made up by two thin signal layers interconnected by vias. Figure 2 shows the copper patterns of the three layers of the studied ball-grid-array (BGA) substrate.

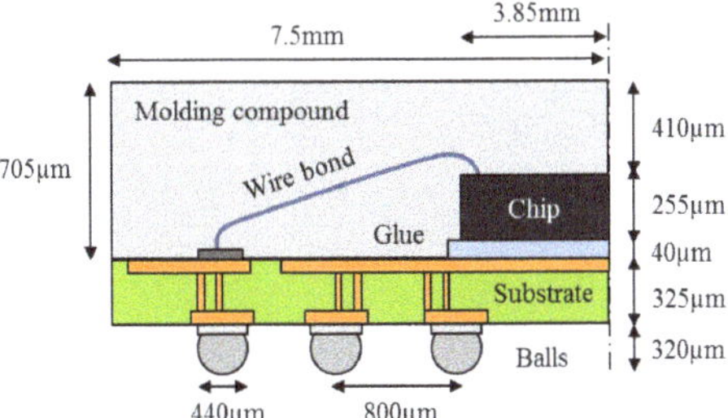

Figure 1. Semi-cross-section of the studied BGA208.

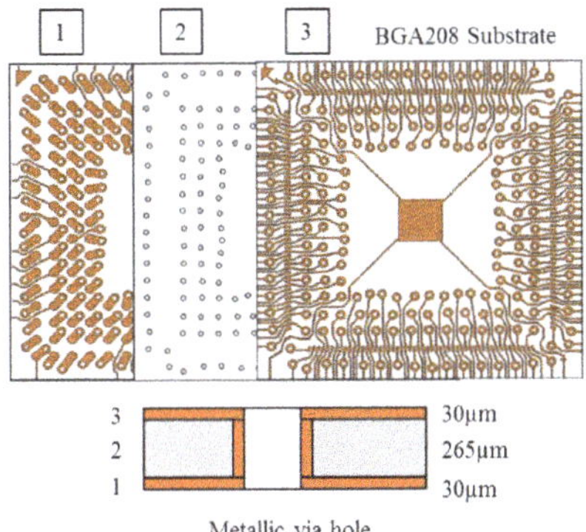

Figure 2. BGA208's laminate description.

Table 1 supplies the thermal properties used to establish the numerical model of this BGA package.

Table 1. Reference material properties.

Constituent	Material	k (W/(m.K))
Molding Compound	Resin	0.66
Wire bond	Gold	320
Chip	Silicon	148
Chip attach adhesive	Silver Glue	2.1
Signal layer and via	Copper	400
Dielectric layer	FR4	0.38
Solder ball	63Sn37pb	51

3. Experimental Measurements

3.1. Experimental Setup

The complex component package cannot be tested independently of a printed circuit board (PCB). Thus, a set of standardized tests [14] was performed for two standardized PCBs, named 2s0p [15] and 2s2p [16], as well as for various stabilized airflow boundary conditions (still air [17] and forced moving air [18]) in order to check the component thermal performances according to JEDEC recommendations.

Figure 3 shows the stack-up of seven layers that alternate between high- (1, 3, 5, 7) and very-low (2, 4, 6)-conductivity layers that are defined for a JEDEC 2s2p thermal test board.

The "*s*" refers to the signal layers and "*p*" to the buried power (or ground plane) layers. The two internal quasi full-covered copper layers act as efficient in-plane heat spreaders. The overall length, width, and thickness of the test board are respectively 102, 112, and 1.6 mm. The typical width of a copper trace is 300 μm.

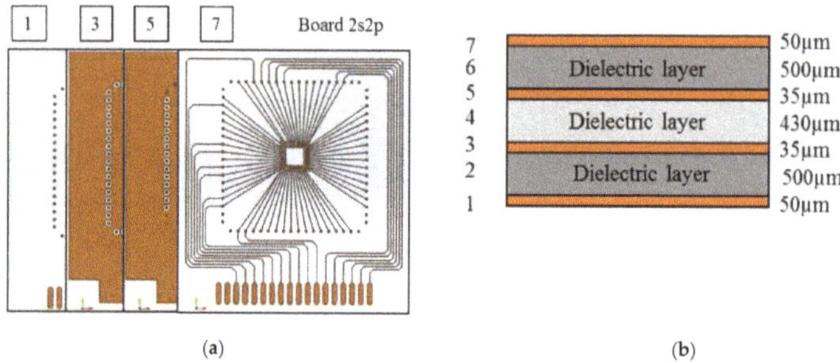

(a)   (b)

Figure 3. JEDEC 2s2p thermal test boards: Copper traces definition (**a**) and stack-up (**b**).

3.2. Measurements

Experimental measurements have been performed by the thermal test services of Analysis Tech following all JEDEC $n°$ 51 requirements. Figure 4 displays the standardized experimental setup used to characterize this BGA208.

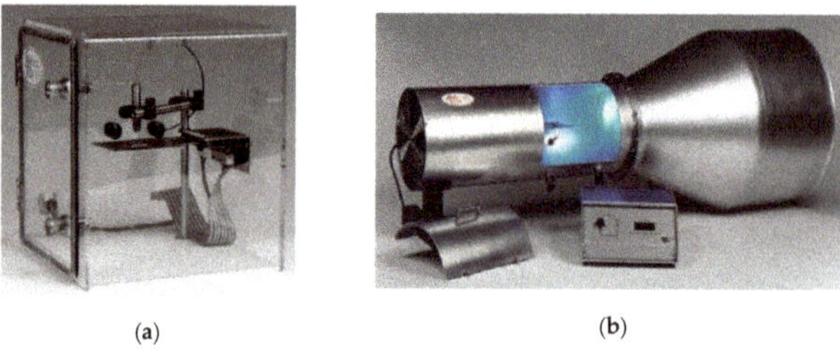

(a)   (b)

Figure 4. Standardized experimental setups for natural convection (**a**) and forced convection (**b**).

Following JEDEC standards, the chip behavior is characterized by a metric, which is called junction-to-ambient thermal resistance, defined by Equation (1):

$$R_{JA}(Q) = \left[T_J - T_\infty\right]/Q \tag{1}$$

That metric indicates the flowing capacity of a uniform power (Q) dissipated in the device through all the thermal paths between the chip junction (T_J) and the ambient air. This parameter can be easily calculated with measured temperatures and power.

Table 2 gives the reference values of R_{JA} used to validate the numerical models.

Table 2. Experimental R_{JA} measurement of BGA208 mounted on the 2s2p board.

Convection Mode	T_∞ (°C)	Q (W)	U (m/s)	R_{JA}^M (K/W)
Natural	22.5	2.001	0	29.21
	21.1	3.037	1	25.37
Forced	20.7	3.02	2	23.91
	20.7	3.06	3	22.87

4. Detailed Numerical Model

4.1. Definition of the Mathematical Model

Let Ω be a domain made of two disjoint sub-domains Ω_1 and Ω_2, as illustrated in Figure 5.

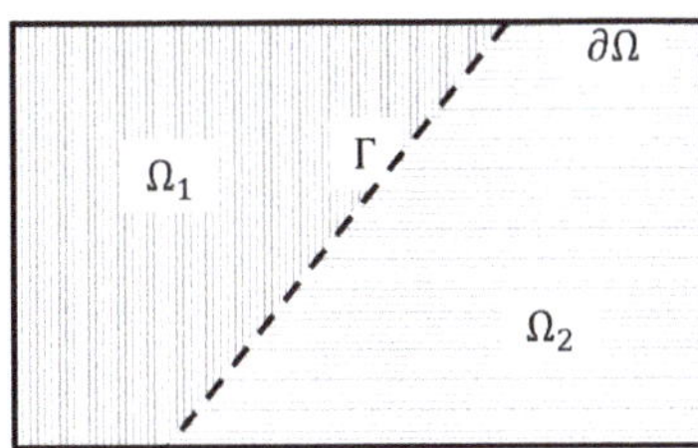

Figure 5. Sub-decomposition of the domain.

The boundary between each sub-domain is referred to as Γ. An interface thermal resistance accounts for imperfect contact (R_c).

Let T be the temperature field of the domain Ω. This latter is subjected to an internal power generation, named ω. The generated heat is exchanged from the outside surfaces $(\partial\Omega)$ considering a Fourier boundary condition. The heat transfer coefficient h gathers convection and radiation phenomena. The thermal conductivity and the thermal capacity are respectively defined as k and C. Heat exchanges are modeled by the heat equation:

$$C\,\dot{T} = \underline{\nabla}\cdot(k\ \underline{\nabla}T) + \omega \quad on\ \Omega \tag{2}$$

$$k\ \underline{\nabla}T_i\cdot\underline{n} = h\ (T_\infty - T) \quad on\ \partial\Omega \tag{3}$$

The heat flux density φ is conserved through Γ, but the imperfect contact creates a temperature discontinuity.

$$k\ \underline{\nabla}T_1\cdot\underline{n_1} = -k\ \underline{\nabla}T_2\cdot\underline{n_2} = \varphi \quad on\ \Gamma \tag{4}$$

$$T_2 - T_1 = \varphi/R_c \quad on\ \Gamma \tag{5}$$

where T_i is the temperature of a given sub-domain Ω_i.

The matrix formulation is established using classic spatial discretization by finite elements:

$$\forall\ i,k \in \{1,2\},\ i \neq k,\quad C_i\,\dot{T}_i = -[K_i + H_i]\ T_i + J_{i,k}\ T_k + U_i \tag{6}$$

Matrix $J_{i,k}$ is a rectangular matrix that ensures the coupling between substructures i and k. U is the vector representing solicitations.

4.2. Experimental Validation of The Numerical Model

To be relevant and adequate, the thermo-fluid simulations were made using a full description of every detail of the laminate structure. As seen in Figure 6, with this fine three-dimensional description of the substrate, meshing the BGA requires around 600,000 degrees of freedom (DoF) and, consequently, high computing resources.

Moreover, the JEDEC test board, described in Figure 3, is completely modeled in 3D to minimize the modeling assumptions, and the experimental setups, displayed in Figure 4, are converted to numeric boundary conditions.

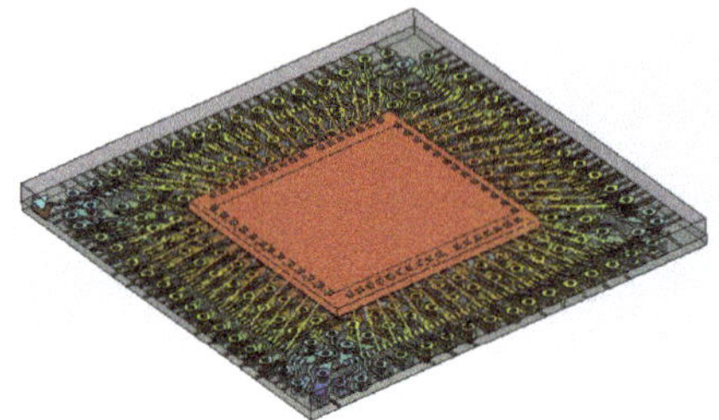

Figure 6. BGA208's numerical model.

Table 3 gives the adjusted thermal properties of the calibrated numerical model of the board substrate.

Table 3. Reference properties for both JEDEC boards.

Constituent	Material	k (W/(m.K))
Signal layer and via	Copper	400
Dielectric layer	FR4	0.38

Numerical simulations were performed using four distinct pieces of computational fluid dynamic software [1,19] and demonstrates, whatever the thermal test board, a very good agreement between the experimental measurements and numerical results on the R_{JA} computation, as reported in Table 4, for the 2s2p PCB.

Table 4. Fitting of the 2s2p thermal metrics (Icepak®).

T_∞ (°C)	Q (W)	U (m/s)	R_{JA}^M (K/W)	R_{JA}^N (K/W)	%E
22.5	2.001	0	29.21	29.33	<1%
21.1	3.037	1	25.37	25.45	<1%
20.7	3.02	2	23.91	24.11	<1%
20.7	3.06	3	22.87	23.07	<1%

It occurs that the discrepancy of the numerical model (R_{JA}^N) in comparison to measurements (R_{JA}^M) is lower than 2%. The 3-D numeric model of the BGA has been validated by experimental measurements, so the first step of model reduction has been achieved.

The mesh size of that realistic numerical model is 28.9 million cells. In the steady state, the convergence for each set of boundary conditions applied to that model is reached in 8h00 using 16 cores and 48 GB of RAM. Cleary, a model order reduction is mandatory to act on the design for overpopulated industrial electronic boards.

5. Reduced-Order Modal Model

Inspired by the classical decomposition in Fourier series, the temperature is searched as a sum of known elementary spatial functions, called the modes, weighted by unknown coefficients. However, the creation of the modal model is more complex and the current study focuses on the substructuring modal method [20]. This latter allows the chip to be handled separately and to reduce it more efficiently.

5.1. Modal Formulation

To ensure the coupling between both sub-domains (Ω_i), the temperature is decomposed on a Dirichlet–Steklov base [11]:

$$T\left(\underline{M}, t\right) = \sum_i x_i^D(t)\ V_i^D\left(\underline{M}\right) + \sum_i x_i^S(t)\ V_i^S\left(\underline{M}\right) \tag{7}$$

Dirichlet's modes are defined as follows:

$$\underline{\nabla} \cdot \left(k\ \underline{\nabla} V_i^D\right) = \lambda_i\ C\ V_i^D\ \ on\ \Omega \tag{8}$$

$$V_i^D = 0\ \ on\ \partial\Omega \tag{9}$$

where λ_i are the eigenvalues.

Temperature fields that can be rebuilt using Dirichlet modes are null on the boundary. Therefore, these fields belong to a subspace of the admissible thermal fields, but smaller. Thus, it is necessary to add a second subspace so that the union of the eigenbasis of the two subspaces gives the space of the admissible thermal fields. This is the role of the Steklov base, whose modes verify the following eigenvalue problem.

$$\underline{\nabla} \cdot \left(k\ \underline{\nabla} V_i^S\right) = 0\ \ on\ \Omega \tag{10}$$

$$k\ \underline{\nabla} V_i^S \cdot \underline{n} = -\lambda_i\ V_i^S\ \ on\ \partial\Omega \tag{11}$$

By construction, the union of the eigenbasis of these two subspaces gives the space of the admissible thermal fields. Thus, temperature fields can be rebuilt on the entire domain.

5.2. Modal Reduction: The Amalgam Method

The modal formulation only shifts the problem: Instead of computing temperature values at the nodes of a mesh, temporal states (or amplitudes) are searched. The next step consists of reducing the size of the model, i.e., reducing the number of degrees of freedom from N to $\widetilde{N}$, where $\widetilde{N} \ll N$. This is done by the amalgam method, where the most prominent modes are selected and the remaining ones are added to them, weighted by a coefficient [7]. These new amalgamated modes are referred to as $\widetilde{V}_i$ and are expressed as a linear combination of the original modes V_i according to

$$\widetilde{V}_i = \sum_p \alpha_{i,p}^\aleph V_{i,p}^\aleph\ \ \text{where}\ \aleph \in \{D, S\} \tag{12}$$

The coefficients $\alpha_{i,p}^\aleph$ are determined by minimizing, in the modal space, the distance between the modal model and a reference model. The quality of the approximation is, thus, dependent of this reference model.

5.3. The State Equation

The state equation is obtained by replacing the temperature field in Equation (6) by its modal decomposition (Equation (7)), while the test functions are the eigenmodes. A simplified version is given here, where it is supposed that the conductivity and capacity used in Equations (2) and (8)–(11) are identical, and where orthogonality properties are used to simplify the problem.

$$\dot{X}_i^D + V_i^D C_i V_i^S \dot{X}_i^S = -\Lambda_i^D X_i^D + V_i^D U_i \tag{13}$$

$$\forall\ \ i,k \in \{1,2\},\ \ i \neq k, V_i^S C_i V_i^D \dot{X}_i^D + V_i^S C_i V_i^S \dot{X}_i^S = -\left(\Lambda_i^D + V_i^S H_i V_i^S\right) X_i^S + V_i^S J V_k^S + V_i^S U_i \tag{14}$$

where Λ_i^D and Λ_i^S are diagonal matrices of eigenvalues such that $\Lambda_i^D(k,k) = \lambda_{i,k}^D$.

6. Utilization of Modal Model to Create a Dynamic Compact Thermal Model

The proposed global hybrid procedure for the creation of the dynamic compact thermal model (DCTM) is outlined in Figure 7. By coupling the model-order-reduction (MOR) technique based on the modal approach [8] and a meta-heuristic optimization [21], that procedure allows us to reduce both creation and simulation times of a suitable model of a sophisticated BGA package. The most relevant benefit is achieved for transient calculations.

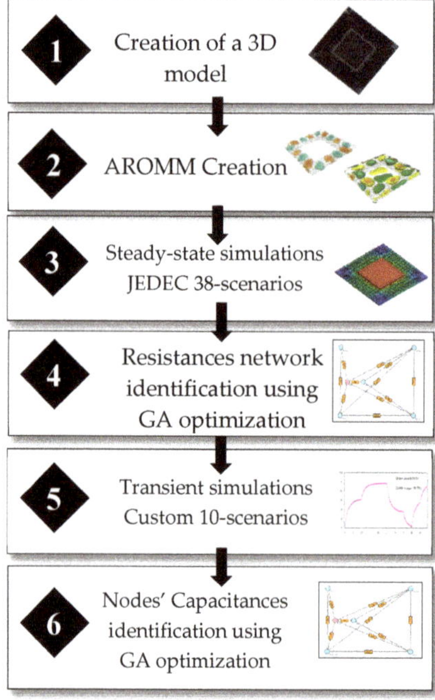

Figure 7. Hybrid dynamic compact thermal model (DCTM) creation flow.

In fine, the developed reduction process enables an amalgam reduced-order modal model to be generated and then a practical dynamic compact thermal model to be derived, both being highly reliable whatever the environmental conditions. In these two cases, models are boundary conditions-independent (BCI) by construction.

The overall DCTM creation time is reduced by 86% using Reduced Order Model mathematical calculations instead of time-consuming Detailed Thermal Model numerical simulations to generate training data required for Genetic Algorithm optimization, as reported in [8].

Example of DCTM Network Definition

The derived BGA208 surrogate model is made to handle multiple thermal paths, so the DCTM network is circumscribed, in this case, to nine nodes corresponding to:

1. One "Junction": Maximum temperature of the chip,
2. One "top inner": Projected chip area on top surface,
3. Two "top outer": The four regrouped corners and four remaining top surfaces,
4. One "Bottom inner": Keep-out ball area
5. Three "Bottom outer" according to ball footprint patterns [8]
6. One "Sides": Regrouped lateral surfaces excluding the balls layer.

The thermal predictions of the deduced DCTM were evaluated for each boundary conditions set, as well as for each thermal board test, and then compared to experimental results, as shown in Table 5.

Table 5. Approximation of 2s2p thermal metrics (Icepak®).

T_∞ (°C)	Q (W)	V (m/s)	R_{JA}^{M} (K/W)	R_{JA}^{DCTM} (K/W)	%E
22.5	2.001	0	29.21	29.73	1.8%
21.1	3.037	1	25.37	25.09	1.1%
20.7	3.02	2	23.91	23.51	1.7%
20.7	3.06	3	22.87	22.16	3.1%

The model agreement is good with a discrepancy lower than 4% while the simulation speed is greatly improved. Thus, the good accuracy of the DCTM permits us to integrate this model inside the system/subsystem simulation to quickly identify thermal issues and optimize cooling solutions.

7. Impact of Chip Power Dissipation Layout

In realistic applications, the functions burnt on the chip are numerous, varied, and dissymmetric, and their activations depend on used or implemented logical functions. Thus, the power distribution of the silicon chip is not uniform and additional thermal analyses need to be carried out to predict the influence of various power dissipation patterns.

As seen in Figure 8, the chip is now partitioned in nine zones (2^9 possible combinations) to model more accurately the heating due to the individual activation of various logical functions. The largest source and the smallest one represent respectively 22.4% and 1.8% of the chip volume.

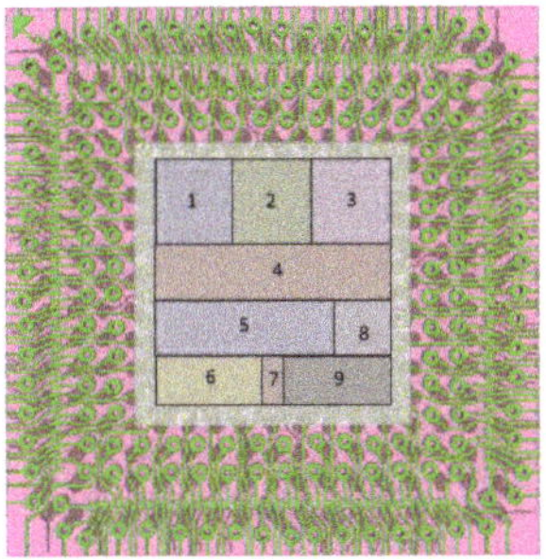

Figure 8. Chip partitioning in nine zones.

In this fictive case, the conventional method used to create dynamic compact thermal models can hardly be applied [22]. First, based on the superposition principle, the number of mandatory simulations (JEDEC 38-set scenarios) to correctly identify the resistances network must be multiplied by ten. Each zone is separately activated, and then all of them.

Second, the meaning of the junction temperature for a nine-heat-sources network is not trivial. For instance, Figure 9 highlights two temperature mappings (only the chip and the copper traces are displayed) when a power dissipation of 2.6 W is uniformly applied on the upper surface of the chip (Figure 9a), or, at the opposite, concentrated on a peculiar zone (Figure 9b), named zone 7 (refer to Figure 8). Both numerical simulations assume similar boundary conditions on package external surfaces, such as $\bar{h}_{TOP} = \bar{h}_{SIDES} = 20$ W/(m^2.K) and $\bar{h}_{BOTTOM} = 800$ W/(m^2.K).

Obviously, for a smaller surface dissipation, the maximum temperature reached by the chip rises significantly (24 °C) and its location is not centered anymore. This phenomenon will be especially exacerbated for dynamic simulations. Indeed, the location of the maximum temperature moves at each time step following the transient power profile applied on each zone. Thus, applying our previous DCTM creation flow seems difficult, and a modal approach is chosen.

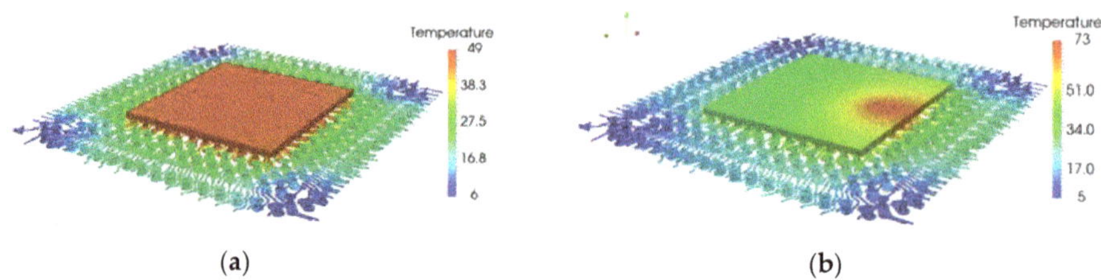

Figure 9. Sensitive temperatures encountered by the chip with uniform power dissipation (**a**) and localized on a peculiar area (**b**).

8. Reduced-Order Modal Model for Multiple Heat Sources

8.1. Multi-Source Numerical Model

As stated in the introduction, this part is not validated experimentally, as the experimental setup is still under development. The reduced-order model is compared to the finite elements model. However, this latter has been validated experimentally in Section 4.2 for a uniform power distribution.

8.2. Computation of the Dirichlet–Steklov Base

The BGA 208 package is split in two substructures: The chip is modeled separately by 8000 DoF. The substructuring technique allows its complete modal base to be deduced in 2.5 min. For the rest of the components (resin, balls, copper tracks), the complete base computation is not feasible, so only reduced percentages of Dirichlet modes and Steklov modes are selected, as commented in [13]. The computation of 17,800 modes (11,200 Dirichlet + 6600 Steklov) is made in 6.5 h.

8.3. Reduction of the Dirichlet–Steklov Base

In the perspective of an industrial application, reference simulations needed by the amalgam procedure should be carried out at a low computational cost and should be easy to conceive. Their objective is not to provide precise temperature fields but to trigger the relevant modes for the amalgam procedure. According to the heat sources number, ten cases are simulated and then concatenated: Each single zone is successively active and, finally, all of them. These reference simulations are obtained via a first-order Euler scheme with constant time steps. The whole process, corresponding to reference simulations and the amalgam procedure, can be performed in 1.5 h. *In fine*, 50 modes are retained for the chip, and 250 for the rest of the package, leading to a reduced modal model of order 300. Consequently, the number of DoF has been reduced by a factor of 2000.

8.4. Steady-State Results

Two cases are presented. They highlight the component thermal behavior when:

1. Test 1: A power dissipation of 2.6 watts is applied on zone 2
2. Test 2: Zones 3, 5, and 8 are respectively submitted to a power dissipation of 0.41, 0.675, and 0.0975 watts.

The mathematical calculations assume the boundary conditions on package external surfaces presented in Table 6.

Table 6. Heat transfer coefficients definition (W/(m^2.K)).

Case	$\bar{h}_{TOP}$	$\bar{h}_{BOTTOM}$	$\bar{h}_{SIDES}$
Test 1	50	250	15
Test 2	1000	40	100

Figure 10 presents the temperature field calculated by the reduced model for the whole package, as well as for the chip on the BGA substrate. The computation time of the derived ROM is lower than 0.5 s for a temporal simulation of 60 s (the time required to reach steady state). The main interest of the modal method lies in its ability to compute, at low cost, the whole temperature field, even for complex geometries such as the BGA packages family. Then, 4.5 s are needed to rebuild the whole temperature field for 101 snapshots. The hot-spot location on the chip, our concern, is properly identified, but fine details are also recovered as the temperature elevation on the copper tracks. The error between the reduced and the finite elements model is also displayed in Figure 10c. In most of the chip, and copper track, the error is below 1 °C (0.8%), which is a very interesting result for this preliminary investigation.

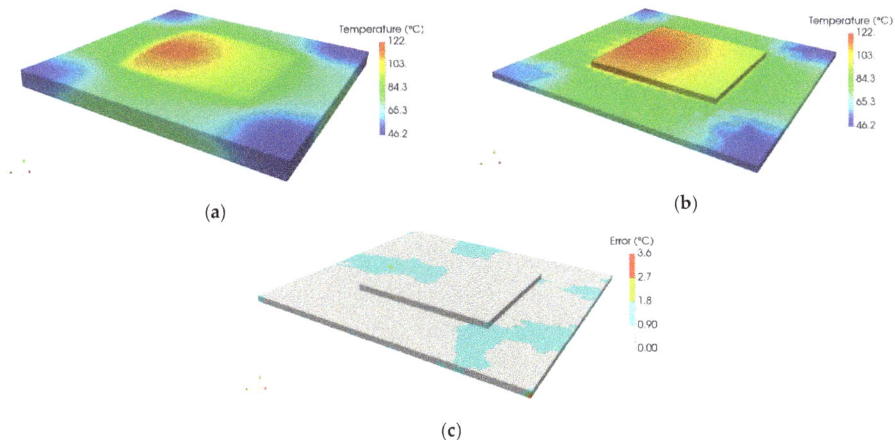

(a)

(b)

(c)

Figure 10. Temperature field with molded resin (**a**), without (**b**), and error field (**c**) at steady state for Test 1.

The temperature distribution at steady state for test 2 is presented in Figure 11 As the boundary conditions differ significantly from test 1 and the dissipated power is reduced, the maximum temperature reached by the chip is much lower and is predicted by the modal model with a maximum error of 0.36 °C (1.6%). Obviously, the hot-spot location moves as the different zones are activated, which is well predicted.

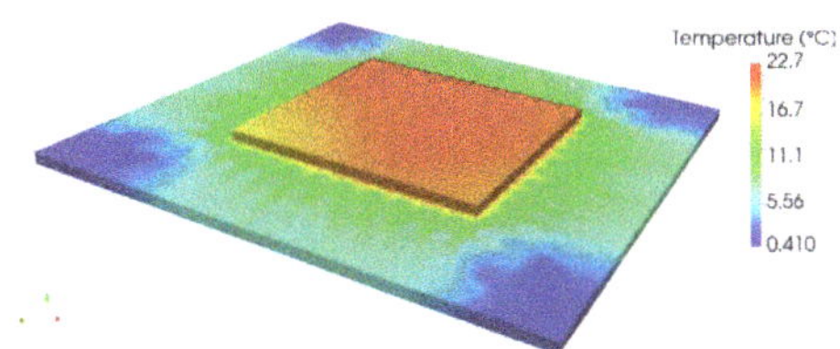

Figure 11. Temperature field for three active heat sources.

Finally, as the temperature field on the chip is different, it substantially affects the heat spreading on the tracks and, thus, the heat distribution on the ball array. The knowledge of the whole temperature field enables the computation of temperature gradients, and opens the way to thermomechanical consideration.

8.5. Transient Results

Dynamic simulations are conducted to compare the thermal prediction of the computed ROM with the FEM simulation (assumed to be the reference) on two test cases. FEM simulations need roughly 47 min to perform a 50 s transient simulation with multi-activations.

Two sets of boundary conditions (different from those presented in Table 6) were chosen and are summarized in Table 7.

Table 7. Heat transfer coefficients definition (W/(m^2.K)).

Case	$\bar{h}_{TOP}$	$\bar{h}_{BOTTOM}$	$\bar{h}_{SIDES}$
Test 1	20	800	10
Test 2	200	500	20

Boundary conditions of Case 1 correspond to a component mounted on a PCB in vertical natural convection plus radiation. Case 2 corresponds to a component sandwiched between the PCB and thermal drain reported on top of the component ($\bar{h}_{TOP}$ integrates all thermal paths: Contact resistances, thermal interface material, and aluminum drain).

Two power transient scenarios are defined: One in which different zones are successively activated, as presented in Table 8, and a second one in which different zones are simultaneously activated, as presented in Table 9. This latter describes a realistic operating case of a BGA. Indeed, power Input-Outputs and firmware are always on and the other functional areas have dynamic activation imposed by software operations.

Table 8. Activation of the different zones of the chip for transient simulation number 1.

Active Zone	φ (W/m^3)	Q (W)	Time (s)
Zone 4	8.2·10^9	0.813	0–10
Zone 7	8.0·10^9	0.162	10–20
Zone 5	2.2·10^9	0.164	20–30
Zone 3	9.6·10^9	0.497	30–40
Zone 2	2.0·10^9	0.103	40–50

Table 9. Activation of the different zones of the chip for transient simulation number 2.

Active Zone	φ (W/m^3)	Q (W)	Time (s)
Zone 1	4.45·10^9	0.23	15–23
Zone 2	9.09·10^9	0.47	23–37 and 46–50
Zone 3	9.09·10^9	0.47	23–37
Zone 4	1.11·10^9	0.11	0–50
Zone 5	3.47·10^9	0.26	40–46
Zone 6	-	-	-
Zone 7	4.44·10^9	0.36	40–50
Zone 8	-	-	-
Zone 9	3.46·10^9	0.14	0–50

The computation time required by the reduced model is 4.5 s, which is 600 times faster than that of the finite elements model.

The maximum temperature reached by the chip computed by the amalgamated reduced-order modal model (AROMM) and by the finite elements model has been compared for both cases. Figure 12a,b present the comparison for boundary conditions case 1 with activation profile 1 (Table 8) and boundary conditions case 2 with activation profile 2 (Table 9), respectively.

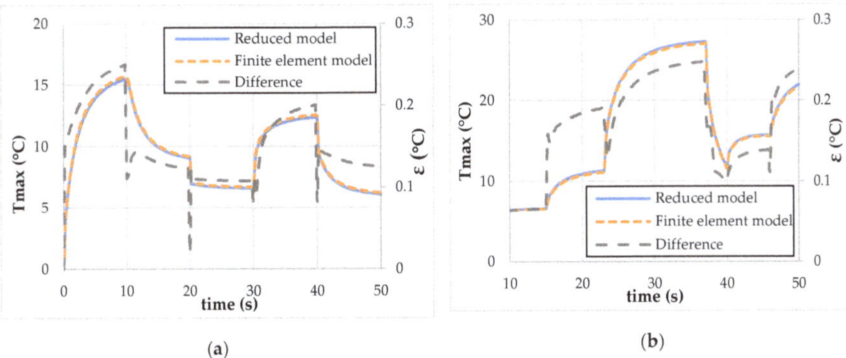

(a) (b)

Figure 12. Temporal evolution of the maximum temperature reached by the ball-grid-array (BGA)–difference on this parameter between the reduced and the finite element model for test case 1 (**a**) and 2 (**b**).

The agreement on this critical parameter is very good, as it never outreaches 0.25 °C, i.e., a relative difference less than 1.6% for the first case and 1% for the second. A sudden rise in the difference between models is noticed when the power changes. This effect is induced by modal reduction as modes with a high time constant have been discarded.

This very good accuracy on the maximal temperature is accompanied by a satisfying precision on the entire chip. Figure 13 presents the temperature field computed by the reduced model at $t = 37$ s, i.e., at the time where the error is the most important. The maximal temperature difference on the chip between the reduced model and the FE one is less than 0.5 °C. On most of the chip (and the copper etches), the error is below 0.2 °C, yielding an average error (in time and space) of 0.08 °C.

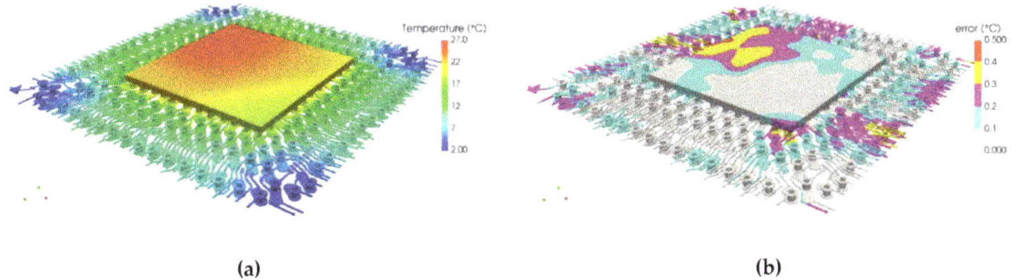

(a) (b)

Figure 13. Temperature field (**a**) computed by the reduced model of order 300 at $t = 37$ s. Error (**b**) with the finite elements simulation at the same time.

The error field is erratic, which is characteristic of modal reduction. Thus, the location of the maximum error cannot be known a priori with this method.

Thus, both test cases using different boundary conditions and power profiles confirm the very good accuracy of the ROM, as the error is below 2% for each time step on the chip. Those results validate the new substructuring model order reduction approach to create an AROMM of complex components.

Moreover, as modal methods compute the temperature field in its integrality, there is no need for an a priori definition of the outputs. Indeed, the localizations of the hottest spot of the chip during the transient simulation (depicted by circles) are highlighted in Figure 14.

Figure 14. Localization of the maximal temperature reached by the chip during the transient simulation.

Obviously, the hot spot moves as the different zones are activated. This simple fact questions the notion of junction temperature. This is confirmed by Figure 15, which compares the maximum temperature reached by the chip to the temperature at the center of the chip: An output at the center of the chip would underestimate the temperature by up to 4 °C, i.e., a relative error of 25%, which is by far greater than the temperature prediction error of the reduced model.

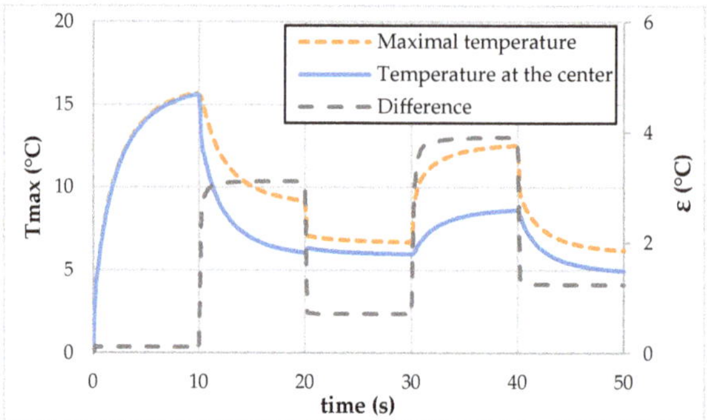

Figure 15. Temporal evolution of the maximum temperature reached by the BGA and the temperature at the center of the chip—Difference between those two quantities.

Another main interest of this substructuring modal approach is to have two distinct models, one for the chip and one for all the other parts of the BGA. If one of them is modified, only the latter must be regenerated. In the case of the component, all constituting parts except the die are imposed by the manufacturer, so this model is realized only once. Then, the model of the chip can be easily regenerated to take into account the new spatial power profile or correction of semiconductor thermal properties.

The substructuring modal approach offers a solution to integrate the real spatial power distribution of the component without additional creation and simulation time. Indeed, this power distribution evolves during the development cycle from the uniform power distribution to the real profile based on electric simulation.

9. Conclusions

This study presents a procedure to validate a numerical thermo-fluid model of a complex electronic component, in this case, a ball grid array package of 208 balls. Then, this detailed thermal model is used to derive a dynamic compact thermal model, inspired by the DELPHI methodology, which can be substituted by the DTM to perform a set of thermo-fluidic simulations while preserving the high level of accuracy. To quicken the reduction process, an amalgamated reduced-order modal model is coupled to meta-heuristic optimizations using genetic algorithms. As a main benefit, the overall DCTM creation time is reduced by 86%.

Further, the AROMM method coupled to a substructuring modal method is applied to a BGA208 with, this time, multiple internal heat sources. This novel method allows the building of reduced models independent of boundary conditions (BCI-AROMM). A reduced-order model of only 300 modes was built and will be improved in future works. However, the time needed to create the model remains important, as 8 h of computation were used. Nevertheless, the first deduced model offers very satisfying results as the error on the maximum temperature never outreaches 2%, as well as for steady-state and transient simulations, for a reduction factor of 600 in computation time. Moreover, this model permits us to study, quickly and accurately, all 3-D thermal phenomena involved by the complex structure of the real component. These numerical results should now be confirmed by experimental data, and an experimental setup is being conceived.

Further, a transient characterization is under investigation. The definition of the adjusted heat capacity parameters is based on one-dimensional network identification using stochastic Bayesian deconvolution [23].

Author Contributions: V.B. proposed this approach, performed the numerical models simulations and contributed to the paper redaction. F.J. ran the computations on the substructured model and contributed to the paper redaction. E.M.-V. designed the test vehicle, provided experimental measurements and contributed to the paper redaction. A.N. elaborated the modal substructuration theory. O.D. implemented and optimized the Genetic Algorithms. All authors have read and agreed to the published version of the manuscript.

Funding: This research received no external funding.

Acknowledgments: The authors would like to thank Vincent Fox for his contribution to this work.

Conflicts of Interest: The authors declare no conflict of interest.

Nomenclature

Latin symbols

C	volumetric heat capacity $\left[\text{K}/\left(\text{m}^3.\text{K}\right)\right]$
h	heat exchange coefficient $\left[\text{W}/\left(\text{m}^2.\text{K}\right)\right]$
k	thermal conductivity $[\text{W}/(\text{m}.\text{K})]$
R_c	contact thermal resistance $\left[\text{m}^2.\text{K}/\text{W}\right)]$
R_{JA}	Junction to ambient thermal resistance $[\text{K/W}]$
Q	Thermal power $[\text{W}]$
T	Temperature $[^{\circ}\text{C}]$
T_{∞}	Air temperature $[^{\circ}\text{C}]$
x	state
V	mode $[\text{K}]$

Greek symbols

λ	eigenvalue
φ	heat flux density $\left[\text{W}/\text{m}^2\right]$
ω	volume power $\left[\text{W}/\text{m}^3\right]$

Superscript

M	measurement
N	numeric
D	Dirichlet
S	Steklov

References

1. Monier-Vinard, E.; Rogié, B.; Bissuel, V.; Laraqi, N.; Daniel, O.; Kotelon, M.-C. State of the art of numerical thermal characterization of electronic component. In Proceedings of the 17th International Conference on Thermal, Mechanical and Multi-Physics Simulation and Experiments in Microelectronics and Microsystems (EuroSimE), Montpellier, France, 18–20 April 2016.
2. JEDEC Standard JESD15-4. *DELPHI Compact Thermal Model Guideline*; Jedec Solid-State Technology Association: Arlington, VA, USA, 2008. Available online: https://www.jedec.org/system/files/docs/JESD15-4.pdf (accessed on 8 June 2020).
3. Atwell, J.A.; King, B.B. Proper orthogonal decomposition for reduced basis feedback controllers for parabolic equations. *Math. Comput. Model.* **2001**, *33*, 1–19. [CrossRef]
4. Zhang, X.; Xiang, H. A fast meshless method based on proper orthogonal decomposition for the transient heat conduction problems. *Int. J. Heat Mass Transf.* **2015**, *84*, 729–739. [CrossRef]
5. Codecasa, L.; D'Alessandro, V.; Magnani, A.; Rinaldi, N.; Zampardi, P.J. Fast Novel Thermal Analysis Simulation Tool for Integrated Circuits (FANTASTIC). In Proceedings of the 20th International Workshop on Thermal Investigations of ICs and Systems, London, UK, 24–26 September 2014.
6. Joly, F.; Quéméner, O.; Neveu, A. Modal Reduction of an Advection-Diffusion Model Using a Branch Basis. *Numer. Heat Transf. Part B Fundam.* **2008**, *53*, 466–485. [CrossRef]
7. Quéméner, O.; Joly, F.; Neveu, A. The generalized amalgam method for modal reduction. *Int. J. Heat Mass Transf.* **2012**, *55*, 1197–1207. [CrossRef]
8. Bissuel, V.; Fox, V.; Monier-Vinard, E.; Neveu, A.; Joly, F.; Daniel, O. Multi-port Dynamic Compact Thermal Models of BGA package using Model Order Reduction and Metaheuristic Optimization. In Proceedings of the 2019 18th IEEE Intersociety Conference on Thermal and Thermomechanical Phenomena in Electronic Systems (ITherm), Las Vegas, NV, USA, 28–31 May 2019.
9. Rogié, B.; Codecasa, L.; Monier-Vinard, E.; Bissuel, V.; Laraqi, N.; Daniel, O.; D'Amore, D.; Magnani, A.; d'Alessandro, V.; Rinaldi, N. Multi-port dynamic compact thermal models of dual-chip package using model order reduction and metaheuristic optimization. *Microelectron. Reliab.* **2018**, *87*, 222–231. [CrossRef]
10. Grosjean, S.; Joly, F.; Vera, K.; Neveu, A.; Monier-Vinard, E. Reduction of an electronic card thermal problem by the modal substructuring method. In Proceedings of the IHTC-16, Beijing, China, 10–15 August 2018.
11. Grosjean, S.; Gaume, B.; Joly, F.; Vera, K.; Neveu, A. A modal substructuring method for non-conformal mesh. Application to an electronic board. *Int. J. Therm. Sci.* **2020**, *152*, 106298. [CrossRef]
12. Codecasa, L.; Bornoff, R.; Dyson, J.; d'Alessandro, V.; Magnani, A.; Rinaldi, N. Versatile MOR-based boundary condition independent compact thermal models with multiple heat sources. *Microelectron. Reliab.* **2018**, *87*, 194–205. [CrossRef]
13. Rogié, B.; Grosjean, S.; Monier-Vinard, E.; Bissuel, V.; Joly, F.; Daniel, O.; Laraqi, N.; Vera, K. Delphi-like dynamical compact thermal models using model order reduction based on modal approach. In Proceedings of the 34th Thermal Measurement, Modeling & Management Symposium (SEMI-THERM), San Jose, CA, USA, 19–23 March 2018.
14. JEDEC Standard JESD51. *Methodology for the Thermal Measurement of Component Packages*; Jedec Solid-State Technology Association: Arlington, VA, USA, 2008. Available online: https://www.jedec.org/system/files/docs/Jesd51.pdf (accessed on 8 June 2020).
15. JEDEC Standard N°51-3. *Low Effective Thermal Conductivity Test Board for Leaded Surface Mount Packages*; Jedec Solid-State Technology Association: Arlington, VA, USA, 1996. Available online: https://www.jedec.org/system/files/docs/JESD51-3.PDF (accessed on 8 June 2020).
16. JEDEC Standard N°51-7. *High Effective Thermal Conductivity Test Board for Leaded Surface Mount Packages*; JEDEC Solid-State Technology Association: Arlington, VA, USA, 1999. Available online: https://www.jedec.org/system/files/docs/jesd51-7.PDF (accessed on 8 June 2020).
17. JEDEC Standard N°51-2a. *Integrated Circuit Thermal Test Method Environmental Conditions—Natural Convection (Still Air)*; Jedec Solid-State Technology Association: Arlington, VA, USA, 2002. Available online: https://www.jedec.org/system/files/docs/JESD51-2A.pdf (accessed on 8 June 2020).
18. JEDEC Standard N°51-6. *Integrated Circuit Thermal Test Method Environmental Conditions—Forced Convection (Moving Air)*; Jedec Solid-State Technology Association: Arlington, VA, USA, 1999. Available online: https://www.jedec.org/system/files/docs/jesd51-6.pdf (accessed on 8 June 2020).

19. Monier-Vinard, E.; Rogié, B.; Bissuel, V.; Laraqi, N.; Daniel, O.; Kotelon, M.-C. State of the Art of Thermal Characterization of Electronic Components using Computational Fluid Dynamic tools. *Int. J. Numer. Methods Heat Fluid Flow* **2017**, *27*. [CrossRef]
20. Laffay, P.O.; Quéméner, O.; Neveu, A.; Elhajjar, B. The Modal Substructuring Method: An Efficient Technique for Large-Size Numerical Simulations. *Num. Heat Transf. Part B Fundam.* **2011**, *60*, 278–304. [CrossRef]
21. Monier-Vinard, E.; Bissuel, V.; Rogié, B.; Laraqi, N.; Daniel, O.; Kotelon, M.-C. Evolution of the DELPHI Compact Thermal Modelling Method: An Investigation on the Boundary Conditions Scenarios. In Proceedings of the THERMINIC XXII, Budapest, Hungary, 21–23 September 2016. [CrossRef]
22. Monier-Vinard, E.; Dia, C.; Bissuel, V.; Laraqi, N.; Daniel, O. Latest developments of Compact Thermal Modeling of System in Package devices by means of Genetic Algorithm. In Proceedings of the ITHERM XIV, Orlando, FL, USA, 27–30 May 2014. [CrossRef]
23. Lai, W.; Liu, X.; Chen, W.; Lei, X.; Tang, X.; Zang, Z. Transient multi-exponential signals analysis using Bayesian deconvolution. *Appl. Math. Comput.* **2015**, *265*, 486–493.

Article

Compact Thermal Model of the Pulse Transformer Taking into Account Nonlinearity of Heat Transfer [†]

Krzysztof Górecki [1,*]**, Kalina Detka** [1] **and Krzysztof Górski** [2]

[1]   Department of Marine Electronics, Gdynia Maritime University, Morska 83, 81-225 Gdynia, Poland;
     k.detka@we.umg.edu.pl

[2]   Command Institute, General Tadeusz Kościuszko Military University of Land Forces,
     Piotra Czajkowskiego 109, 51-147 Wrocław, Poland; k.gorski@we.am.gdynia.pl

*   Correspondence: k.gorecki@we.umg.edu.pl

†   This paper is an extended version of our paper published in Proceedings of 25th International Workshop on
     Thermal Investigations of ICs and Systems Therminic 2019, Lecco, Italy, 27–29 September 2019,
     doi: 10.1109/THERMINIC.2019.8923510.

Received: 14 May 2020; Accepted: 29 May 2020; Published: 1 June 2020

Abstract: This paper presents a compact nonlinear thermal model of pulse transformers. The proposed model takes into account differentiation in values of the temperatures of a ferromagnetic core and each winding. The model is formulated in the form of an electric network realising electrothermal analogy. It consists of current sources representing power dissipated in the core and in each of the windings, capacitors representing thermal capacitances and controlled current sources modelling the influence of dissipated power on the thermal resistances in the proposed model. Both self-heating phenomena in each component of the transformer and mutual thermal couplings between each pair of these components are taken into account. A description of the elaborated model is presented, and the process to estimate the model parameters is proposed. The proposed model was verified experimentally for different transformers. Good agreement between the calculated and measured waveforms of each component temperature of the tested pulse transformers was obtained. Differences between the results of measurements and calculations did not exceed 9% for transformers with a toroidal core and 13% for planar transformers.

Keywords: nonlinear thermal model; SPICE; pulse transformer; thermal phenomena; self-heating; modelling; measurements

---

1. Introduction

Pulse transformers are an important component of switched-mode power converters [1–3]. These transformers have simple structure, and they consist of two kinds of components, i.e., a ferromagnetic core and at least two windings. During the operation of the considered device, an increase in temperature of each transformer component is observed [4–6]. This increase is a result of thermal phenomena occurring in the pulse transformer, such as self-heating in each component of the transformer and mutual thermal interaction between each pair of these components [7–9].

Knowing the core temperature and the windings temperatures is important from the point of view of electrical and magnetic properties of a pulse transformer. As is shown in [10,11], the temperature significantly changes the characteristics of ferromagnetic materials used to make a transformer core and causes a change in the resistances of the windings. In particular, an excessive increase in temperature can lead to damage in the insulation of the windings or can reduce the magnetic permeability of the core [8,11,12]. Additionally, an increase in the temperatures of electronic components causes a decrease in their lifetime [13,14].

In order to calculate the temperature waveforms of electronic components with thermal phenomena taken into account, a thermal model of these components is indispensable [15,16]. In many papers [10,11,17–25], thermal models of transformers are described, but they have disadvantages. One group of thermal models is microscopic models, which make it possible to calculate temperature distribution in the considered component. For example, in [11,18], the finite element method is used to determine the temperature distribution in a transformer, but at the same time the distribution of the wasted power per unit of volume in the transformer is assumed. Another group of thermal models is compact thermal models, which take into account only one temperature characterising the whole device [17].

Reference [26] presents three-dimensional (3-D) numerical compact thermal models of planar transformers. The DC thermal network of the cited model is inspired by the Delphi method, which allows obtaining shorter time of calculations than the finite volume method. The presented results do not illustrate the influence of dissipated power on the temperatures of the core and the windings. Additionally, this model is dedicated to the ANSYS software, and it is difficult to implement it in other software, e.g., in SPICE.

The similar approach to modelling thermal properties of electronic components is presented in [27,28]. The model presented in [27] uses homogenisation techniques to reduce calculation requirements. The advantage of this model is the reduction of the number of thermal factors to 6–8, and good agreement between the results of calculations and measurements can still be obtained. Unfortunately, the mentioned model could be used in the software dedicated only to a 3-D thermal analysis. In turn, the approach to the thermal modelling of components of electric machines presented in [28] requires time-consuming measurements of the tested prototype. In the paper [28], some temperature waveforms of the tested machine are presented.

Reference [29] describes a simplified form of an electrothermal model dedicated to planar magnetic components (inductors and transformers), which operate in the space industry. An important part of this model is the thermal model dedicated to the ANSYS program. Due to the fact that a lot of factors are taken into account, the network representation of the proposed model is complex and contains three subcircuits representing the thermal network of the transformer windings, the thermal network of the transformer core and the thermal network of the connection between the windings and the Printed Circuit Board (PCB).

References [30,31] are dedicated to parameters estimation of thermal models of electronics devices in the Delphi-inspired form. To this end, genetic algorithms are used. Unfortunately, for multiple heat sources in such models, calculations are time-consuming, and the total simulation time can reach 800 h.

Reference [32] presents a thermal model of a planar transformer. The form of this model is obtained on the basis of computation fluid dynamics. The structure of this model is adequate for planar transformers only. Unfortunately, in the paper [32], no results of experimental verification of this model are presented.

Compact thermal models of transformers are described in [10,11,17,18,26]; however, differences between the core and windings temperatures are typically not taken into account in the mentioned models. In addition, in the models described in [24,25], the dependence of the dissipation efficiency of heat generated in the device on the power dissipated in the transformer is omitted [33].

It is widely known [7,15,16,34–36] that some factors, such as ambient temperature, cooling systems or power dissipated in electronic devices, influence the efficiency of heat removal from the devices. Reference [37] presents a nonlinear thermal model of a planar transformer. In this model, the influence of power dissipated in particular components of the transformer on the efficiency of heat removal is taken into account.

This paper, which is an extended version of Reference [38], proposes a compact nonlinear thermal model (CNTM) of a pulse transformer based on a thermal model of a planar transformer described in [19]. In comparison to [19], which presents a linear thermal model, the nonlinearity of the heat transfer process is taken into account. In comparison to [38], a detailed description of the nonlinear thermal model of the transformer and the new results of measurements and calculations illustrating the effectiveness of this model for transformers including different ferromagnetic cores are presented in this paper. The model proposed by the authors' takes into account self-heating phenomena in all components of the transformer and mutual thermal couplings between each pair of these components.

The selected thermal models of transformers given in the literature are discussed in Section 2. The form of a nonlinear thermal model is presented in Section 3. The obtained results of measurements and calculations proving the effectiveness of the elaborated model are shown in Section 4. The advantage of this nonlinear thermal model over a linear thermal model is experimentally confirmed for the selected transformers containing cores with different shapes and made of different ferromagnetic materials. For all models described in the following sections, thermal resistance and thermal capacitance are given in K/W and J/K, respectively, whereas all temperatures are expressed in Celsius degrees.

2. Selected Thermal Models of Transformers in the Literature

A network representation of classical linear thermal models of transformers [10,39] is shown in Figure 1. It can be seen that only one internal temperature of the whole transformer is used. In this model, differences in the temperatures of the core and the windings are not taken into account.

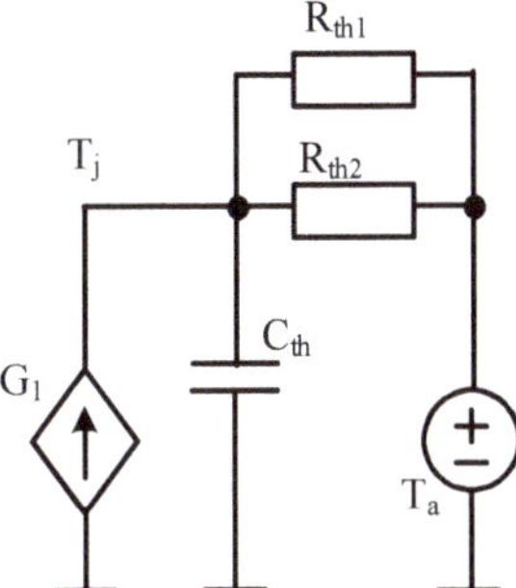

Figure 1. Network representation of the thermal model described in [10,39].

In the presented model, the controlled current source G_1 represents the sum of the power dissipated in the core and in the windings of the transformer. Thermal capacitance is represented by capacitor C_{th}, while R_{th1} is the thermal resistance characterising heat convection and R_{th2} is thermal resistance characterising the heat radiation from the surface of the examined device. Voltage source T_a is the ambient temperature, and the voltage in node T_j is the temperature of the transformer.

In [17,33], a thermal model of magnetic components (transformers and inductors) is proposed. This model enables calculating the temperature difference between the core and ambient temperature (ΔT_C) and the temperature difference between the windings and ambient temperature (ΔT_W) by taking into account self-heating and mutual thermal couplings between the core and the windings. The network representation of this model is shown in Figure 2.

However, in this model, only single thermal time constants for the windings (R_{thW} and C_{thW}) and for the core (R_{thC} and C_{thC}) are taken into account. Current sources G_{PC} and G_{PW} describe power losses in the core and in the windings, respectively; whereas current sources G_{PC1} and G_{PW1} model the influence of mutual thermal couplings between the core and the winding on the temperature differences ΔT_W and ΔT_C.

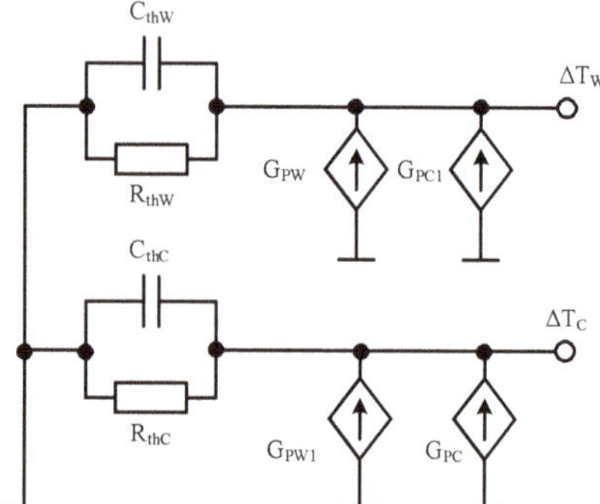

Figure 2. Network representation of the thermal model of magnetic components from [18].

The thermal model of a transformer proposed in [17], of which the diagram is presented in Figure 3, makes it possible to calculate the temperature of the core T_C and the temperatures of both windings T_W by taking into account self-heating and mutual thermal couplings between the core and the windings. This model has the form of RC Foster networks excited by current sources representing the powers dissipated in the core (P_{thC} and P_{thWC1}) and in the windings (P_{thW} and P_{thWC1}).

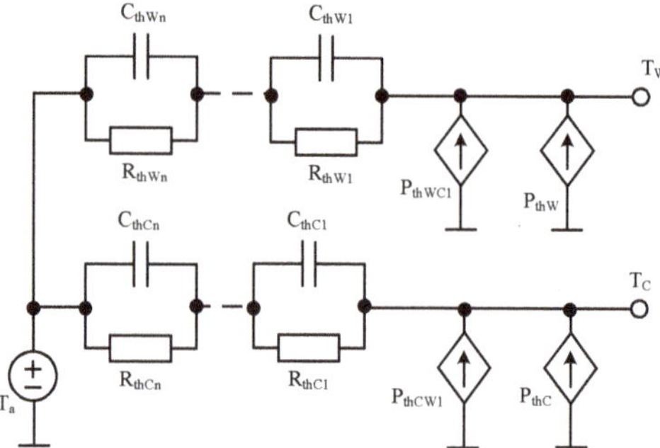

Figure 3. Network representation of the thermal model of a transformer proposed in [17].

In order to take into account thermal couplings between the core and the windings, controlled current sources P_{thWC1} and P_{thCW1} are applied. The voltage source T_a represents ambient temperature. In the described model, common RC networks are used to model self-heating and mutual thermal couplings between the components of the transformer.

A disadvantage of thermal models of transformers described in this section is they do not take into account differentiations of windings temperatures, nonlinearities of thermal properties and thermal couplings occurring between the components of the transformer. Therefore, in the following Section, the authors' nonlinear thermal model of the transformer is proposed. In this model, nonlinearities of thermal phenomena and thermal couplings between the components of the transformer are taken into account.

3. Proposed Nonlinear Thermal Model of Pulse Transformers

As is shown in [17,19], temperature distributions on each of the windings and on the core in pulse transformers are practically uniform. Therefore, their thermal properties can be described with the use of a compact thermal model [7,19]. On the other hand, the temperatures of the core and the windings can be significantly different from each other [7]. Therefore, in such models, the differences of the core and windings temperatures should be taken into account. In each transformer component, a self-healing phenomenon occurs, and additionally, thermal interaction between each pair of transformer components is observed.

The nonlinear thermal model of the transformer worked out by the authors has the form of a subcircuit dedicated for SPICE. This form is inspired by the linear thermal model of a planar transformer proposed in [19] and by the nonlinear thermal model of semiconductor devices described in [37,40]. In the new model, the differences in the temperatures of components in the transformer (the core and all the windings) are taken into account. These temperatures result from self-heating phenomena in every component and mutual thermal couplings between each pair of the components of the transformer. The presented model makes it possible to calculate the temperature waveforms of the core and each winding by taking into account both the mentioned phenomena. When formulating the considered model, based on the results of measurements shown in [7], changes in the powers dissipated in each component influence only the values of transformer thermal resistances in the thermal model, whereas thermal capacitances do not depend on power. The network representation of the proposed model is presented in Figure 4.

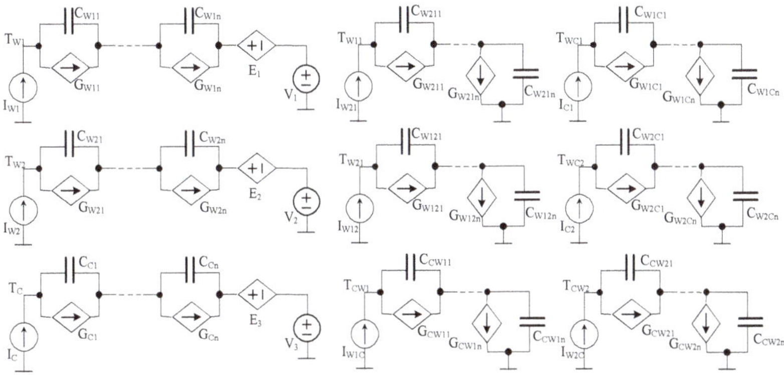

Figure 4. Network representation of a compact nonlinear thermal model (CNTM) of a pulse transformer.

In this network, nine circuits can be distinguished. Three of them located on the left-hand side of Figure 4 are used to calculate the temperature waveforms of particular components of the transformer—the primary winding T_{W1}, the secondary winding T_{W2} and the core T_C. These temperatures (given in °C) correspond to the voltages (given in V) in the nodes and are denoted as T_{W1}, T_{W2} and T_C, respectively. Current sources I_C, I_{W1} and I_{W2} model power losses in the core and the two windings, respectively.

Circuits including capacitors and controlled current sources model self-transient thermal impedances of the primary winding ($C_{W11}, \ldots, C_{W1n}$ and $G_{W11}, \ldots, G_{W1n}$), of the secondary winding ($C_{W21}, \ldots, C_{W2n}$ and $G_{W21}, \ldots, G_{W2n}$) and of the core ($C_{C1}, \ldots, C_{Cn}$ and $G_{C1}, \ldots, G_{Cn}$), respectively. The voltages on these circuits (given in V) correspond to differences (given in °C) between the temperatures of the transformer components and ambient temperature, resulting from self-heating phenomena.

The controlled voltage sources E_1, E_2 and E_3 represent the influence of mutual thermal couplings between the components of the transformer on the temperatures of these components. Capacitors represent the thermal capacitances of all the elements in the heat flow path, whereas the controlled current sources represent nonlinear thermal resistances between the elements in the heat flow path.

The output voltage of the controlled voltage source E_1 is equal to the sum of voltages in nodes T_{W11} and T_{WC1}, the output voltage of the controlled voltage source E_2 is equal to the sum of voltages in nodes T_{W21} and T_{WC2}, and the output voltage of the controlled voltage source E_3 is equal to the sum of voltages in nodes T_{CW1} and T_{CW2}. Voltage sources V_1, V_2 and V_3 represent ambient temperature.

Six other subcircuits are used to model mutual thermal couplings between each pair of the transformer components. Current sources in these subcircuits represent power (given in W) dissipated in particular components of the transformer (I_{W12} and I_{W1C} in the primary winding, I_{W21} and I_{W2C} in the secondary winding and I_{C1} and I_{C2} in the core). Networks containing capacitors and controlled current sources connected to the above-mentioned current sources model mutual transient thermal impedances between each pair of components of the transformer.

All self-transient and mutual transient thermal impedances can be described by Equation (1) [7,15]:

$$Z_{th}(t) = R_{th} \cdot \left[1 - \sum_{i=1}^{N} a_i \cdot \exp\left(-\frac{t}{\tau_{thi}}\right) \right], \tag{1}$$

where R_{th} is the thermal resistance, a_i is the coefficient (without unit) corresponding to thermal time constant τ_{thi} (given in s), and N is the number of thermal time constants.

The dependence of R_{th} on the dissipated power is described as:

$$R_{th} = R_{th0} \cdot \left[1 + \alpha \cdot \exp\left(-\frac{p - p_0}{b}\right) \right], \tag{2}$$

where R_{th0} denotes the minimum value of the thermal resistance, p denotes the power dissipated in a heating component of the transformer, α is the parameter without unit, and p_0 and b are model parameters given in W.

Changes in values of thermal resistance are modelled by the controlled current source G_i. The output current of the controlled source is described as:

$$G_i = V_{Gi} / (a_i \cdot R_{th}), \tag{3}$$

where V_{Gi} denotes the voltage on the current source G_i.

Thermal capacitances are given as:

$$C_i = \tau_{thi} / (a_i \cdot R_{th}). \tag{4}$$

The values of the model parameters are estimated using the results of measurements of self- and mutual transient thermal impedances existing in the transformer thermal model. Measurements of such parameters at different powers dissipated in the core and in the windings of the tested transformer are realised by the method described in [7]. The values of parameters in Equation (1) are estimated for each transient thermal impedance using the method described in [15,41]. Next, parameters α, p_0 and b in Equation (2) are estimated for self-thermal and mutual thermal resistance in the transformer model by using local estimation [15].

4. Results

In order to verify the presented model and its practical use, measurements and calculations of the temperature waveform of each component of the tested transformers, which contained ferromagnetic cores with different shapes and sizes and were made of different materials, were performed. A planar transformer with a ferrite core and transformers with ring cores made of different materials were measured and modelled as examples.

The planar transformer, of which the cuboidal core dimension was 22 mm × 16 mm × 9 mm, was made of ferrite material 3F3 and contained windings in the form of printed paths on laminate FR-4, which was 1 mm thick. The primary winding contained three turns with a width of 2.5 mm, and the secondary winding contained four turns with a width of 1 mm. Transformers with a toroidal core contained identical primary and secondary windings. On each of them, 20 turns of copper wire in the enamel with a diameter of 0.8 mm were wound. The toroidal core had an external diameter equal to 26 mm, the internal diameter equal to 16 mm and a width equal to 11 mm. Cores made of toroidal powdered iron (RTP), ferrites (RTF) and nanocrystals (RTN) were used for toroidal transformers.

The values of the parameters in the thermal model were estimated for transformers containing ferromagnetic cores with different shapes and dimensions and made of different ferromagnetic materials.

The values of the parameters in Equation (2), which represents the proposed compact nonlinear thermal model of the transformer (CNMT), are summarized in Table 1 (for a toroidal transformer) and Table 2 (for a planar transformer). The values of the parameters presented in Tables 1 and 2 describe the model of the network representation shown in Figure 4.

Table 1. Values of the parameters in Equation (2) for a toroidal transformer with a toroidal powdered iron core (RTP).

Thermal Resistance	R_{th0} [K/W]	α	p_0 [W]	b [W]
R_{thW1}	12.75	10	0	1.13
R_{thW1W2}	8.6	10	0	1.3
R_{thW1C}	9.9	10.5	0	1.2

Table 2. Values of the parameters in Equation (2) for a planar transformer.

Thermal Resistance	R_{th0} [K/W]	α	p_0 [W]	b [W]
R_{thW1}	26	0.27	1	2
R_{thW1W2}	15	0.733	1	5.5
R_{thW1C}	11	0.636	0.5	3
R_{thC}	11.5	0.435	10	10
R_{thCW1}	5.1	0.96	10	9.9
R_{thCW2}	2.4	2.125	10	10

Comparing the values of the parameters describing thermal resistances in the thermal model of the toroidal transformer, it is obvious that the values of parameters α, p_0 and b were nearly the same for all considered thermal resistances. Differences were observed in the values of the parameter R_{th0}. This meant that courses $R_{thW1}(p_{W1})$, $R_{thW1W2}(p_{W1})$ and $R_{thW1C}(p_{W1})$ were nearly parallel. In contrast, big differences were observed between the values of the parameters describing the dependences of thermal resistances in the thermal model of the planar transformer on powers P_{W1} and P_C.

The values of thermal capacitances were estimated with the use of the method described in [15]. As an example, in Table 3, the values of these parameters obtained for a planar transformer are summarised.

Table 3. Values of the parameters a_i and τ_{thi} of the selected self-transient and mutual transient thermal impedances in a thermal model of a planar transformer.

Parameter	Z_{thW1} (t)	Z_{thW1C} (t)	Z_{thW1W2} (t)	Z_{thC} (t)	Z_{thCW1} (t)	Z_{thCW2} (t)
a_1	0.274	0.271	0.626	0.297	0.224	0.18
a_2	0.448	0.456	0.374	0.676	0.776	0.82
a_3	0.225	0.273		0.027		
a_4	0.053					
τ_{th1} [s]	350.33	498.84	350.72	432.04	678.02	1067.99
τ_{th2} [s]	60.22	103.66	192.79	139.67	194.39	221.5
τ_{th3} [s]	14.31	16.26		17.11		
τ_{th4} [μs]	40					

It can be clearly seen that, in the considered self-transient and mutual transient thermal impedances, different numbers of thermal time constants occurred. In self-transient thermal impedances, three or four thermal time constants were used, whereas mutual transient thermal impedances were described with two or three thermal time constants. The values of the considered thermal time constants were in a range from 40 µs to over 1000 s.

For example, by means of Equation (2), the measured (indicated by points) and modelled (indicated by lines) dependences of thermal resistances in the thermal model of the transformer on the power dissipated in one of the components of the transformer are shown in Figures 5 and 6.

Figure 5 illustrates the influence of power P_{W1} dissipated in the primary winding of the transformer containing a toroidal core made of powdered iron on the thermal resistance of the winding R_{thW1} (blue colour) and on mutual thermal resistances between this winding and the secondary winding R_{thW1W2} (black colour), as well as those between the core and the primary winding R_{thW1C} (red colour).

Figure 6a illustrates the influence of power dissipated in the primary winding of the planar transformer on the thermal resistances R_{thW1} (blue colour), R_{thW1W2} (black colour) and R_{thW1C} (red colour). Figure 6b illustrates the influence of power dissipated in the core of the transformer on the thermal resistance of the core R_{thC} (blue colour) and the mutual thermal resistances between the core and both windings R_{thCW1} (red colour) and R_{thCW2} (green colour).

As it is obvious in Figures 5 and 6, it is possible to accurately model the measured dependences of the considered thermal resistance on the dissipated power using Equation (2). Visible differences between self-thermal and mutual thermal resistances were observed for both considered transformers. These thermal resistances differed from one another. It is worth noticing that the considered differences were bigger for the planar transformer than for the toroidal transformer. Changes in dissipated power can cause changes in thermal resistance even by 25%.

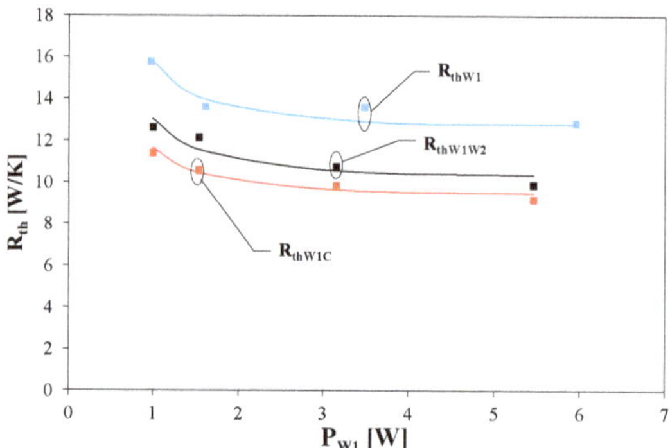

Figure 5. Measured (indicated by points) and modelled (indicated by lines) dependences of the selected thermal resistances in the thermal model of the transformer with a toroidal core made of powdered iron on the power dissipated in the primary winding.

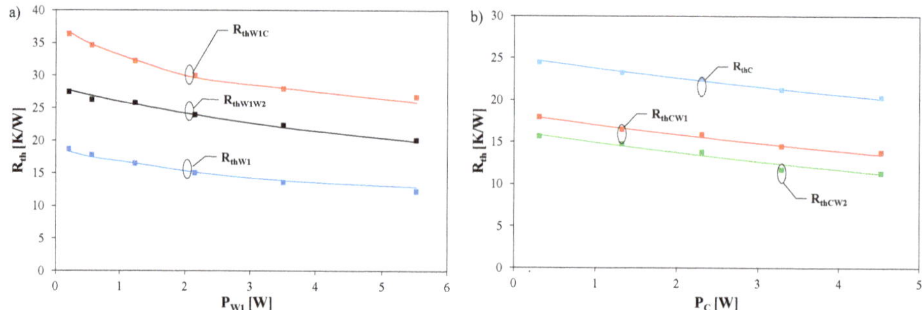

Figure 6. Measured (indicated by points) and modelled (indicated by lines) dependences of the selected thermal resistances in the thermal model of the planar transformer on the power dissipated in the primary winding (**a**) and in the core (**b**).

Calculations and measurements were performed at power dissipation in only one of the components of the tested transformers and at simultaneous dissipation of power in different components of the tested devices. Power dissipated in each component of the transformer always had a shape of a single rectangular impulse with a long duration time. The results of calculations obtained by means of the nonlinear thermal model were compared to the results of calculations performed by means of the linear thermal model described in [19] and the results of measurements performed with the use of a pyrometer. The windings and the core were excited with different powers. The successive figures (Figures 7–12) present the calculated and measured waveforms of the temperatures of the primary winding T_{W1}, the secondary winding T_{W2} and the core T_C of the tested transformers. In these figures, the results of measurements is represented by points, the results of calculations performed using the CNTM is represented by solid lines, and the results of calculations using the compact linear thermal model (CLTM) is represented by dashed lines [19].

Calculations were performed for the values of parameters describing the nonlinear thermal model of transformers according to the principles shown in Section 3. On the other hand, for the model from [19], the values of parameters estimated at the lowest measured values of power dissipated in each component of the transformers were used.

Figure 7 presents the measured and calculated waveforms of temperatures T_{W1} and T_C in the toroidal transformer with the RTP core at a dissipated power P_{W1} of 2.83 W in the primary winding.

It can be observed that the nonlinear thermal model makes it possible to obtain very good agreement between the results of calculations and measurements. In contrast, the values of the considered temperatures obtained with the use of calculations performed with the linear thermal model were higher than the results of measurements by even 40 °C.

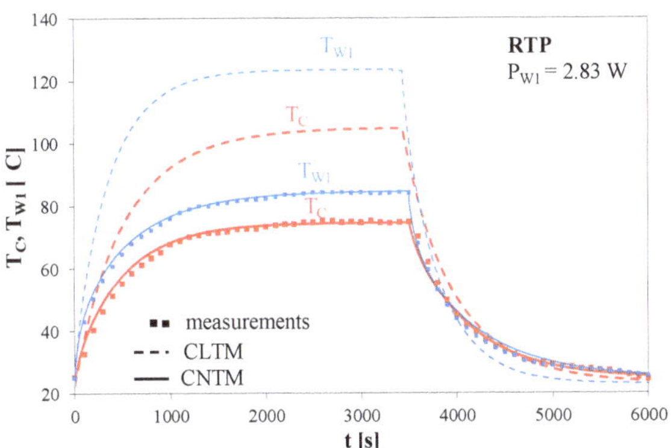

Figure 7. Measured and calculated temperature waveforms of the primary winding T_{W1} and the core T_C of a toroidal transformer with an RTP at a dissipated power P_{W1} of 2.83 W in the primary winding.

Figure 8 shows the measured and calculated temperature waveforms of the primary winding T_{W1}, the secondary winding T_{W2} and the core T_C for a transformer containing an RTN. These waveforms were obtained, while the primary winding of the tested transformer was excited by a single rectangular pulse with duration times equal to 4000 s (Figure 8a) and 5000 s (Figure 8b). P_{W1} was 1.05 W in the case presented in Figure 8a, and P_{W1} was equal to 2.82 W in the case presented in Figure 8b.

As it is seen, very good agreement between the results of calculations performed with the use of the CNTM and measurements was achieved for both the values of power dissipated in this transformer. In contrast, for the linear thermal model, an excess of temperature of each component of the transformer over ambient temperature was overestimated by 10% at lower values of the considered powers and even 50% overestimated for higher values of the considered dissipated power.

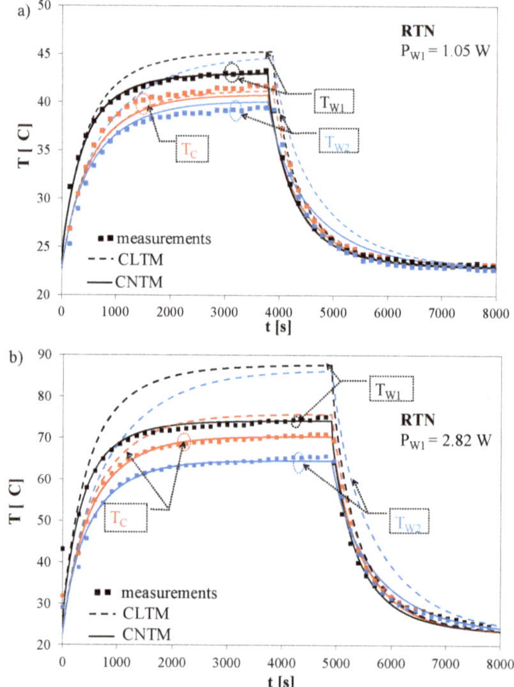

Figure 8. Measured and calculated temperature waveforms of the components of a transformer with a toroidal nanocrystalline core (RTN) at dissipated powers P_{W1} of 1.05 W (**a**) and 2.82 W (**b**) in the primary winding.

Figure 9 illustrates the measured and calculated temperature waveforms of the primary winding T_{W1}, the secondary winding T_{W2} and the core T_C for the transformer with an RTF. The primary winding was stimulated by a single rectangular impulse with a duration time equal to 7000 s at P_{W1} of 0.77 W.

By analysing the temperature waveforms of the core T_C, the primary winding T_{W1} and the secondary winding T_{W2} presented in Figure 9, it can be observed that the results of calculations performed by means of the nonlinear thermal model assured better agreement with the results of measurements than the results of calculations obtained by means of the model described in [19]. The obtained difference between the results of calculations performed with the use of the linear thermal model and the measurements results was 10%.

Figure 10 presents the measured and calculated waveforms of the temperature of the primary winding T_{W1}, the temperature of the secondary winding T_{W2} and the core temperature T_C of the planar transformer by stimulating the primary winding with a single rectangular impulse having a duration time equal to 5500 s and P_{W1} of 1 W (Figure 10a) and with a single rectangular impulse having a duration time of 5000 s and P_C of 2 W (Figure 10b).

As one can notice, the results of calculations by means of the nonlinear thermal model assured better agreement with the results of measurements than the results of calculations performed by means of the linear thermal model. In Figure 10a, it is visible that the difference between the waveforms of temperatures T_{W1} and T_C obtained with the use of the considered models was more than 2 °C. The biggest difference between the results of calculations was observed for temperature T_{W2}. In Figure 10b, the considered differences were bigger than in the case presented in Figure 10a. These differences reached even 20 °C. When power was dissipated in the core only, the temperatures of both windings were nearly the same.

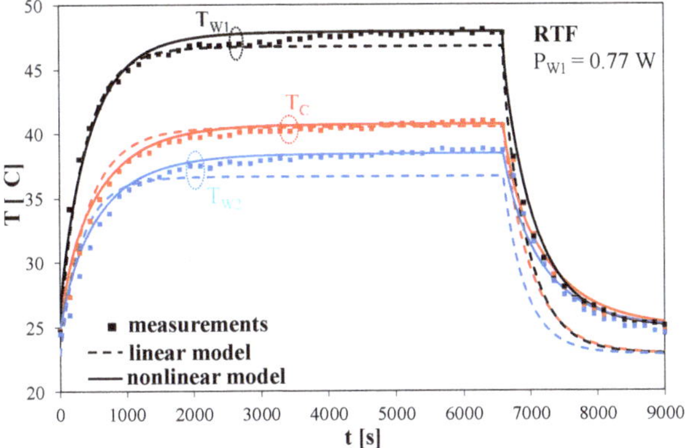

Figure 9. Measured and calculated waveforms of the temperature of the primary winding T_{W1}, the temperature of the secondary winding T_{W2} and the temperature of the core T_C at a dissipated power P_{W1} of 0.77 W in the primary winding.

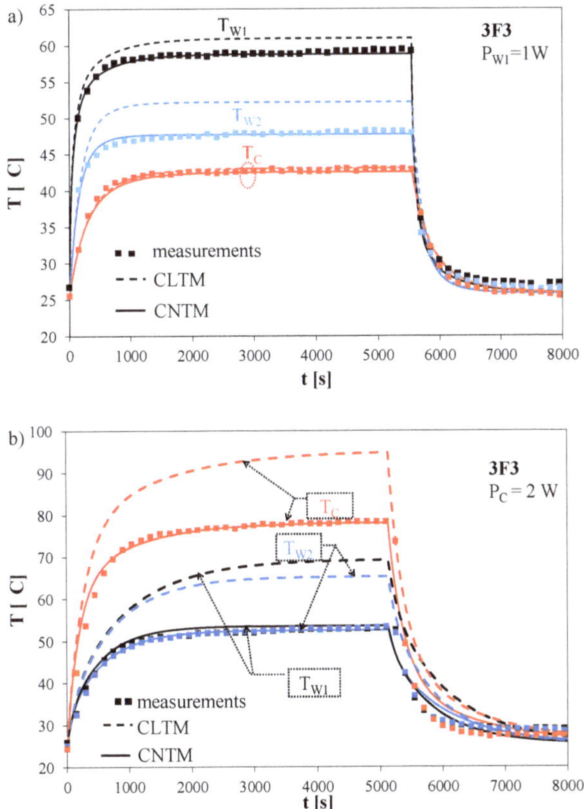

Figure 10. Measured and calculated waveforms of the temperature of the primary winding T_{W1}, the temperature of the secondary winding T_{W2} and the temperature of the core T_C of the planar transformer at P_{W1} of 1 W in the primary winding (**a**) and P_C of 2 W in the core (**b**).

The results of measurements and calculations presented above corresponded to untypical situations, when power was dissipated in one of the transformer components only. Figures 11 and 12 present the results of measurements and calculations of temperature waveforms of components of the selected transformers obtained in the case when power was dissipated in both the components of the selected transformers.

Figure 11 shows the measured and calculated temperature waveforms of the primary winding T_{W1}, the secondary winding T_{W2} and the core T_C for the transformer containing the toroidal ferrite core (RTF) when it was excited by power dissipated simultaneously in the primary and secondary windings with a single rectangular pulse having duration times equal to 4500 s (Figure 11a) and 5500 s (Figure 11b). P_{W1} was equal to 0.88 W and P_{W2} was equal to 1 W in the case presented in Figure 11a, whereas P_{W1} was equal to 2.4 W and P_{W2} was equal to 2.7 W in the case presented in Figure 11b.

By analysing temperature waveforms of the core and the windings of the transformer containing an RTF (Figure 11), it is easy to observe that the difference between the results of calculations performed using the linear thermal model and the results of measurements was up to 20 °C. In contrast, the results of calculations made using the nonlinear thermal model showed very good agreement with the results of measurements. Differences between the results of measurements and calculations performed with the linear thermal model were bigger at higher powers dissipated in the components of the transformer.

Figure 12 presents the calculated and measured waveforms of the temperatures of each component of the planar transformer by simultaneously stimulating the core and the primary winding with a single rectangular impulse having a duration time equal to 4500 s, P_{W1} of 2.3 W and P_C of 2 W.

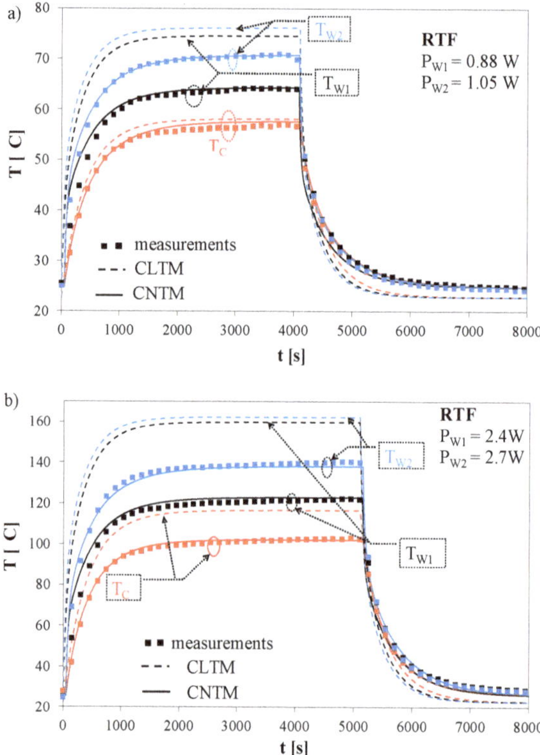

Figure 11. Measured and calculated temperature waveforms of the primary winding T_{W1}, the secondary winding T_{W2} and the core T_C with simultaneous power dissipation in the windings: (**a**) P_{W1} = 0.88 W and P_{W2} = 1.05 W; (**b**) P_{W1} = 2.4W and P_{W2} = 2.7 W.

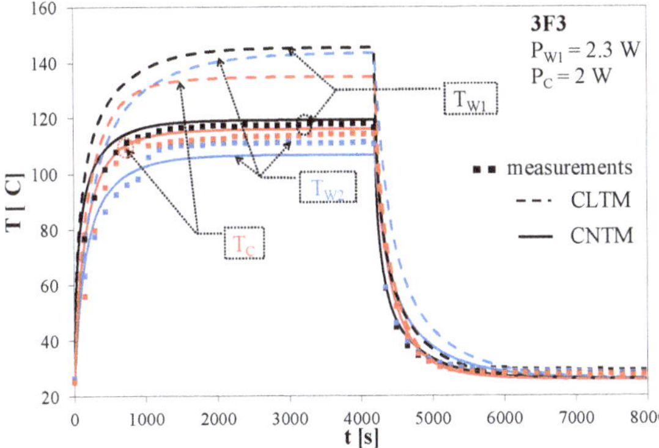

Figure 12. Measured and calculated waveforms of the temperature of the primary winding T_{W1}, the temperature of the secondary winding T_{W2} and the temperature of the core T_C of the planar transformer with an impulse power P_C of 2 W in the core and an impulse power P_{W1} of 2.3 W in the primary winding.

As one can notice, the results of calculations by means of the nonlinear thermal model assured better agreement with the results of measurements than the results of calculations performed by means of the linear thermal model. The observed differences between the results of measurements and the results of calculations performed by means of the linear thermal model exceeded even 40 °C, whereas the difference between the results of calculations performed by means of the nonlinear thermal model and the results of measurements was no more than about 5 °C. It is also worth noticing that in the considered operation conditions the temperatures of the transformer components were low and did not exceed 10 °C.

Table 4 contains the values of the maximum errors of the temperatures of transformers components δ_{TW1}, δ_{TW2} and δ_{TC} obtained using the CLTM and the CNTM of the transformer analysed in this section.

Table 4. Values of the maximum relative error of the temperatures of transformers components calculated with the CNTM and the compact linear thermal model (CLTM).

Core Type	Power Dissipated	Model Type	δ_{TW1} [%]	δ_{TW2} [%]	δ_{TC} [%]
RTP	$P_{W1} = 2.83$ W	CLTM	48.8%	-	40.83%
		CNTM	8.16%	-	3.22%
RTN	$P_{W1} = 1.05$ W	CLTM	5.49%	13.37%	2.86%
		CNTM	1.33%	5.22%	2.86%
	$P_{W1} = 2.82$ W	CLTM	19.2%	31.34%	8.4%
		CNTM	2.78%	9.05%	5.9%
Toroidal ferrite core (RTF)	$P_{W1} = 0.88$ W	CLTM	15%	13%	4%
	$P_{W2} = 1.05$ W	CNTM	4%	3.5%	2%
	$P_{W1} = 2.4$ W	CLTM	33%	16.5%	16%
	$P_{W2} = 2.7$ W	CNTM	2%	2%	1%
	$P_{W1} = 0.77$ W	CLTM	10.5%	13.02%	10.74%
		CNTM	1.53%	6.25%	7%
3F3	$P_{W1} = 1$ W	CLTM	4.18%	10.78%	4.20%
		CNTM	3.18%	3.13%	4.16%
	$P_C = 2$ W	CLTM	30.6%	25%	20.7%
		CNTM	7.4%	6%	17.5%
	$P_{W1} = 2.3$ W	CLTM	24%	29.09%	47.32%
	$P_c = 2$ W	CNTM	13.14%	4.47%	35.67%

It is clearly seen in Table 4 that the considered relative error of calculations of temperatures obtained using the CLTM achieved 48.8% for the primary winding temperature of the transformer with an RTP core whereas such an error of calculations of the primary winding temperature obtained using the CNTM was equal to 8.16% for the same transformer. It is also worth noticing that the value of the relative error of the temperatures of transformers components obtained using the CNTM did not exceed 9% for all the considered transformers containing a toroidal core made of different materials. It was also observed that an increase in power dissipated in the transformer components caused an increase of the value of the relative error of calculations for both the considered models.

5. Conclusions

This paper describes a new CNTM of a pulse transformer elaborated by the authors. It allows determining the waveforms of the temperatures of the core and each winding. Calculations were performed by taking into account a self-heating phenomenon and mutual thermal interaction between the transformer components. In our model, the dependences of self-thermal and mutual thermal resistances depend on power dissipated in the transformer components. The proposed model was verified experimentally for transformers including ferromagnetic cores with different shapes and made of different materials. High accuracy of this model and its advantage over the CLTM were also demonstrated. The results of calculations and measurements presented in this paper also confirmed that power dissipated in the transformer components influenced mostly thermal resistance and its influence on thermal capacity was neglected.

It is also worth noticing that the obtained differences between the values of the temperatures of the transformer components can be equal to 50 °C. Such big differences justify the use different temperatures of the windings and the core in the thermal model. The linear thermal model makes it possible to obtain good agreement between the results of calculations and measurements only for very low powers dissipated in each component of the examined transformers. The nonlinear thermal model makes it possible to obtain good agreement between the results of calculations and measurements over a wide range of power dissipated in the windings or in the core of the examined transformers by considering the dependence of R_{th} on power.

The presented results of calculations and measurements proved that the nonlinear thermal model of a pulse transformer proposed in this paper is able to accurately describe dynamic thermal properties of the transformer including cores with different dimensions and shapes and made of different ferromagnetic materials. It was also shown that using this model one can obtain accurate results of calculations at different conditions of power dissipation in the tested transformers.

It is also worth noticing that the proposed model is universal and it can be useful for different type of transformers such as planar transformers or transformers containing a toroidal core. Additionally, this model takes into account properties of materials used to make the core. The biggest advantage of the proposed model is its simple structure and that easy implementation, e.g., in the SPICE program, which is widely used by designers of electronic circuits. In addition, analyses made with the proposed model are not time-consuming, which is very important from the economic point of view. The proposed model can be used to formulate electrothermal models of transformers, which take into account the nonlinearity of a heat removal process.

The CNTM of the pulse transformer proposed in this paper can be useful in designing switch-mode power supplies, and it can be used as a component of an electrothermal model of transformers.

Author Contributions: Conceptualization (K.G. (Krzysztof Górecki)); methodology (K.G. (Krzysztof Górecki), K.G. (Krzysztof Górski)); validation (K.G. (Krzysztof Górecki), K.D., K.G. (Krzysztof Górski)); investigation (K.G. (Krzysztof Górecki), K.G. (Krzysztof Górski)); resources (K.D.); writing—original draft preparation (K.G. (Krzysztof Górecki), K.D.); writing—review and editing (K.G. (Krzysztof Górecki), K.D.); visualization (K.G. (Krzysztof Górecki), K.G. (Krzysztof Górski)); supervision (K.G. (Krzysztof Górecki)). All authors have read and agreed to the published version of the manuscript.

Funding: The project is financed in the framework of the program by the Ministry of Science and Higher Education called "Regionalna Inicjatywa Doskonałości" in the years 2019–2022 (project number: 006/RID/2018/19; the sum of financing: 11,870,000 PLN).

Conflicts of Interest: The authors declare no conflict of interest.

Abbreviations and Notations

R_{th}	thermal resistance
C_{th}	thermal capacitance
T_{W1}	the temperature of the transformer primary winding
T_{W2}	the temperature of the transformer secondary winding
T_C	the temperature of the transformer core
T_a	ambient temperature
$Z_{th}(t)$	transient thermal impedance
τ_{thi}	thermal time constant
a_i	the dimensionless coefficient corresponding to a thermal time constant
N	the number of thermal time constants
RTP	toroidal powdered iron core
RTN	toroidal nanocrystalline core
RTF	toroidal ferrite core
SPICE	Simulation Program with Integrated Circuits Emphasis
CNTM	compact nonlinear thermal model
CLTM	compact linear thermal model
ANSYS	engineering simulation and 3D design software

References

1. Rashid, M.H. *Power Electronic Handbook*; Academic: New York, NY, USA, 2007.
2. Erickson, R.; Maksimović, D. *Fundamentals of Power Electronics*; Springer Science and Business Media LLC: Berlin/Heidelberg, Germany, 2001.
3. Basso, C. *Switch-Mode Power Supply SPICE Cookbook*; McGraw-Hill: New York, NY, USA, 2001.
4. Barlik, R.J.; Nowak, K.M. *Energoelektronika. Elementy Podzespoły, Układy*; Oficyna Wydawnicza Politechniki Warszawskiej: Warszawa, Poland, 2014.
5. Kazimierczuk, M.K. *Pulse-Width Modulated DC-DC Power Converters*; Wiley: Hoboken, NJ, USA, 2008.
6. Magambo, J.S.N.T.; Bakri, R.; Margueron, X.; Le Moigne, P.; Mahe, A.; Guguen, S.; Bensalah, T. Planar Magnetic Components in More Electric Aircraft: Review of Technology and Key Parameters for DC–DC Power Electronic Converter. *IEEE Trans. Transp. Electrif.* **2017**, *3*, 831–842. [CrossRef]
7. Górecki, K.; Górski, K.; Zarębski, J. Investigations on the Influence of Selected Factors on Thermal Parameters of Impulse-Transformers. *Inf. MIDEM J. Microelectron. Electron. Compon. Mater.* **2017**, *47*, 3–13.
8. Detka, K.; Górecki, K.; Zarębski, J. Modeling Single Inductor DC–DC Converters with Thermal Phenomena in the Inductor Taken Into Account. *IEEE Trans. Power Electron.* **2017**, *32*, 7025–7033. [CrossRef]
9. Maksimovic, D.; Stankovic, A.; Thottuvelil, V.; Verghese, G. Modeling and simulation of power electronic converters. *Proc. IEEE* **2001**, *89*, 898–912. [CrossRef]
10. Wilson, P.; Ross, J.; Brown, A. Simulation of magnetic component models in electric circuits including dynamic thermal effects. *IEEE Trans. Power Electron.* **2002**, *17*, 55–65. [CrossRef]
11. Tsili, M.A.; Amoiralis, E.I.; Kladas, A.; Souflaris, A.T. Power transformer thermal analysis by using an advanced coupled 3D heat transfer and fluid flow FEM model. *Int. J. Therm. Sci.* **2012**, *53*, 188–201. [CrossRef]
12. Valchev, V.C.; Bossche, A.V.D. *Inductors and Transformers for Power Electronics*; Informa UK Limited: London, UK, 2018.
13. Castellazzi, A.; Gerstenmaier, Y.; Kraus, R.; Wachutka, G. Reliability analysis and modeling of power MOSFETs in the 42-V-PowerNet. *IEEE Trans. Power Electron.* **2006**, *21*, 603–612. [CrossRef]
14. Narendran, N.; Gu, Y. Life of LED-Based White Light Sources. *J. Disp. Technol.* **2005**, *1*, 167–171. [CrossRef]
15. Górecki, K.; Zarębski, J.; Górecki, P.; Ptak, P. Compact thermal models of semiconductor devices–a review. *Int. J. Electron. Telecommun.* **2019**, *65*, 151–158.

16. Janicki, M.; Sarkany, Z.; Napieralski, A. Impact of nonlinearities on electronic device transient thermal responses. *Microelectron. J.* **2014**, *45*, 1721–1725. [CrossRef]

17. Górecki, K.; Rogalska, M. The compact thermal model of the pulse transformer. *Microelectron. J.* **2014**, *45*, 1795–1799. [CrossRef]

18. Penabad-Duran, P.; López-Fernández, X.; Turowski, J. 3D non-linear magneto-thermal behavior on transformer covers. *Electr. Power Syst. Res.* **2015**, *121*, 333–340. [CrossRef]

19. Górecki, K.; Gorski, K. Compact thermal model of planar transformers. In Proceedings of the 2017 MIXDES-24th International Conference Mixed Design of Integrated Circuits and Systems, Bydgoszcz, Poland, 22–24 June 2017; Institute of Electrical and Electronics Engineers (IEEE): Piscataway, NJ, USA, 2017; pp. 345–350.

20. Gamil, A.; Al-Abadi, A.; Schatzl, F.; Schlucker, E. Theoretical and Empirical-Based Thermal Modelling of Power Transformers. In Proceedings of the 2018 IEEE International Conference on High Voltage Engineering and Application (ICHVE), Athens, Greece, 10–13 September 2018; IEEE: Piscataway, NJ, USA, 2018; pp. 1–4. [CrossRef]

21. Souza, L.; Lemos, A.; Caminhas, W.; Boaventura, W. Thermal modeling of power transformers using evolving fuzzy systems. *Eng. Appl. Artif. Intell.* **2012**, *25*, 980–988. [CrossRef]

22. Tang, W.H.; Wu, Q.H.; Richardson, Z.J. A simplified transformer thermal model based on thermal–electric analogy. *IEEE Trans. Power Deliv.* **2004**, *19*, 1112–1119. [CrossRef]

23. Tang, W.; Wu, Q.H.; Richardson, Z. Equivalent heat circuit based power transformer thermal model. *IEE Proc.-Electr. Power Appl.* **2002**, *149*, 87. [CrossRef]

24. Swift, G.; Molinski, T.S.; Bray, R. A fundamental approach to transformer thermal modeling. Part I: Theory and equivalent circuit. *IEEE Trans. Power Deliv.* **2001**, *16*, 171–175. [CrossRef]

25. Haritha, V.; Rao, T.; Amit Jain Ramamoorty, E. Thermal Modeling of Electrical Transformers. In Proceedings of the 16th National Power Systems Conference, Hyderabad, India, 15–17 December 2010; pp. 597–602.

26. Bissuel, V.; Codecasa, L.; Monier-Vinard, E.; Rogie, B.; Olivier, A.; Mahe, A.; Laraqi, N.; Dralessandro, V.; Gougis, C. Novel Approach to the Extraction of Delphi-like Boundary-Condition-Independent Compact Thermal Models of Planar Transformer Devices. In Proceedings of the 2018 24rd International Workshop on Thermal Investigations of ICs and Systems (THERMINIC), Stockholm, Sweden, 26–28 September 2018; Institute of Electrical and Electronics Engineers (IEEE): Piscataway, NJ, USA, 2018; pp. 1–7.

27. Lopez, G.S.; Exposito, A.D.; Muñoz-Antón, J.; Ramirez, J.A.O.; Lopez, R.P.; Delgado, A.; Oliver, J.A.; Prieto, R. Fast and Accurate Thermal Modeling of Magnetic Components by FEA-Based Homogenization. *IEEE Trans. Power Electron.* **2020**, *35*, 1830–1844. [CrossRef]

28. Vansompel, H.; Yarantseva, A.; Sergeant, P.; Crevecoeur, G. An Inverse Thermal Modeling Approach for Thermal Parameter and Loss Identification in an Axial Flux Permanent Magnet Machine. *IEEE Trans. Ind. Electron.* **2018**, *66*, 1727–1735. [CrossRef]

29. De-La-Hoz, D.; Salinas, G.; Svikovic, V.; Alou, P. Simplification of thermal networks for magnetic components in space power electronics. *Energies* **2020**, in press.

30. Monier-Vinard, E.; Bissuel, V.; Laraqi, N.; Dia, C. Latest developments of compact thermal modeling of system-in-package devices by means of Genetic Algorithm. In Proceedings of the Fourteenth Intersociety Conference on Thermal and Thermomechanical Phenomena in Electronic Systems (ITherm), Orlando, FL, USA, 27–30 May 2014; Institute of Electrical and Electronics Engineers (IEEE): Piscataway, NJ, USA, 2014; pp. 998–1006.

31. Rogie, B.; Codecasa, L.; Monier-Vinard, E.; Bissuel, V.; Laraqi, N.; Daniel, O.; D'Amore, D.; Magnani, A.; Dralessandro, V.; Rinaldi, N. Delphi-like dynamical compact thermal models using model order reduction. In Proceedings of the 2017 23rd International Workshop on Thermal Investigations of ICs and Systems (THERMINIC), Amsterdam, The Netherlands, 27–29 September 2017; pp. 1–6. [CrossRef]

32. Shen, Z.; Shen, Y.; Liu, B.; Wang, H. Thermal Coupling and Network Modeling for Planar Transformers. In Proceedings of the 2018 IEEE Energy Conversion Congress and Exposition (ECCE), Portland, OR, USA, 23–27 September 2018; pp. 3527–3533. [CrossRef]

33. Górecki, K.; Godlewska, M. Modelling characteristics of the impulse transformer in a wide frequency range. *Int. J. Circuit Theory Appl.* **2020**, *48*, 750–761. [CrossRef]

34. Yener, Y.; Kakac, S. *Heat Conduction*; Taylor & Francis: Abingdon, UK, 2008.

35. Górecki, K.; Zarębski, J. Modeling the Influence of Selected Factors on Thermal Resistance of Semiconductor Devices. *IEEE Trans. Compon. Packag. Manuf. Technol.* **2014**, *4*, 421–428. [CrossRef]

36. Gorecki, K.; Gorecki, P. A new form of the non-linear compact thermal model of the IGBT. In Proceedings of the 2018 IEEE 12th International Conference on Compatibility, Power Electronics and Power Engineering (CPE-POWERENG 2018), Doha, Qatar, 10–12 April 2018; Institute of Electrical and Electronics Engineers (IEEE): Piscataway, NJ, USA, 2018; pp. 1–6.

37. Górecki, K.; Górski, K. Non-linear thermal model of planar transformers. In Proceedings of the 2017 21st European Microelectronics and Packaging Conference (EMPC) & Exhibition, Warsaw, Poland, 10–13 September 2017; Institute of Electrical and Electronics Engineers (IEEE): Piscataway, NJ, USA, 2017.

38. Gorecki, K.; Detka, K. The Nonlinear Compact Thermal Model of the Pulse Transformer. In Proceedings of the 2019 25th International Workshop on Thermal Investigations of ICs and Systems (THERMINIC), Lecco, Italy, 25–27 September 2019; Institute of Electrical and Electronics Engineers (IEEE): Piscataway, NJ, USA, 2019.

39. Andreu, D.; Boucher, J.; Maxim, A. New SPICE behavioural macromodelling method of magnetic components including the self-heating process. In Proceedings of the IEEE Annual Power Electronics Specialist Conference PESC, Charleston, SC, USA, 1 July 1999; Volume 2, pp. 735–740.

40. Górecki, K.; Gorecki, P.; Zarebski, J. Measurements of Parameters of the Thermal Model of the IGBT Module. *IEEE Trans. Instrum. Meas.* **2019**, *68*, 4864–4875. [CrossRef]

41. Górecki, K.; Rogalska, M.; Zarębski, J. Parameter estimation of the electrothermal model of the ferromagnetic core. *Microelectron. Reliab.* **2014**, *54*, 978–984. [CrossRef]

Article

Electro-Thermal Simulation of Vertical VO$_2$ Thermal-Electronic Circuit Elements

Mahmoud Darwish, Péter Neumann, and János Mizsei and László Pohl *

Department of Electron Devices, Budapest University of Technology and Economics, 1117 Budapest, Hungary; mahmoud@eet.bme.hu (M.D.); neumann@eet.bme.hu (P.N.); mizsei@eet.bme.hu (J.M.)
* Correspondence: pohl@eet.bme.hu

Received: 15 May 2020; Accepted: 1 July 2020; Published: 3 July 2020

Abstract: Advancement of classical silicon-based circuit technology is approaching maturity and saturation. The worldwide research is now focusing wide range of potential technologies for the "More than Moore" era. One of these technologies is thermal-electronic logic circuits based on the semiconductor-to-metal phase transition of vanadium dioxide, a possible future logic circuits to replace the conventional circuits. In thermal-electronic circuits, information flows in a combination of thermal and electronic signals. Design of these circuits will be possible once appropriate device models become available. Characteristics of vanadium dioxide are under research by preparing structures in laboratory and their validation by simulation models. Modeling and simulation of these devices is challenging due to several nonlinearities, discussed in this article. Introduction of custom finite volumes method simulator has however improved handling of special properties of vanadium dioxide. This paper presents modeling and electro-thermal simulation of vertically structured devices of different dimensions, 10 nm to 300 nm layer thicknesses and 200 nm to 30 μm radii. Results of this research will facilitate determination of sample sizes in the next phase of device modeling.

Keywords: beyond CMOS; VO$_2$; thermal-electronic circuits; electro-thermal simulation; vertical structure

---

1. Introduction

Miniaturization of electronic devices' feature size is the primary driver of computing development. Scaling down has been described by Moore's law for many years. Recently, complimentary metal–oxide–semiconductor (CMOS) scaling down has been slower than ever before owing to scaling limits that appear on very small dimensions. The technological advancement described by Moore's Law for CMOS technology since the past 50 years is expected to flatten out completely by 2025 [1,2]. Finding "Beyond CMOS" solutions has become more essential to keep the computing advancement flourishing. It takes about 10 years for a new technology to move from the laboratory to mass production, for example, FinFET [1]. There are several potential "More than Moore" devices in laboratories [3], but currently, none is in a phase to state that this will be the successor.

A report commissioned by the Defense Technical Information Center (DTIC) [4] proposes four basic computational models: classical digital computing, analog computing, neuro-inspired computing and quantum computing. The latter three areas are suitable for solving special problems much more effectively than before; however, they are generally not able to replace traditional tasks where classical digital computing performs well, such as 3D graphics, mobile devices and similar others.

Architecture specialization can help to increase computing power for a while [5], for example, the use of tensor processing units (TPUs) developed for artificial intelligence (AI) [6] or the wider use of field-programmable gate arrays (FPGAs) in data centers [7]. The degree of specialization is well illustrated by the fact that there were 4 separate accelerators in the Apple A4 system on a chip (SoC), 28 in the A8, and more than 40 in the A12 [8]. Another way to increase the performance of CMOS technology-based systems is to use design and technological developments such as chip stacking in 3D using through-silicon vias (TSVs) [9], advanced energy management, near-threshold voltage (NTV) operation and application of an increased number of metal layers [3].

Following are some more promising future devices compatible with CMOS technology [10]:

- In a tunneling FET, conduction occurs through band-to-band tunneling. Gate voltage shifts the energy bands and changes the probabilities of tunneling [11].
- In a graphene pn-junction, transmission or a total internal reflection of electrons occurs by switching the electrostatic p and n doping of graphene by applying voltage to electrodes. The current routes to one or other output of the device [12].
- The bilayer pseudospin FET is an orbitronic device. Holes are injected into one monolayer of graphene and electrons into another monolayer and they may bind into excitons. The excitons may relax into a Bose–Einstein condensate (BEC) state. The current between source and drain first grows with the increase of voltage and then decreases as the carrier imbalance destroys BEC causing negative differential resistance [13].
- Spintronic devices are based on magnetic dipoles represented by electrons with polarized spins or ferromagnetic elements. Spintronic devices are nonvolatile (preserve the state when the power is turned off). Some possible types are as follows: The SpinFET combines a MOSFET and a switchable magnetic element [14]. The spin transfer domain wall device operates by the motion of a domain wall in a ferromagnetic wire [15]. The spintronic majority gate uses ferromagnetic wires and majority of the input currents' directions sets the direction of magnetization [16]. In the all spin logic device, nanomagnets are placed over a copper wire and a diffusion spin current exerts a torque on a nanomagnet to switch its polarization [17].

In this article, we discuss a potential device that is compatible with CMOS integration and in which information transmits by two physical processes: electrical and thermal. This device is the phonsistor (a portmanteau of "phonon transistor") [18]. A digital circuit can be built from this device, which is called thermal-electronic logic circuits (TELC) [19,20]. TELC can be implemented using materials that have semiconductor-to-metal transition (SMT). Vanadium dioxide (VO_2) has shown properties as a good candidate for implementation of TELC structures with its SMT at around 67 °C [21].

Vanadium dioxide has some special properties whose investigation is not yet complete. Its optical properties [22], the relationship between SMT and structural phase transition [23,24], the nature of SMT [20,25] or the monoclinic metallic phase [26] are studied. Its applications cover many areas such as thermal rectification [27,28], high performance electromechanical switches [29], smart window coatings [30], neuromorphic devices [31,32] and non-volatile memory arrays [32]. Phonsistor and TELC are not the only transistor-type applications, we also find examples of purely thermal transistors in the literature [33,34].

VO_2 encounters electrical resistivity change when it exposes to heat. The material shows much lesser ability to conduct electrical current at room temperature than at higher temperatures. Moreover, the resistivity drops steeply around 67 °C (340 K), changing the material electrical properties to a conductor. The transition from a low-conducting phase to a high-conducting phase is called "semiconductor-to-metal transition" (SMT). Phase transition occurs due to the sudden change in the structure of VO_2 from the tetragonal structure (at low temperatures) to the monoclinic structure (at high temperatures) [19]. Figure 1 illustrates the electric resistivity change by 3 to 4 orders of magnitude versus temperature. The electro-thermal behavior of VO_2 is highly technology-dependent. It also depends on the manufacturing process and the substrate [35–37].

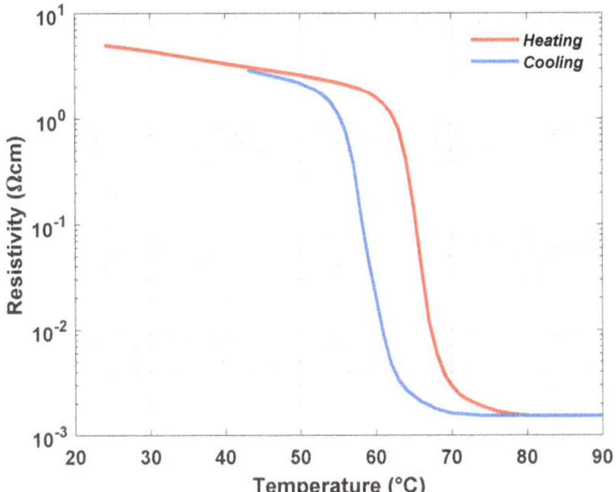

Figure 1. Temperature dependence of the resistivity of vanadium dioxide for (100) VO_2; based on [20].

A simple phonsistor consists of two components: a heating resistor and an SMT resistor. The heat generated by Joule heating in the resistor transports by phonons to the SMT resistor by diffusion. This process is similar to the diffusion of minority charge carriers at the base of a bipolar transistor. The switching voltage and holding current of the SMT resistor decrease due to external heating; this is the role of the heating resistor [38]. The abrupt variation in the VO_2 electric resistivity makes it capable of switching on and off electrically depending on its thermal state. Hence the term "thermal-electronic logic circuits" (TELC) is used.

Figure 2 shows a sample TELC structure where R1, R2, R3, R4, and R5 are conventional resistors and V1, V2, and V3 are VO_2 SMT resistors. For instance, V1 switches ON if R1 AND R2 are both ON. V2 switches ON if R3 OR R4 is ON. Combinations that are more sophisticated can be implemented by altering the ambient temperature, driving current, elements sizes, distances between the resistors and their locations. When the driving current flows in the conventional resistors, the Joule-heating will heat them and the generated heat will diffuse to the surrounding by convection. With enough heating delivered, VO_2 SMT resistors will reach the desired temperature to drop the resistivity value and transition to the conducting phase. This action allows current to pass through VO_2 SMT resistors. Furthermore, heat dissipation layers inserted between each section of the structure is important to prevent heat overlapping.

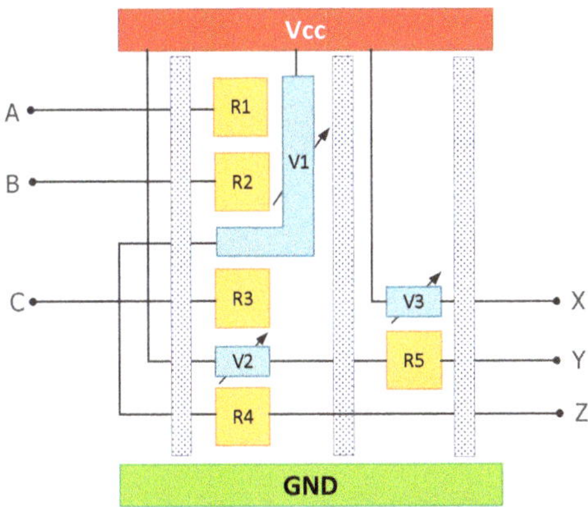

Figure 2. Implementation of VO_2 thermal switching in a TELC circuit.

Many problems still need solutions to implement the circuit shown. For example, how the dimensions of an SMT resistor affect the electro-thermal characteristics of the device needs clarification. This paper investigates that how the characteristics of vertical VO_2 resistors depend on their sizes by simulation. Previously, we studied lateral structures by performing high-resolution electro-thermal simulations using a range of different dimensions [39]. Measurement results for lateral TELC structures are also available in [40].

The goal of our long-term study is designing and creating real TELC structures that require a comprehensive study of VO_2 thin films' behavior along with how these structures act at different dimensions, arrangements and ambient temperatures. In this article, we examine VO_2 based vertical structures at different dimensions.

2. Experimentation

Standard CMOS technology steps were used to prepare the samples for vertical thermo-electrical device test in Figure 3. The substrate was an n-type Si with 3.5×10^{17} cm^{-3} dopant concentration for better spreading resistance of the vertical structure. The 100 nm SiO_2 insulator layer was grown by thermal oxidation. A window was etched into the insulator layer that will be the active conductive channel shape of the vertical structure. The VO_2 layer (approximately 50 nm thin) was deposited by RF sputtering at 650 °C high temperature. The radii of the window changed between 10 μm up to 100 μm.

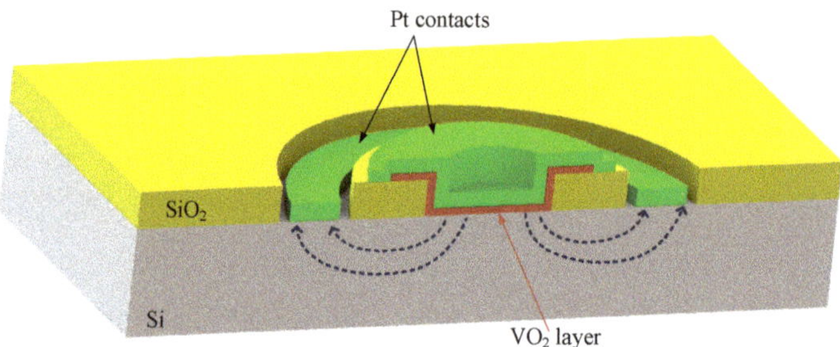

Figure 3. Cross-sectional view of the vertical VO_2 structure prepared in the lab.

For the tailoring of SMT layer, wet chemistry was applied to remove the surround cover area of the window. The top and collector Pt contacts (approx. 10 nm thin) of devices were also prepared by RF sputtering. Standard optical lithography was applied to transfer the patterns during technology steps. During the vertical structure preparation, lateral structures were prepared to check the SMT layer quality and functional operation. The dimensions of VO_2 channels in lateral structures were 100 μm × 700 μm.

Probe station was used for electrical measurement of samples. The electrical measurement was carried out by Keithley SourceMeter and samples were elevated at different environmental temperatures by a Cole-Parmer Thermostat System.

3. Electro-Thermal Simulation

To study thermal-electrical circuits by simulation, a distributed parametric simulator is required, typically with some finite algorithm (FDM, FEM, FVM). The difficulty of simulation is illustrated by the fact that the resistivity of VO_2 changes rapidly as a function of temperature (see Figure 1) and the characteristic has a hysteresis.

Steep characteristic is the bigger problem. One consequence of this is developing of more stable working points. Figure 4 shows a VO_2 resistor. If a small current is applied to the resistor, whose temperature has increased (because of Joule heating), it does not reach the level required for the phase change, the current density will be of same magnitude throughout the resistor volume. At higher driving currents, the temperature begins to rise. In the part of resistor that first reaches the required temperature for phase change, the resistivity suddenly decreases and the current density increases. In the remaining resistor, however, the current density decreases and thus the temperature also decreases. The positive feedback results in vast majority of current flowing in a hot channel between the anode and cathode. If the material of resistor, structure and environment are homogeneous, the channel forms in the centerline of resistor. However, if this is not the case, the channel may create elsewhere, or even more channels may create. In consequence, a steady-state simulation that takes into account the process of current and temperature distribution must be performed. This example is achieved by time-domain simulation.

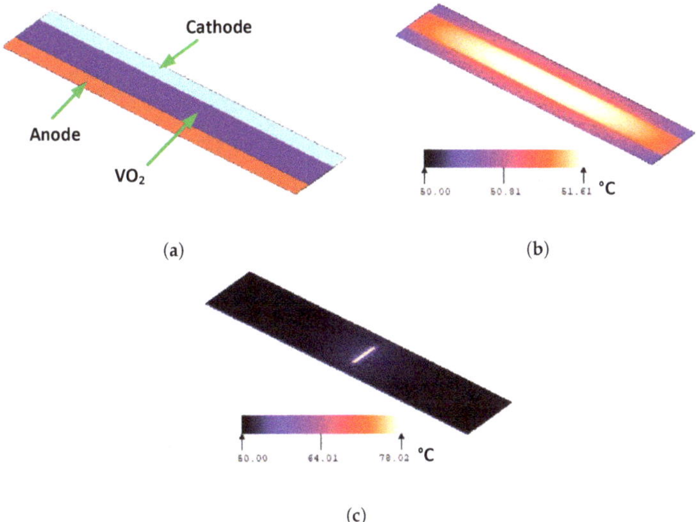

Figure 4. A 200 µm × 20 µm × 0.5 µm VO$_2$ resistor: (**a**) Structure; (**b**) Temperature distribution at low current (2 mA); (**c**) Temperature distribution at high current (8 mA).

The structure of the device shown in Figure 4 and the boundary conditions of the simulation are the same as those of the device shown in [41]. The 200 µm × 20 µm × 0.5 µm VO$_2$ resistor is directly connected to the platinum anode and cathode. There was 25 °C air above the device, and the bottom was connected to a cold plate at 50 °C. In the simulation, the driving current was increased in 2 mA increments up to 20 mA and then decreased in the same increments to 0. The temperature distribution is shown in Figure 4b is for the increasing 2 mA, while the distribution is shown in Figure 4c is for the decreasing 8 mA. In [41], the process of channel formation and termination can be viewed in a series of images.

Another problem arising from the steep characteristic is that the simulation is difficult to converge. In the literature, we find that either electro-thermal simulation results for very specific VO$_2$ structures [29,42], or they achieve the convergence by applying special boundary conditions [43]. In the case of thermal-electrical circuits, these solutions cannot be applied. Our attempts with ANSYS ended in failure, so we made the custom, finite volumes method based simulator developed at the department suitable for handling the special properties of vanadium dioxide [41].

3.1. Computational Method (SUNRED)

In the custom simulator, the examined structure divides by a grid into elementary rectangular cells and each cell is filled with a homogeneous material see Figure 5a. The real structure is three-dimensional as the figure shows a layer of it for clarity. The elementary cells are modeled by a circuit. Adjacent elementary cells are connected by an electrical and a thermal line (the ground is not marked), as shown in Figure 5b. The model obtained from the finite volume method gives the internal structure of elementary cells [44].

A typical structure is shown in Figure 5c. Dissipation caused by current flowing in the electrical part is realized by the dependent power source of the thermal model. Electrical and thermal circuit elements are temperature dependent. The computational task is to determine the voltages and temperatures with knowledge of the excitations and boundary conditions. The simulator performs the solution by successive network reduction method (SUNRED). Figure 5d, which merges cell pairs in each step, knocks out the internal nodes from the equations until finally, a single cell remains. The boundary conditions are applied to this cell, and then the voltages and temperatures of the internal nodes are calculated in a series of successive backward substitution steps.

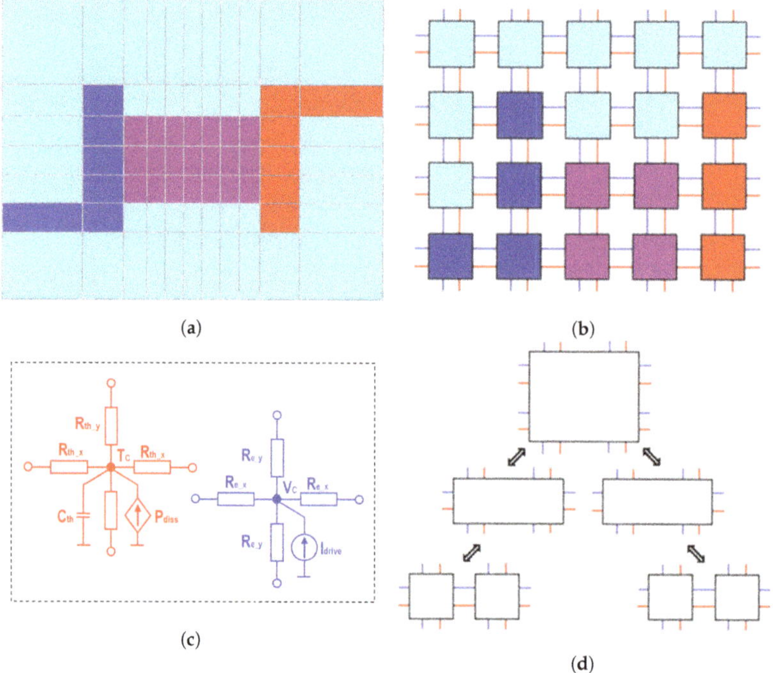

Figure 5. Operation of the custom solver (for clarity in 2D): (**a**) Dividing the model into elementary cells; (**b**) Model network built from elementary cells; (**c**) Electro-thermal elementary cell, red: thermal, blue: electrical; (**d**) Successive network reduction.

In the time domain, the solution is done by the backward Euler method, and the nonlinear calculation is by successive approximation. A heuristic method is used to deal with the extreme nonlinearity of VO_2, details of which are described in [41]. Having more than one value of resistivity at one temperature value, in addition to the memory effect in VO_2, make the hysteresis loop essential in the materials resistivity function. The hysteresis of the resistivity of VO_2 is modeled by a parallelogram, see Figure 6.

3.2. Hysteresis Model

The resistivity function consists of two domains in SUNRED, normal domain and hysteresis domain. In the hysteresis domain, resistivity is multi-valued and any resistance-temperature combination can occur, not just the edges. For instance, when the structure temperature is raised starting from the room temperature 25 °C, resistivity drops linearly in the temperature range

(T0 = 25 °C, T1 = 58 °C). In the following temperature range (T1 = 58 °C, T2 = 64 °C), resistivity remains constant in the hysteresis domain. The material phase change begins above T2 and hence triggers the semiconductor-to-metal transition. SMT leads to a sharp drop in resistivity in the temperature range (T2 = 64 °C, T4 = 73 °C). At 73 °C, the material encounters a full phase transition and resistivity above this temperature is always constant. If the temperature between T2 and T4 starts to decrease, the resistivity remains constant until it reaches the T3-T1 section. It then goes to section T3-T1 to T1 and then T0. During the cooling phase, resistivity function does not follow the same pattern as the heating phase because of the hysteresis. It remains constant until reaching T3 = 67 °C. A reverse phase change occurs in the temperature range (T3 = 67 °C, T1 = 58 °C), bringing VO$_2$ back to the nonconductive state. Resistivity moves up linearly on further reduction of temperature [41].

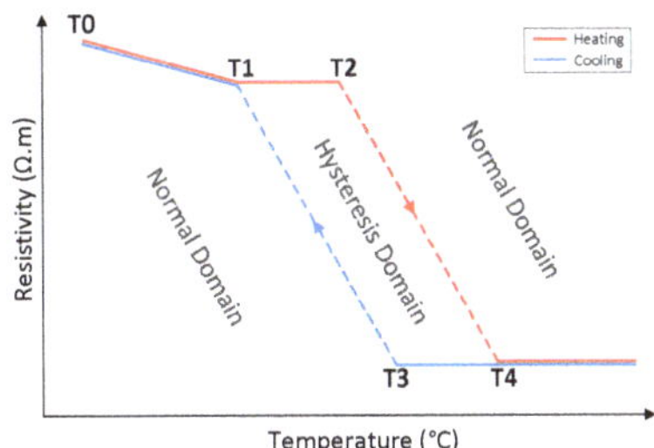

Figure 6. Hysteresis model in the custom simulator.

3.3. Geometry of the Simulated Structure

An equivalent geometry of our vertical structure in Figure 7 is built in the SUNRED simulator ($r = 10\,\mu m$, $h = 50\,nm$). The VO$_2$ layer is sandwiched between platinum and silicon layers. The silicon substrate layer is 6.2 mm × 6.2 mm and 0.3 mm thickness. For faster simulations and simplicity, the quarter of the entire geometry is simulated. In this case, electric current must be multiplied by four while describing the full structure and the cavity length (r) represents the radius of the whole cavity. The bottom of the substrate (the bottom boundary condition) is a heat transfer coefficient (HTC) specified in Section 4. All other sides have a boundary condition of $10\,W/m^2\,K$ *HTC* (the value for air).

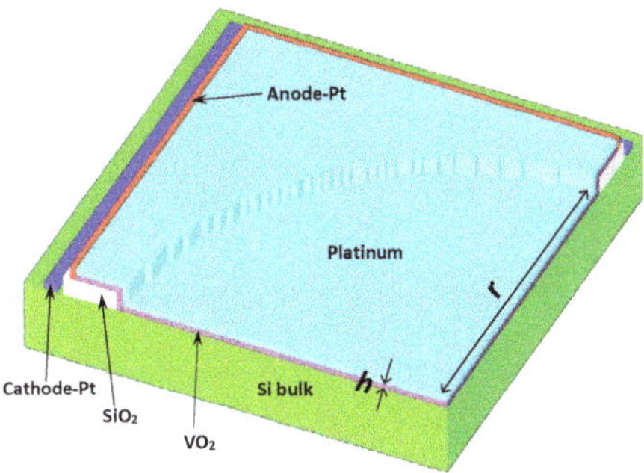

Figure 7. SUNRED model for VO$_2$ vertical resistor.

3.4. Excitation Required for the Simulation

SUNRED capabilities allow different methods for creating excitation in the simulated structure. Electrical current, power (heat flow), voltage and temperature are the available options. For our study, we use electric current on the anode as excitation for the structure. The principal is to assign a different current value in each simulation step. Setting the cathode as the electric ground of the geometry (electric potential = zero) initiates the direction of the electric current from anode to cathode. Optimal path for the current is to pass through VO_2 and Si, reaching the cathode and creating the current flow because air and SiO_2 have high resistivity (refer to Figure 3).

The simulation starts with a low excitation current and gradual increase in each step. Driving current warms up VO_2 layer above the ambient temperature (50 °C) due to the generation of Joule-heat. Higher excitation current leads to a higher temperature of VO_2 resistor and further elevation of temperature is required to simulate the phase-change of VO_2. A similar excitation current profile is used for the second half of the simulation but with a falling direction instead of rising, Figure 8. The peak value and difference of excitation current in each step can be modified to improve the resolution. Smaller current steps give higher resolution.

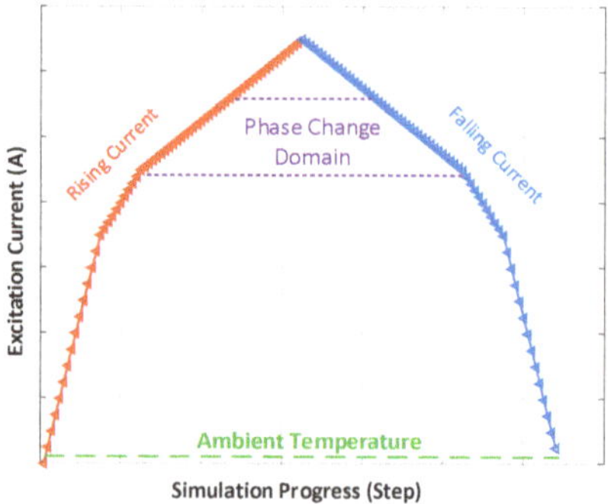

Figure 8. Excitation current per simulation step.

4. Results and Discussion

Each flow of current accompanies an electric voltage developed between anode and cathode. (Voltage, Current) pair in each step is gathered to draw the V-I characteristic curve of our structure. For instance, the V-I curve for one of the performed simulations on vertical VO_2 resistors is shown in Figure 9a. The simulated geometry is ($r = 10\,\mu m$, $h = 60\,nm$) and the ambient temperature is 50 °C. After analyzing several V-I characteristics, a general visualization of the V-I curve of vertical VO_2 geometries can be formalized as illustrated in Figure 9b.

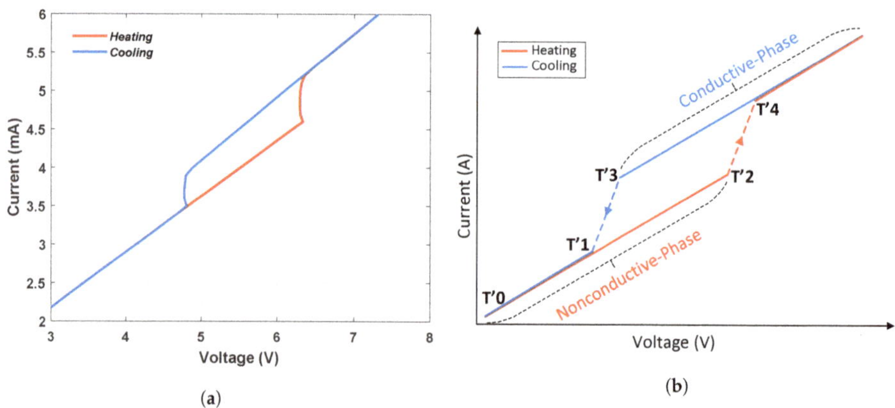

Figure 9. V-I curve of VO_2 vertical structures: (**a**) simulation example, $r = 10\,\mu m$, $h = 60\,nm$, and ambient temperature is 50 °C and (**b**) the general shape for all other simulations.

The starting point of the study is to simulate the geometry in Figure 7 using the real size of produced sample in laboratory (r = 10 µm, h = 50 nm). The low-current section of the measured curve is strongly nonlinear, which did not occur in case of lateral SMT resistors [40,41]. The main difference between the current vertical structure and the previously studied lateral structures is that in lateral structures the platinum anode and cathode were directly attached to VO_2, while in the vertical structure the silicon substrate was inserted between the cathode and VO_2.

More than 90% of the voltage applied to the entire structure drops in the region between the cathode and the VO_2 resistor even when the VO_2 resistor is OFF. The three main parts of the region are the platinum-silicon transition at the cathode, the substrate, and the silicon-VO_2 transition. Since the doping of the substrate is high, its resistivity is around $0.05\,\Omega$ cm, the voltage on it is only a fraction of the voltage applied to the region. In addition, the temperature-dependent nonlinearity of the substrate is very low [45], it cannot be responsible for the measured nonlinearity. The Pt-Si interface can show a Shottky like barrier which will depend on the Si dopant concentration [46]. Annealing could create a platinum-silicide transition that would give an ohmic contact, however, annealing cannot be applied due to VO_2. Based on the band structure of VO_2 [47], it is likely that a barrier will also be created at the junction of the two semiconductor materials. Most of the voltage in the region, therefore, drops at the two transitions.

Based on the measured results, the current dependence on resistivity of the transitions is mainly responsible for the nonlinearity, the temperature dependence is not significant, because at higher currents the curve is almost linear. Since the purpose is the investigation of size dependence of VO_2 resistor at higher currents, the simulation is greatly simplified if nonlinearity is not taken into account and the transitions are modeled with constant resistivity. Accordingly, in the simulated model, a 100 nm thick resistor layer has been placed between both the Pt-Si and Si-VO_2 interfacing surfaces.

Three cases were examined: (a) if a barrier is assumed only at the Pt-Si transition, (b) if a barrier is assumed only at the Si-VO_2 transition and (c) if barriers are assumed at both transitions with same voltage drops. In case (a) the resistivity of the layer between Pt-Si is $1.33 \times 10^5\,\Omega$ cm and the *HTC* at the bottom of the substrate is $46\,W/m^2\,K$; in case (b), the resistivity of the layer between Si-VO_2 is $377\,\Omega$ cm and the *HTC* at the bottom of the substrate is $100\,W/m^2\,K$; and in case (c), the resistivity of the layer between Pt-Si is $6.7 \times 10^4\,\Omega$ cm, between Si-VO_2 it is $189\,\Omega$ cm and the HTC is $60\,W/m^2\,K$. Results of the three simulations are shown in Figure 10b.

Both curves show an acceptable degree of validity. However, it is difficult to reach a near-absolute accuracy while neglecting the nonlinearity of the transitions. The location of the resistor defines the dissipation point. From circuitry point of view, the three cases are equivalent. We assume that there is a barrier at both transitions, so in the rest of the article, we perform the simulations with the model corresponding to case (c).

The experimental data in Figure 10a show a "noisy" behavior in the ascending current phase. When the current increases through the SMT layer, some conductive channels are switched ON, which as a consequense gives the noisy path. When the current decreases, the SMT conductive channel collapse faster. This phenomenon can also be observed with decreasing current, but it is much more moderated than with the increasing current case. The origin of the noisy signal is that a lot of parallel conductive channels are in the SMT layer (depends on the crystal structure). The real VO_2 layer is not a homogenous crystal, it has a spiky structure [48]. While input power increases, some of the conductive channels are "switched ON", which are connected in parallel with the original SMT phase. If the dissipated power decreases (the resistance decreases) then it can switch back, which shows a jump back to the original path. This phenomenon also appears in other structures, for example in [40], the lateral structure with two Pt contacts are directly attached to VO_2, no Si substrate is interposed.

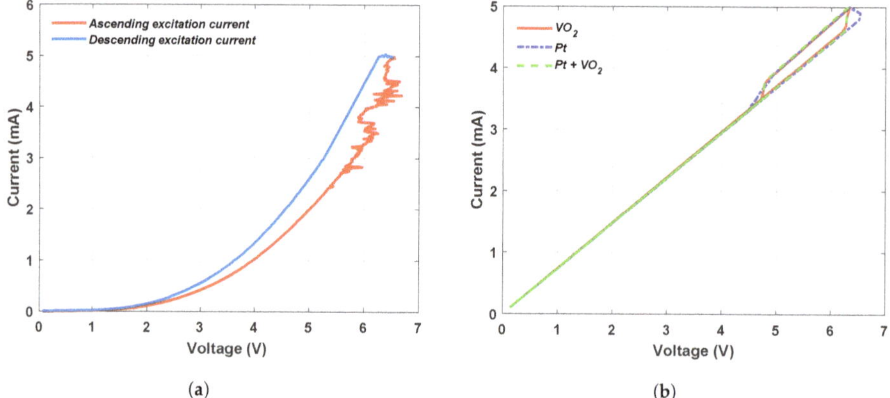

Figure 10. V-I curve of ($r = 10\,\mu m$, $h = 50\,nm$) vertical VO_2 resistor based on: (**a**) Laboratory measurement. (**b**) Simulation results.

4.1. Dependence on Thickness

In the first set of simulations, we investigated how the VO_2 structure behaves on different thicknesses and what is the effect of changing h on the characteristics of our structure. The range of h used in the simulations is (10 nm to 300 nm) which is investigated at different radii. Figure 11 shows how the resistance (R) of VO_2 layer changes with thickness. Resistance during the nonconductive and conductive phases is the point of interest here (refer to Figure 9b). Resistance is calculated as the $1/slope = (\Delta V / \Delta I)$ of the V-I curve. It can be concluded from Figure 11a that the nonconductive phase resistance increases as thickness increases. Furthermore, R dependence on h becomes lower at larger radii. On the other hand, conductive phase resistance shows no direct relation with thickness at these h and r ranges, Figure 11b.

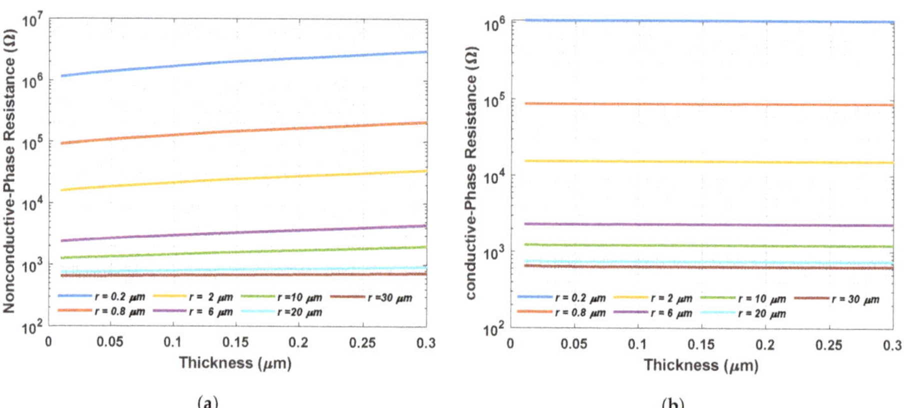

Figure 11. VO_2 resistance during (**a**) the nonconductive phase and (**b**) during the conductive phase at different thicknesses.

The current at which VO_2 resistor is turned on (when resistivity drops significantly) is called the opening current. Similarly, the current at which the VO_2 resistor is turned off (during the cooling phase) is called the closing current. Voltages associated with these currents are opening voltage and closing voltage, respectively. Figure 12 illustrates the dependence of opening and closing currents on the variation of thickness. Since the opening current is associated with the nonconductive phase resistance and the closing current is associated with the conductive phase resistance, the opening current falls as thickness increases due to the elevation of R versus thickness (in the nonconductive phase), Figure 12a. On the other hand, the current remains unchanged as thickness increases because resistance is not affected by thickness (in the conductive phase), Figure 12b. A high degree of compatibility can be noticed between Figures 11 and 12.

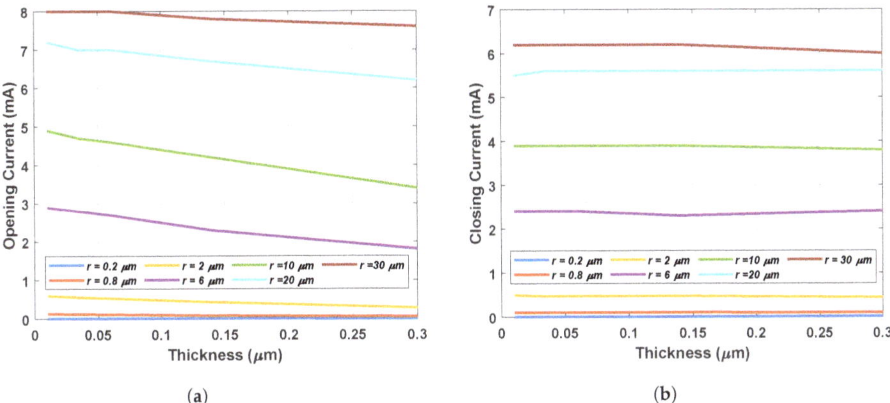

(a)

(b)

Figure 12. The dependence of the (**a**) opening and (**b**) closing currents on the thickness of the VO$_2$ resistor.

4.2. Dependence on Radius

The second set of simulations is to study how the structure responds to change in radius. Simulations performed with different radii r in the range of (200 nm to 30 μm) and are investigated at different thicknesses. Figure 13 concludes that radius has a strong effect on the VO$_2$ resistance, and the resistance drops nonlinearly for both the nonconductive phase (Figure 13a) and the conductive phase (Figure 13b) as radius increases. It is explained with the relation $R = \rho \frac{l}{A}$, ρ is the resistivity, l is length (which is h here) and A is the cross-sectional area (which is here πr^2).

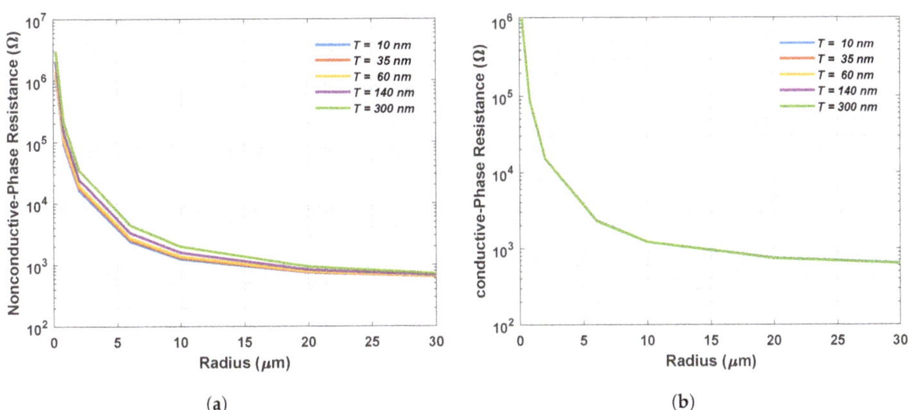

(a)

(b)

Figure 13. VO$_2$ resistance during (**a**) the nonconductive phase and (**b**) during the conductive phase at different radii.

As a consequence of the resistance dependence on radius, opening and closing currents increase as radius increases. Figure 14 illustrates the effect of changing the radius on the opening current (Figure 14a) and the closing current (Figure 14b). For low current applications, it is worth to consider decreasing the radius since it allows the phase change to occur at much lower currents.

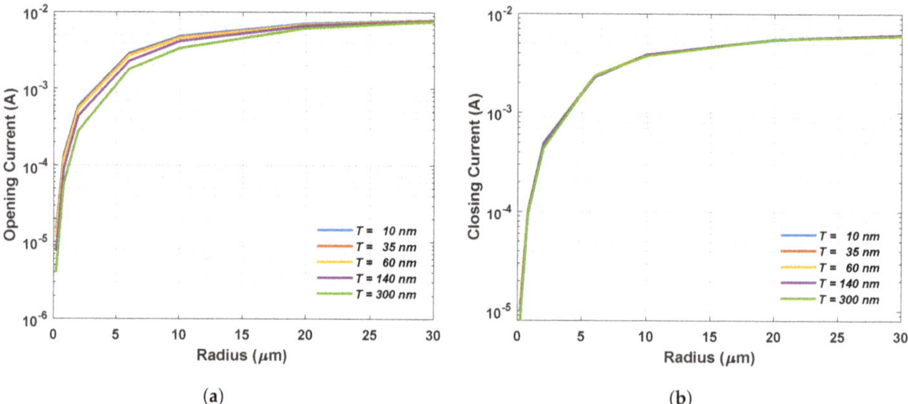

Figure 14. The dependence of the (a) opening and (b) closing currents on the radius of the VO$_2$ resistor.

During the hysteresis domain, the resulting V-I curve also has hysteric nature (refer to Figure 9b). Maximum difference between the resistance of VO$_2$ resistor during heating and cooling phases is called the maximum switching resistance. It can be observed that the maximum switching resistance varies linearly versus thickness Figure 15a, in contrast, it has a nonlinear dependence on the radius Figure 15b.

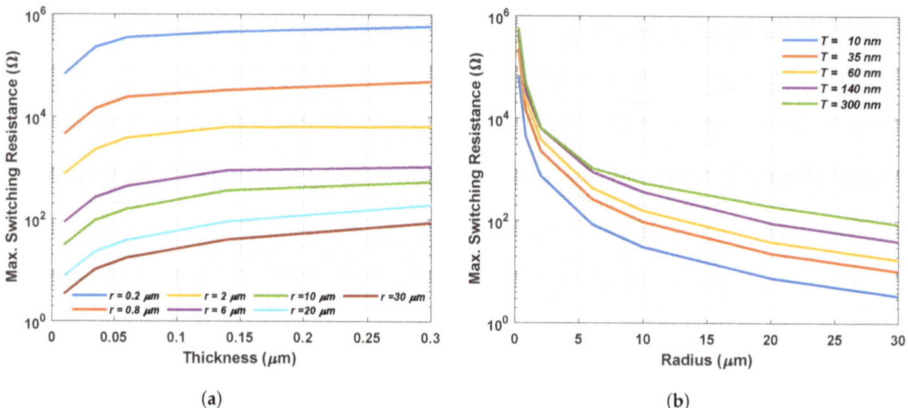

Figure 15. The dependence of the maximum switching resistance on (a) the thickness and (b) the radius of the VO$_2$ resistor.

5. Conclusions

The results of this article offer a good base of sample sizes estimates for further studies in submicron modeling of VO$_2$ resistors. A valid operating model will pave the way for designing thermal-electronic logic circuits. The results provided in our previous work [39] about the lateral structures and results of this article about the vertical structures will straiten the range of samples sizes for producing and studying. This will save both time and effort.

Author Contributions: Conceptualization, P.N. and M.D.; methodology, M.D.; software, L.P.; validation, M.D.; technology, J.M.; sample creation and measurement, P.N.; writing—original draft preparation, M.D., L.P., P.N.; writing—review and editing, L.P., M.D.; visualization, M.D.; supervision, L.P.; project administration, J.M.; funding acquisition, P.N. All authors have read and agreed to the published version of the manuscript.

Funding: The research reported in this paper and carried out at the Budapest University of Technology and Economics was supported by the "TKP2020, National Challenges Program" of the National Research Development and Innovation Office (BME NC TKP2020). This research also received fund by the Higher Education Excellence Program of the Ministry of Human Capacities in the frame of Nanotechnology research area of the Budapest

University of Technology and Economics (BME FIKP-NANO) and the Science Excellence Program at BME under the grant agreement NKFIH-849-8/2019 of the Hungarian National Research, Development and Innovation Office is also acknowledged. The research reported in this paper was partially supported by the Stipendium Hungaricum Scholarship Programme of the Hungarian Government.

Abbreviations

The following abbreviations are used in this manuscript:

VO_2	Vanadium dioxide
CMOS	Complementary metal–oxide–semiconductor
TELC	Thermal-electronic logic circuits
SMT	Semiconductor-to-metal transition
Si	Silicon
SiO_2	Silicon dioxide
RF	Radio frequency
Pt	Platinum
FDM	Finite difference method
FEM	Finite element method
FVM	Finite volume method
HTC	Heat transfer coefficient
SUNRED	Successive network reduction method (also the simulator name)

References

1. Shalf, J. The future of computing beyond Moore's Law. *Philos. Trans. R. Soc. A Math. Phys. Eng. Sci.* **2020**, *378*, 20190061. [CrossRef] [PubMed]
2. Markov, I.L. Limits on fundamental limits to computation. *Nature* **2014**, *512*, 147–154. [CrossRef]
3. Shalf, J.M.; Leland, R. Computing beyond Moore's Law. *Computer* **2015**, *48*, 14–23. [CrossRef]
4. Joneckis, L.; Koester, D.; Alspector, J. *An Initial Look at Alternative Computing Technologies for the Intelligence Community*; Technical report; Institute for Defense Analyses: Alexandria, VA, USA, 2014.
5. Thompson, N.; Spanuth, S. The Decline of Computers As a General Purpose Technology: Why Deep Learning and the End of Moore's Law are Fragmenting Computing. *SSRN Electron. J.* **2018**. [CrossRef]
6. Jouppi, N.P.; Young, C.; Patil, N.; Patterson, D.; Agrawal, G.; Bajwa, R.; Bates, S.; Bhatia, S.; Boden, N.; Borchers, A.; et al. In-Datacenter Performance Analysis of a Tensor Processing Unit. In *Proceedings of the 44th Annual International Symposium on Computer Architecture*; ACM: New York, NY, USA, 2017; pp. 1–12. [CrossRef]
7. Caulfield, A.M.; Chung, E.S.; Putnam, A.; Angepat, H.; Fowers, J.; Haselman, M.; Heil, S.; Humphrey, M.; Kaur, P.; Kim, J.Y.; et al. A cloud-scale acceleration architecture. In Proceedings of the 2016 49th Annual IEEE/ACM International Symposium on Microarchitecture (MICRO), Taipei, Taiwan, 15–19 October 2016; pp. 1–13. [CrossRef]
8. Hill, M.D.; Reddi, V.J. Accelerator-level Parallelism. *arXiv* **2019**, arXiv:1907.02064.
9. Ingerly, D.B.; Enamul, K.; Gomes, W.; Jones, D.; Kolluru, K.C.; Kandas, A.; Kim, G.S.; Ma, H.; Pantuso, D.; Petersburg, C.; et al. Foveros: 3D Integration and the use of Face-to-Face Chip Stacking for Logic Devices. In Proceedings of the 2019 IEEE International Electron Devices Meeting (IEDM), San Francisco, CA, USA, 7–11 December 2019; pp. 19.6.1–19.6.4. [CrossRef]
10. Nikonov, D.E.; Young, I.A. Overview of Beyond-CMOS Devices and a Uniform Methodology for Their Benchmarking. *Proc. IEEE* **2013**, *101*, 2498–2533. [CrossRef]
11. Seabaugh, A.C.; Zhang, Q. Low-Voltage Tunnel Transistors for Beyond CMOS Logic. *Proc. IEEE* **2010**, *98*, 2095–2110. [CrossRef]
12. Low, T.; Appenzeller, J. Electronic transport properties of a tilted graphene p-n junction. *Phys. Rev. B* **2009**, *80*, 155406. [CrossRef]

13. Banerjee, S.; Register, L.; Tutuc, E.; Reddy, D.; MacDonald, A. Bilayer PseudoSpin Field-Effect Transistor (BiSFET): A Proposed New Logic Device. *IEEE Electron Device Lett.* **2009**, *30*, 158–160. [CrossRef]

14. Sugahara, S.; Tanaka, M. A spin metal–oxide–semiconductor field-effect transistor using half-metallic-ferromagnet contacts for the source and drain. *Appl. Phys. Lett.* **2004**, *84*, 2307–2309. [CrossRef]

15. Currivan, J.A.; Jang, Y..; Mascaro, M.D.; Baldo, M.A.; Ross, C.A. Low Energy Magnetic Domain Wall Logic in Short, Narrow, Ferromagnetic Wires. *IEEE Magn. Lett.* **2012**, *3*, 3000104–3000104. [CrossRef]

16. Nikonov, D.E.; Bourianoff, G.I.; Ghani, T. Proposal of a Spin Torque Majority Gate Logic. *IEEE Electron Device Lett.* **2011**, *32*, 1128–1130. [CrossRef]

17. Behin-Aein, B.; Datta, D.; Salahuddin, S.; Datta, S. Proposal for an all-spin logic device with built-in memory. *Nat. Nanotechnol.* **2010**, *5*, 266–270. [CrossRef]

18. Mizsei, J.; Lappalainen, J.; Bein, M.C. Thermal-electronic integrated logic. In Proceedings of the 19th International Workshop on Thermal Investigations of ICs and Systems (THERMINIC), Berlin, Germany, 25–27 September 2013; pp. 128–134. [CrossRef]

19. Zylbersztejn, A.; Mott, N.F. Metal-insulator transition in vanadium dioxide. *Phys. Rev. B* **1975**, *11*, 4383–4395. [CrossRef]

20. Limelette, P. Universality and Critical Behavior at the Mott Transition. *Science* **2003**, *302*, 89–92. [CrossRef] [PubMed]

21. Lappalainen, J.; Heinilehto, S.; Jantunen, H.; Lantto, V. Electrical and optical properties of metal-insulator-transition VO_2 thin films. *J. Electroceram.* **2009**, *22*, 73–77. [CrossRef]

22. Chain, E.E. Optical properties of vanadium dioxide and vanadium pentoxide thin films. *Appl. Opt.* **1991**, *30*, 2782. [CrossRef]

23. Rozgonyi, G.A.; Hensler, D.H. Structural and Electrical Properties of Vanadium Dioxide Thin Films. *J. Vac. Sci. Technol.* **1968**, *5*, 194–199. [CrossRef]

24. Budai, J.D.; Hong, J.; Manley, M.E.; Specht, E.D.; Li, C.W.; Tischler, J.Z.; Abernathy, D.L.; Said, A.H.; Leu, B.M.; Boatner, L.A.; et al. Metallization of vanadium dioxide driven by large phonon entropy. *Nature* **2014**, *515*, 535–539. [CrossRef]

25. Qazilbash, M.M.; Brehm, M.; Chae, B.G.; Ho, P.C.; Andreev, G.O.; Kim, B.J.; Yun, S.J.; Balatsky, A.V.; Maple, M.B.; Keilmann, F.; et al. Mott Transition in VO_2 Revealed by Infrared Spectroscopy and Nano-Imaging. *Science* **2007**, *318*, 1750–1753. [CrossRef]

26. Kim, H.T.; Lee, Y.W.; Kim, B.J.; Chae, B.G.; Yun, S.J.; Kang, K.Y.; Han, K.J.; Yee, K.J.; Lim, Y.S. Monoclinic and correlated metal phase in VO_2 as evidence of the Mott transition: coherent phonon analysis. *Phys. Rev. Lett.* **2006**, *97*, 266401. [CrossRef]

27. Gomez-Heredia, C.L.; Ramirez-Rincon, J.A.; Ordonez-Miranda, J.; Ares, O.; Alvarado-Gil, J.J.; Champeaux, C.; Dumas-Bouchiat, F.; Ezzahri, Y.; Joulain, K. Thermal hysteresis measurement of the VO_2 emissivity and its application in thermal rectification. *Sci. Rep.* **2018**, *8*, 8479. [CrossRef]

28. Hamaoui, G.; Horny, N.; Gomez-Heredia, C.L.; Ramirez-Rincon, J.A.; Ordonez-Miranda, J.; Champeaux, C.; Dumas-Bouchiat, F.; Alvarado-Gil, J.J.; Ezzahri, Y.; Joulain, K.; et al. Thermophysical characterisation of VO_2 thin films hysteresis and its application in thermal rectification. *Sci. Rep.* **2019**, *9*, 8728. [CrossRef]

29. Hu, B.; Zhang, Y.; Chen, W.; Xu, C.; Wang, Z.L. Self-heating and External Strain Coupling Induced Phase Transition of VO_2 Nanobeam as Single Domain Switch. *Adv. Mater.* **2011**, *23*, 3536–3541. [PubMed]

30. Soltani, M.; Chaker, M.; Haddad, E.; Kruzelesky, R.V. Thermochromic vanadium dioxide smart coatings grown on Kapton substrates by reactive pulsed laser deposition. *J. Vac. Sci. Technol. A Vac. Surf. Films* **2006**, *24*, 612–617. [CrossRef]

31. Datta, S.; Shukla, N.; Cotter, M.; Parihar, A.; Raychowdhury, A. Neuro Inspired Computing with Coupled Relaxation Oscillators. In *Proceedings of the The 51st Annual Design Automation Conference on Design Automation Conference—DAC'14*; ACM Press: New York, NY, USA, 2014; pp. 1–6. [CrossRef]

32. Zhou, Y.; Ramanathan, S. Mott Memory and Neuromorphic Devices. *Proc. IEEE* **2015**, *103*, 1289–1310. [CrossRef]

33. Ordonez-Miranda, J.; Ezzahri, Y.; Drevillon, J.; Joulain, K. Dynamical heat transport amplification in a far-field thermal transistor of VO_2 excited with a laser of modulated intensity. *J. Appl. Phys.* **2016**, *119*, 203105. [CrossRef]

34. Prod'Homme, H.; Ordonez-Miranda, J.; Ezzahri, Y.; Drevillon, J.; Joulain, K. VO$_2$ -based radiative thermal transistor in the static regime 2017. *arXiv* **2017**, arXiv:1710.10332.

35. Chen, C.; Zhao, Y.; Pan, X.; Kuryatkov, V.; Bernussi, A.; Holtz, M.; Fan, Z. Influence of defects on structural and electrical properties of VO$_2$ thin films. *J. Appl. Phys.* **2011**, *110*, 023707. [CrossRef]

36. Jian, J.; Chen, A.; Zhang, W.; Wang, H. Sharp semiconductor-to-metal transition of VO$_2$ thin films on glass substrates. *J. Appl. Phys.* **2013**, *114*, 244301. [CrossRef]

37. Taha, M.; Walia, S.; Ahmed, T.; Headland, D.; Withayachumnankul, W.; Sriram, S.; Bhaskaran, M. Insulator—Metal transition in substrate-independent VO$_2$ thin film for phase-change devices. *Sci. Rep.* **2017**, *7*, 17899. [CrossRef] [PubMed]

38. Mizsei, J.; Bein, M.; Lappalainen, J.; Juhász, L.; Plesz, B. The Phonsistor—A Novel VO$_2$ Based Nanoscale Thermal-electronic Device and Its Application in Thermal-electronic Logic Circuits (TELC). *Mater. Today Proc.* **2015**, *2*, 4272–4279. [CrossRef]

39. Pohl, L.; Darwish, M.; Mizsei, J. Electro-Thermal Investigation of SMT Resistors for Thermal-Electrical Logic Circuits by Simulation. In Proceedings of the 2019 25th International Workshop on Thermal Investigations of ICs and Systems (THERMINIC), Lecco, Italy, 25–27 September 2019; pp. 1–4. [CrossRef]

40. Mizsei, J.; Lappalainen, J. Microelectronics, Nanoelectronics: step behind the red brick wall using the thermal domain. *Mater. Today Proc.* **2019**, *7*, 888–893. [CrossRef]

41. Pohl, L.; Ur, S.; Mizsei, J. Thermoelectrical modelling and simulation of devices based on VO$_2$. *Microelectron. Reliab.* **2017**, *79*, 387–394. [CrossRef]

42. Gopalakrishnan, G.; Ruzmetov, D.; Ramanathan, S. On the triggering mechanism for the metal–insulator transition in thin film VO$_2$ devices: electric field versus thermal effects. *J. Mater. Sci.* **2009**, *44*, 5345–5353. [CrossRef]

43. Kumar, S.; Pickett, M.D.; Strachan, J.P.; Gibson, G.; Nishi, Y.; Williams, R.S. Local Temperature Redistribution and Structural Transition During Joule-Heating-Driven Conductance Switching in VO$_2$. *Adv. Mater.* **2013**, *25*, 6128–6132. [CrossRef]

44. Nagy, G.; Horváth, P.; Pohl, L.; Poppe, A. Advancing the thermal stability of 3D ICs using logi-thermal simulation. *Microelectron. J.* **2015**, *46*, 1114–1120. [CrossRef]

45. PV Lighthouse: Resistivity calculator, Available online: https://www.pvlighthouse.com.au (accessed on 21 June 2020).

46. Muta, H. Electrical Properties of Platinum-Silicon Contact Annealed in an H 2 Ambient. *Jpn. J. Appl. Phys.* **1978**, *17*, 1089–1098. [CrossRef]

47. Martens, K.; Radu, I.P.; Mertens, S.; Shi, X.; Nyns, L.; Cosemans, S.; Favia, P.; Bender, H.; Conard, T.; Schaekers, M.; et al. The VO$_2$ interface, the metal-insulator transition tunnel junction, and the metal-insulator transition switch On-Off resistance. *J. Appl. Phys.* **2012**, *112*, 124501. [CrossRef]

48. Rao Popuri, S.; Artemenko, A.; Labrugere, C.; Miclau, M.; Villesuzanne, A.; Pollet, M. VO$_2$ (A): Reinvestigation of crystal structure, phase transition and crystal growth mechanisms. *J. Solid State Chem.* **2014**, *213*, 79–86. [CrossRef]

Article

Investigation and Modeling of the Magnetic Nanoparticle Aggregation with a Two-Phase CFD Model

Péter Pálovics *, Márton Németh and Márta Rencz

Department of Electron Devices, Budapest University of Technology and Economics,
Magyar tudósok krt. 2, Bld. Q, H-1117 Budapest, Hungary; nemeth.marton@vik.bme.hu (M.N.);
rencz.marta@vik.bme.hu (M.R.)
* Correspondence: palovics.peter@vik.bme.hu

Received: 4 August 2020; Accepted: 7 September 2020; Published: 17 September 2020

Abstract: In this paper the magnetic nanoparticle aggregation procedure in a microchannel in the presence of external magnetic field is investigated. The main goal of the work was to establish a numerical model, capable of predicting the shape of the nanoparticle aggregate in a magnetic field without extreme computational demands. To that end, a specialized two-phase CFD model and solver has been created with the open source CFD software OpenFOAM. The model relies on the supposed microstucture of the aggregate consisting of particle chains parallel to the magnetic field. First, the microstructure was investigated with a micro-domain model. Based on the theoretical model of the particle chain and the results of the micro-domain model, a two-phase CFD model and solver were created. After this, the nanoparticle aggregation in a microchannel in the field of a magnet was modeled with the solver at different flow rates. Measurements with a microfluidic device were performed to verify the simulation results. The impact of the aggregate on the channel heat transfer was also investigated.

Keywords: magnetic nanoparticle; microfluidics; CFD; OpenFOAM; two-phase solver; rheology

---

1. Introduction

Magnetic nanoparticles (MNPs) are magnetizable nanosized (1–500 nm) objects with various shapes. These particles are utilized broadly in biomedical applications, including drug delivery; hyperthermia treatment; MRI as contrast agents; chemical reaction enhancement with microreactors; contaminant removal for sewage treatment; and microfluidic cooling. These applications utilize various advantageous features of the MNPs, e.g., the high surface-to-volume ratio, the controllability with external magnetic field, the non-toxicity of the coating and the selective heat absorption against the alternating magnetic field. In this section, we briefly review the most relevant applications and summarize the problems we are facing during the modeling of systems using MNPs.

One of the most promising applications of MNPs is to enhance organic chemical reactions. In this case, catalyst is attached to the coated MNP. Due to the huge surface-to-volume ratio, high reaction rates can be achieved, while thanks to the MNP's magnetic core, the catalyst remains on the surface of the MNP and can be recovered [1]. In other applications the catalyst coated MNP is magnetically anchored in a flow-through reactor [2–4]. In this setting the chemical reaction is continuous; therefore, it can be parallelized and scaled according to demands. While in the former application, the reaction happens similarly to a traditional chemical reaction, in the flow-through reactor unexpected effects are observed. The most interesting among them is the catalytic reaction rate dependency on the flow rate [4]. In [5] it is presented that when the microreactor is not fully packed with MNPs, the flow goes around the chemically active aggregate, which creates a bypass for the reagents outside the MNP aggregated phase. This is an undesired effect, as only a part of the reagents participate in the reaction.

To design and apply such MNP-based flow-through reactors, a modeling method is required, as the shape and the quantity of the anchored MNP aggregate matters while the reaction rates are calculated. The simulation cannot be done by modeling the motion of each particle, because of their prohibitively high number for computing (the estimated number of particles can be in the range of 10^9–10^{10} even in a microreactor). Consequently, we need to find a way to simulate the MNP aggregate in macro scale, as a continuous phase. If we can determine the shape and the size of the MNP aggregate, we are able to also create temperature control on the reaction (with distant heating through a magnetic field), which is a key to realizing complex reactions such as PCR (polymerase chain reaction) in flow-through MNP reactors [6].

Although the MNPs are very useful for creating drugs or diagnosing diseases trough chemical reactions, they can be also used inside the body for diagnostic purposes or for therapy. The MNPs are most commonly used in the MRI (magnetic resonance imaging) diagnostic tool as contrast agents [7]. The functionalized MNPs can accumulate in specific tissues, creating a great contrast on the MRI results. From the clinical research and practice the MNP seems to be a good candidate for drug delivery too, as in a small dose it is non-toxic, and the degradation time is long enough [8].

Another biomedical application evolves when the MNPs are aggregated in blood vessels with the help of an external magnetic field. The aggregate blocks the blood flow; therefore, no oxygen and nutrients can go through. This is important in cancer treatment, where the malignant cells should be killed. The method can be combined with heating with alternating magnetic field; that is called hyperthermia therapy. Using magnetic nanoparticles for hyperthermia cancer treatment is in pre-clinical state [9].

In these applications the accumulation in flow and the thermal effects have key importance. The thermal behaviour of the MNP suspension is also influenced by the magnetic field. The MNPs are arranged into chains in a magnetic field and the heat conductivity of the suspension depends on the lengths of these evolved particle chains. On the other hand, the aggregate changes the flow path; therefore, a higher wall-to-middle heat transfer occurs, which may elevate the Nusselt number. Moreover, according to the attracting force on the MNP, the heat transported by the nanoparticles from the suspension to the aggregate is also important [10]. This leads to a heat transfer system, where the heat transfer coefficient can be influenced by an external magnetic field. Reports on elevated Nusselt numbers with the magnetic field intensity were published in [11,12], where slight elevation was observed in the alternating and constant magnetic field, but others reported lower local Nusselt numbers [13]; they presumed it was because of the increased viscosity. As these examples show, calculating the energy transport in a MNP suspension is very challenging. To design a cooler, hyperthermia treatment or temperature controlled MNP microreactor a model is needed, wherein the MNPs are handled as being continuous, and not individually.

Numerical investigations of the magnetized particles in a fluid are presented in several works in the literature. In [14,15] the individual particles are modeled separately, which offers a detailed view of the particle aggregation in a micro-scale. The disadvantage of this particle-based method is its high computational demand, which was discussed above. In other papers the nanoparticle aggregation is investigated with continuum-based approaches, enabling one to simulate the particle suspensions [16,17]. It should be noted that in these continuum-based works, although the force of the external magnetic field on the particles is taken into account, the particle–particle magnetic interactions are not investigated, which can be a relevant phenomenon during the MNP aggregation.

In this article we present a new two-phase CFD (computational fluid dynamics) model capable of predicting the shape of the magnetic nanoparticle aggregate in case of a given fluid flow and a magnetic field. The details of this model and the micro-structure of the aggregate are the main topics of this paper. Our model's novelty with respect to the similar works in the literature is that it relies on the suspected micro-structure of the aggregate in which the particle–particle interactions are included.

This is relevant, as it will be shown that these interactions play a major role in the aggregation of the nanoparticles in fluid flow. The balance between the particle–particle forces, the drag force and the external magnetic force determines the shape of the formed aggregate in the magnetic field.

The structure of the paper is as follows. First the theoretical model of the nanoparticle aggregate in the case of the magnetic field is presented. Then the specialized two-phase CFD model and solver are discussed, which relies on the micro-domain model. After the theoretical section, the experimental results are presented. A microfluidic chip containing a straight channel was prepared in order to investigate the nanoparticle aggregation in the field of a neodymium magnet. Besides the observation of the aggregation, thermal measurements were also performed. Finally the numerical results are presented. First the particle aggregates were investigated in a micro-domain at different particle concentrations, which were needed to establish the two-phase model. Then the nanoparticle aggregation in the channel was simulated with the two-phase solver at the different flow rates. The simulated and measured results are compared.

2. Micro-Domain Model

In this section the theoretical model for the magnetic nanoparticle dynamics is presented briefly. This model is not the main scope of the paper, but its description is necessary to understand the build-up of the two-phase model. A detailed discussion of the micro-domain model is going to be published in another paper.

In this model each magnetic nanoparticle is considered separately (discrete particle method). Detailed description and material parameters of the MNP can be found in the experimental section. Here we only summarize the details necessary for the modeling. The nanoparticles have a magnetite core with $d_{core} = 210$ nm, and the core is coated with a 20 nm thick silica shell. The particle size therefore is $d_p = 250$ nm. The core is treated as a linear magnetic material with the magnetic susceptibility of $\chi_p = 2.8$.

2.1. Magnetization

Suppose that we place one particle in a uniform magnetic field $\mathbf{H}_0$. Therefore, the particle's magnetite core will be magnetized. The magnetization of the spherical core is

$$\mathbf{M} = \frac{3\chi_p}{3 + \chi_p}\mathbf{H}_0,\tag{1}$$

where χ_p is the magnetic susceptibility of the magnetite core [18]. This means that the magnetic moment of the core will be

$$\mathbf{m} = V_{core}\mathbf{M};\tag{2}$$

i.e., $\mathbf{m} = \frac{4\pi}{3}r_{core}^3\frac{3\chi_p}{3+\chi_p}\mathbf{H}_0$. The spherical magnetic core has its own magnetic field, which is the field of a dipole

$$\mathbf{H}(\mathbf{r}) = \frac{1}{4\pi}\frac{1}{r^3}\left[3(\mathbf{m}\hat{\mathbf{r}})\hat{\mathbf{r}} - \mathbf{m}\right],\tag{3}$$

where $\mathbf{r}$ is the position vector from the particle center (and $r = |\mathbf{r}|$), and $\hat{\mathbf{r}}$ is the unit vector in that direction, $\hat{\mathbf{r}} = \mathbf{r}/r$ [18].

When two particles are close, they modify each other's magnetic moment. This means that Equation (2) should be modified to take into account the effect of the surrounding particles. The corrected magnetic moment for the particle i is

$$\mathbf{m}_i = V_{core}\frac{3\chi_p}{\chi_p + 3}\left(\mathbf{H}_0 + \sum_{j\neq i}\mathbf{H}_j\right),\tag{4}$$

where $\sum_j\mathbf{H}_j$ notes the field of the surrounding particles j.

2.2. Forces Effecting the Particle

Suppose now that the particle is placed in the fluid flow in a microchannel. The fluid is water and its relative permeability is assumed to be 1. Next we put a neodymium magnet over the channel. The non-homogeneous field of the magnet acts with a force on the particle:

$$\mathbf{F}_{M\_field} = \mu_0(\mathbf{m}\nabla)\mathbf{H}_0 = \mu_0 \mathbf{m}_i \cdot \nabla \mathbf{H}_0 \tag{5}$$

where μ_0 is the vacuum permeability. The equality of the last two phrases in Equation (5) can be done, as $\nabla \times \mathbf{H}_0 = 0$.

The drag force on the particle is the Stokes force:

$$\mathbf{F}_S = -6\pi\mu_f r_p(\mathbf{U}_p - \mathbf{U}_f), \tag{6}$$

where μ_f is the dynamic viscosity of the fluid, r_p is the particle radius and the last term is the velocity difference between the fluid and the particle. Besides this, if the particle rotates in the fluid, a stopping torque from the viscous fluid appears

$$\boldsymbol{\tau} = -8\pi r_p^3 \mu_f \cdot \boldsymbol{\omega}_p, \tag{7}$$

where ω_p is the particle's angular velocity [19].

If two particles are close, a magnetic force appears, as one particle is at the other particle's non-uniform magnetic dipole field. The magnetic force between the close particles i and j is

$$\mathbf{F}_{M\_ij} = \frac{3\mu_0 m_i m_j}{4\pi r_{ij}^4}[\hat{\mathbf{r}}_{ij}(\hat{\mathbf{m}}_i\hat{\mathbf{m}}_j)+$$
$$\hat{\mathbf{m}}_i(\hat{\mathbf{r}}_{ij}\hat{\mathbf{m}}_j) + \hat{\mathbf{m}}_j(\hat{\mathbf{r}}_{ij}\hat{\mathbf{m}}_i) - 5\hat{\mathbf{r}}_{ij}(\hat{\mathbf{r}}_{ij}\hat{\mathbf{m}}_i)(\hat{\mathbf{r}}_{ij}\hat{\mathbf{m}}_j)], \tag{8}$$

where m_i and m_j are the magnetic moment's of the particles, r_{ij} is the distance between their centers and $\hat{\mathbf{r}}_{ij}$ is the unit vector of the distance [14]. This force is assumed to be the main reason for the aggregation. It causes the particles to arrange into chains; the direction of these is parallel to the main magnetic field. The particles in the chain are attracted each other. It can be shown that in the field of the neodymium magnet, which is presented in the numerical results section, the attractive force between two magnetized particles in the chain is approximately three orders of magnitude higher than the force of the neodymium magnet on the particle. If the chain is bent as a result of other forces, it tries to rotate back to be parallel with the magnetic field.

The gravitational and buoyancy forces at this size range are negligible compared to the magnitude of the previously discussed forces.

2.3. Chain Formulation

As was mentioned in the previous section, in a magnetic field the magnetic nanoparticles are magnetized and due to the magnetic force between the particles they aggregate into a chain, which is parallel to the magnetic field. The chain formulation of the magnetized particles was also observed experimentally; see, e.g., [20]. The chain formulation is also widely investigated in magnetorheological (MR) fluids [21].

In the following we focus on the particle chain which is placed in a fluid flow with homogeneous strain rate. The idea originated from the electrorheological (ER) particle chain investigations in [22]. Our goal was to find the net impact of the magnetic interactions of the particles on the fluid dynamics.

The arrangement is shown in Figure 1a. The fluid flows to the direction x, and has a homogeneous strain rate of $\dot{\gamma}_f$; i.e., $\frac{dv_x}{dz} = \dot{\gamma}_f$. The magnetic field is vertical and uniform, noted as $\mathbf{H}_0$, which causes the particles to arrange into a vertically oriented chain.

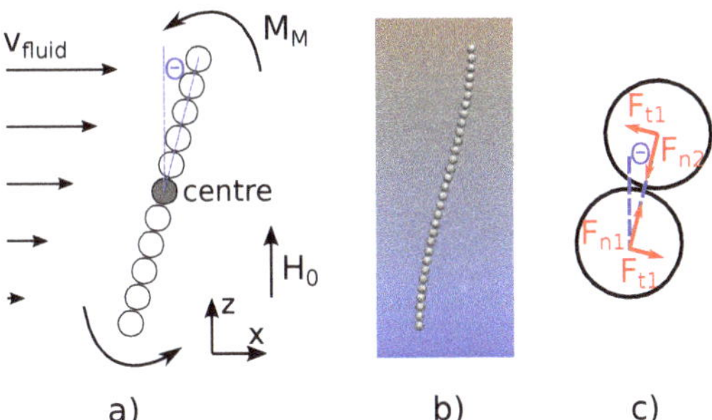

Figure 1. (**a**) Linear particle chain model in a fluid flow with homogeneous strain rate $\dot{\gamma}_f$. (**b**) Simulation result of the same problem with our OpenFOAM based solver. (**c**) Magnetic force between two neighboring particles on the bent chain.

Suppose that initially the chain is vertical and the particles have zero velocity. The fluid then starts to move the chain to the direction x due to the drag force. However, as the fluid has a strain rate, the drag force at the top of the chain is higher than at the bottom section, which causes the chain to start to rotate. As the chain rotates, a counter magnetic torque between the neighboring particle pairs appears due to the magnetic particle–particle interaction. This can be seen more clearly if Equation (8) is rewritten with its tangential and normal components:

$$\mathbf{F}_{M\_ij} = -\frac{3\mu_0 m_i m_j}{4\pi r_{ij}^4} \left[(3\cos^2\Theta - 1)\cdot \mathbf{e}_n + \sin 2\Theta \cdot \mathbf{e}_t \right] \tag{9}$$

where Θ is the angle between the magnetic field and the particle-pair direction; see Figure 1c. The magnetic moments can be expressed with the magnetic field. Based on Equation (4) the moments are proportional to the magnetic field; therefore, the pair force is $F_{M_{ij}} \propto \mathbf{H}_0^2$.

An excellent description of this force and the angle-dependency is presented in [14]. Here only the main consequences are collected based on [14]:

- The normal force $\mathbf{F}_{M\_ij\_n}$ between the particles is attractive until $\Theta \leq 54.74°$. This attractive force is the main reason for the chain formulation.
- The tangential force $\mathbf{F}_{M\_ij\_t}$ is 0 at $\Theta = 0$, and it approximately linearly increases with the increasing Θ at small angles.

The chain rotates in the non-homogeneous flow until the torque from the magnetic forces counter-balances the torque of the fluid. Finding the exact rotated shape which is shown in Figure 1b is complicated; therefore, in the follow the chain is considered linear, similarly to a model in [22]. Now the torque of the flow is calculated on the linear chain. First consider the case when the number of particles in the chain is odd. In this case the axis is fixed to the center particle; see Figure 1. We will concentrate to the effect of the upper section of the particles on the center particle. The torque of the ith particle's drag from the center particle is:

$$\mathbf{M}_{d_i} = \mathbf{F}_{drag} \times \mathbf{r} = (6\pi\mu_f r_p v_{f\_rel}) \cdot (i \cdot 2r_p) \cdot \cos\Theta \tag{10}$$

where $v_{f\_rel} = (2\cdot r_p i)\cdot \dot{\gamma}_f \cdot \cos\Theta$ as all particles have the same velocity, which is equal to the fluid velocity at the center due to the symmetry. In other words, the relative velocity between the particle and the fluid phase is zero at the center. The total torque of the fluid on the chain based on the last two equations is:

$$M_d = 2 \cdot \sum_{i=1}^{\frac{N-1}{2}} 6\pi\mu_f \dot{\gamma}_f r_p^3 \cdot (2i)^2 \cos^2\Theta. \tag{11}$$

The multiplication by 2 has to be done as the bottom section of the chain is also rotated.

The counter torque from the magnetic pair interactions between the neighboring particles can be calculated with Equation (9). For one particle pair the torque is caused by the tangential magnetic force pair, which is shown in Figure 1c. As there are $N - 1$ pairs in the chain, the total magnetic torque is:

$$M_m = (N - 1) \cdot \frac{3\mu_0 m_i m_j}{4\pi(2r_p)^4} \sin 2\Theta \cdot 2r_p \tag{12}$$

In the steady rotated chain the net torque is zero; therefore, $M_m = M_d$. By comparing Equation (11) with Equation (12) the tangent of the angle of the steady chain can be calculated as:

$$\tan \Theta = \frac{64\pi^2 \mu_f \dot{\gamma}_f r_p^6}{\mu_0 m_i m_j} \cdot \frac{1}{N - 1} \sum_{i=1}^{\frac{N-1}{2}} (2i)^2 \tag{13}$$

The presented equation shows that an increasing strain rate increases the angle. Increasing the magnetic field in turn decreases the angle, as $m_i m_j \propto \mathbf{H}_0^2$, and in this case the magnetic counter torque becomes higher. Longer chains have a higher rotation angle.

If N is even, i.e., there is no center particle, the torque of the drag in Equation (11) slightly changes, and the summation for i changes to $\sum_{i=1}^{\frac{N}{2}} (2i - 1)^2$. This has to be applied also for Equation (13). It should be noted that the long chains can be broken over a critical angle; that problem is presented in detail in [22]. In the current work this phenomenon is not investigated.

Using the approximation of $\sin 2\Theta \approx 2\Theta \approx 2 \tan \Theta$, in Equation (13) the magnetic torque for the chain can be rewritten as:

$$M_{chain} \approx 12\pi\mu_f \dot{\gamma}_f r_p^3 \sum_{i=1}^{\frac{N-1}{2}} (2i)^2 \tag{14}$$

when N is odd, and for the even case, the summation should be changed as described above. This equation shows that we can determine the magnetic torque of the chain from the chain length. The equation can be used to calculate the increased viscosity of the magnetic nanoparticle phase (see later). It can be noted that because of $\sum_{i=1}^{N} i^2 = \frac{1}{6}(2N^3 + 3N^2 + N)$ with the increasing chain length, the steady-state magnetic torque is increasing rapidly.

Finally, we also investigated the case when the chain is close to a horizontal wall and one of the chain ends is fixed to it. The fluid velocity is parallel to the wall and its profile is $v_x(z) = \dot{\gamma}_f \cdot z$, where z is the distance from the wall. The chain becomes bent due to the drag and the total magnetic torque can be identified with a similar derivation, which is shown above. In this case the total magnetic torque is approximately four times higher for a chain with N particles compared to Equation (14).

3. The Two-Phase CFD Model

A two-phase CFD model was elaborated to handle the nanoparticle aggregation procedure in the magnetic field. The main scope of the work was to create an Euler–Euler-based model, which is able to predict the shape of a macroscopic-sized nanoparticle aggregate in the magnetic field in a fluid flow.

The Euler–Euler-based model considers two-phases: the fluid (phase *b*) and the nanoparticle phase (phase *a*). The main issue is the presented micro-chain structure of the nanoparticles, i.e., how the particle–particle magnetic interactions can be treated in a continuous model. Our solution relies on the idea that the effect of the rotated chains in a non-homogeneous flow can be managed as an increased viscosity of the nanoparticle phase. The idea originated from [22]. It should be emphasized that the macroscopic treatment of the micro-chain structure is relevant, as the chain formulation plays the major role in the aggregation. Neglecting the particle–particle interaction in the two-phase model would mean that we do not see any aggregation in the simulation at all, or only at really slow flow rates. In the following section this novel model is presented in detail.

3.1. Governing Equations

The created numerical model is based on the `twoPhaseEulerFoam` solver in the open source CFD software OpenFOAM v5x [23]. The background of an earlier version of the solver is presented in [24].

In the model, each phase in a computational cell has its own volumetric fraction: α_a and α_b, i.e., $\alpha_a + \alpha_b = 1$ in every cell. The volumetric conservation equation for the mixture is

$$\nabla \cdot (\alpha_a \mathbf{U}_a + \alpha_b \mathbf{U}_b) = 0. \tag{15}$$

The conservation of the momentum for the fluid is:

$$\rho_b \alpha_b \frac{\partial \mathbf{U}_b}{\partial t} + \rho_b \alpha_b (\mathbf{U}_b \cdot \nabla) \mathbf{U}_b = $$
$$-\alpha_b \nabla p - \nabla \cdot \alpha_b \boldsymbol{\tau}_b + \mathbf{M}_b + \mathbf{S}_b. \tag{16}$$

where p is the pressure, $\boldsymbol{\tau}_b$ is the stress tensor, $\mathbf{M}_b$ represents the momentum transfer from the other phase and $\mathbf{S}_b$ denotes any other source terms. For the particle phase a similar equation can be applied. The stress tensor $\boldsymbol{\tau}$ for the fluid is

$$\boldsymbol{\tau}_b = \mu_b \left[(\nabla \mathbf{U}_b) + (\nabla \mathbf{U}_b)^T \right] - \frac{2}{3} \mu_b (\nabla \cdot \mathbf{U}_b) \mathbf{I}, \tag{17}$$

where $\mathbf{I}$ is the identity matrix. For the particle phase a similar form of the stress tensor is applied; however, the value of μ_a will be set by a specialized viscosity model considering the effect of the particle chains (see later).

The momentum transfer between the phases is based on the Wen–Yu drag model [25]. This model can be found originally in OpenFOAM for the Eulerian–Lagrangian libraries as a drag model for the particles. The drag force on one particle is calculated as

$$\mathbf{F}_{drag} = \frac{3}{4} V_p \cdot \mathrm{CdRe}\{\alpha_b \cdot Re\} \frac{\mu_b \alpha_b^{-3.65}}{d_p^2} (\mathbf{U}_b - \mathbf{U}_{particle}), \tag{18}$$

where V_p is the particle volume; Re is the Reynolds-number which is calculated as $Re = \rho_f \cdot \mathrm{mag}\{\mathbf{U}_{particle} - \mathbf{U}_b\} \cdot d_p / \mu_b$ and the function $\mathrm{CdRe}\{x\} = 24(1 + 0.15 x^{0.687})$. As in our case $Re << 1$, the CdRe term in Equation (18) will be approximately 24. It should be noted that the force is equal to the Stokes force when $\alpha_a \rightarrow 0$. Converting the particle force to a volumetric force density is done using the simplifications above:

$$\mathbf{M}_a = 18 \frac{\alpha_a}{\alpha_b^{3.65}} \frac{\mu_b}{d_p^2} (\mathbf{U}_b - \mathbf{U}_a), \tag{19}$$

which represents the momentum transfer between the two phases. The magnetic force density of the neodymium magnet on the aggregate phase can be determined from the particle magnetic force, which was shown in Equation (5). Based on this equation and the magnetic moment formula in Equation (2) the magnetic force density is

$$\mathbf{f}_m = \mu_0 \left(\frac{d_{core}}{d_p} \right)^3 \frac{3 \chi_p}{\chi_p + 3} \cdot \alpha_a \cdot \mathbf{H}_0 \cdot \nabla \mathbf{H}_0, \tag{20}$$

where the $(d_{core}/d_p)^3$ term is caused by the fact that only the core of the particle can be magnetized.

Finally, the phase fraction α_a field is updated by solving the following equation:

$$\frac{\partial \alpha_a}{\partial t} + \nabla \cdot (\mathbf{U}_a \alpha_a) = 0, \tag{21}$$

and $\alpha_b = 1 - \alpha_a$ [24].

3.2. Viscosity Model

In the micro-domain model, the rotated chains represent a magnetic torque-density in the aggregate. This phenomenon can be handled as an increased viscosity of the nanoparticle phase. To calculate the viscosity increase, first the magnetic torque density should be determined in the domain. For one chain, the sum of the magnetic torque was presented in Equation (12). The formula can be approximated by Equation (14). The latter provides the magnetic torque if the chain length is known. The problem is that in the two-phase model only the volumetric fraction α_a of the nanoparticle phase is known; i.e., there is no information about the chain lengths in the aggregate. It can be assumed that the chain length distribution depends on the volume fraction of the particles, and also on the magnetic field **H**. To overcome this problem, several micro-domain simulations were prepared with different nanoparticle concentrations in a given homogeneous magnetic field. In these simulations the particles initially were placed randomly, and then the self-aggregation appeared in the simulations due to the magnetic particle–particle forces. As a result of each simulation, the chain length distribution could be identified at the different concentrations; see more details in the Results section.

Knowing the chain length distribution in a domain, the magnetic torque density can be calculated as

$$\tau_m = \frac{1}{V_{domain}} \sum_{i=1}^{n_{chains}} M_{chain} \tag{22}$$

where the summation goes over all chains, and M_{chain} is calculated based on Equation (14). By performing micro-domain simulations at different nanoparticle concentrations, the $\tau_m\{\alpha_a\}$ torque density dependence can be identified. In the next step the magnetic torque density is embedded into the viscosity of the nanoparticle phase. Our numerical implementation is based on OpenFOAM's Herschel–Bulkey model. First, the strain rate of the nanoparticle phase is calculated explicitly from the previous time step as:

$$\dot{\gamma}_a = \sqrt{2\mathbf{D}:\mathbf{D}}, \tag{23}$$

where $\mathbf{D} = \frac{1}{2}(\nabla \mathbf{U}_a + \nabla \mathbf{U}_a^T)$ is the strain rate tensor, and $\mathbf{D}:\mathbf{D} = D_{ij}D_{ij}$; i.e., it notes the sum of the product of the tensor components. Then the viscosity of the nanoparticle phase is calculated as

$$\mu_a = \frac{\tau_m}{\dot{\gamma}_a}. \tag{24}$$

This model follows a relatively simple approach, as the effects of the chains are considered to be isotropic. This method was found to be numerically stable. However, the discussed torque density derivation is only valid when the direction of the chains corresponds to the dominant elements of the strain rate tensor. In our case this condition is roughly true, as at the aggregate-fluid boundary the flow is generally parallel to the boundary. Nevertheless, we intend to improve the model in this aspect in the future.

3.3. Numerical Model

In the original `twoPhaseEulerFoam` solver the PIMPLE (merged SIMPLE + PISO) segregated algorithm is used to solve the conservation equations. The description of the SIMPLE and PISO algorithms can be found, e.g., in [26]. The phase fraction Equation (21) is solved with OpenFOAM's MULES (multi-dimensional limiter for explicit solution) algorithm, which is tailored to guarantee the boundedness of the phase fractions.

In viscoelastic problems the SIMPLEC algorithm is preferred [27]. A detailed explanation of the algorithm is presented for one phase in [26,27], and our following description relies on these works. Although the algorithm exists in OpenFOAM, in the two-phase solver `twoPhaseEulerFoam` it has not been implemented yet.

In our work the SIMPLEC algorithm is embedded into the two-phase solver as follows. Based on the conservation of the momentum, in a new time step the predicted velocity $\mathbf{U}^*$ for an arbitrary cell in the mesh can be calculated as:

$$A_p\mathbf{U}_p^* + \sum_l A_l\mathbf{U}_l^* = \mathbf{Q} - \alpha\nabla p^{old}, \tag{25}$$

where $\mathbf{U}_p$ is the cell-center velocity value, and $\mathbf{U}_l$ denotes the neighboring cell velocities. $\mathbf{Q}$ means any other explicit source terms, while p^{old} is the pressure from the previous time step. The equation can be written for both phases. As in OpenFOAM the following notations are used:

$$\mathbf{H} = \mathbf{Q} - \sum_l A_l\mathbf{U}_l^*, \quad H_1 = -\sum_l A_l, \quad A = A_p; \tag{26}$$

with those notations the equation can be expressed as

$$AU_p^* = \mathbf{H} - \alpha\nabla p^{old}$$
$$\mathbf{U}_p^* = \frac{\mathbf{H}}{A} - \frac{\nabla p^{old}}{A}. \tag{27}$$

The problem with the guessed velocity fields $\mathbf{U}_a^*$, $\mathbf{U}_b^*$ is that they do not fulfill the volumetric continuity Equation (15). Therefore, a correction for the velocities $\mathbf{U}_a^{new} = \mathbf{U}_a^* + \mathbf{U}_a'$ and $\mathbf{U}_b^{new} = \mathbf{U}_b^* + \mathbf{U}_b'$ is needed with a new pressure field $p^{new} = p^{old} + p'$, whose fields satisfy both Equations (15) and (25):

$$A_p(\mathbf{U}_p^* + \mathbf{U}_p') + \sum_l A_l(\mathbf{U}_l^* + \mathbf{U}_l') = \mathbf{Q} - \alpha\nabla(p^{old} + p') \tag{28}$$

and

$$\nabla \cdot \left[\alpha_a(\mathbf{U}_a^* + \mathbf{U}_a') + \alpha_b(\mathbf{U}_b^* + \mathbf{U}_b')\right] = 0. \tag{29}$$

subtracting from Equation (25) to (28) the velocity correction can be expressed with the pressure correction as

$$A_p\mathbf{U}_p' + \sum_l A_l\mathbf{U}_l' = -\alpha\nabla p'. \tag{30}$$

In the SIMPLEC algorithm the following approximation is used

$$\mathbf{U}_p' = \frac{1}{\sum_l A_l}\sum_l A_l\mathbf{U}_l'; \tag{31}$$

i.e., the cell velocity correction is approximated to be the weighted mean of the neighbor corrections. Substituting it into Equation (30):

$$\mathbf{U}' = -\frac{\alpha}{A_p + \sum_l A_l}\nabla p' = -\frac{\alpha}{A - H_1}\nabla p'. \tag{32}$$

The volumetric continuity Equation (29) can be rewritten by expressing the velocity corrections with the pressure correction, resulting in the pressure equation. By solving this equation, the pressure correction, i.e., the new pressure field, can be determined. Having these values the velocity corrections for each phase can be calculated using Equation (32). With these the continuity equation, Equation (29), can be rewritten using OpenFOAM's notation as

$$\nabla \cdot \left[\alpha_a\left(\frac{\mathbf{H}_a}{A_a} - \frac{\alpha_a\nabla p^{old}}{A_a}\right) + \alpha_a\mathbf{U}_a'\right]$$
$$+\nabla \cdot \left[\alpha_b\left(\frac{\mathbf{H}_b}{A_b} - \frac{\alpha_b\nabla p^{old}}{A_b}\right) + \alpha_b\mathbf{U}_b'\right] = 0. \tag{33}$$

by expanding the velocity corrections based on Equation (32):

$$
\nabla \cdot \left[\alpha_a \left(\frac{\mathbf{H}_a}{A_a} - \frac{\alpha_a \nabla p^{old}}{A_a} \right) - \frac{\alpha_a}{A_a - H_{1_a}} \nabla p' \right]
$$
$$
+ \nabla \cdot \left[\alpha_b \left(\frac{\mathbf{H}_b}{A_b} - \frac{\alpha_b \nabla p^{old}}{A_b} \right) - \frac{\alpha_b}{A_b - H_{1_b}} \nabla p' \right] = 0.
\tag{34}
$$

Finally by expressing the pressure correction as $p' = p^{new} - p^{old}$ the final form of the pressure equation is

$$
\nabla \cdot \left[\alpha_a \left(\frac{\mathbf{H}_a}{A_a} \right) + \alpha_b \left(\frac{\mathbf{H}_b}{A_b} \right) \right] - \nabla \cdot \left[\left(\frac{\alpha_a^2}{A_a} + \frac{\alpha_b^2}{A_b} \right) \nabla p^{old} \right] + \nabla \cdot \left[\left(\frac{\alpha_a^2}{A_a - H_{1_a}} + \frac{\alpha_b^2}{A_b - H_{1_b}} \right) \nabla p^{old} \right] =
$$
$$
\nabla \cdot \left[\left(\frac{\alpha_a^2}{A_a - H_{1_a}} + \frac{\alpha_b^2}{A_b - H_{1_b}} \right) \nabla p^{new} \right]
\tag{35}
$$

As it was discussed previously, after the new pressure field is calculated, the corrections for the velocities $\mathbf{U}'_a$, $\mathbf{U}'_b$ are obtained by using Equation (32).

In the numerical code the momentum equations (Equation (25)) for both phases were constructed, from which the $\mathbf{H}_{phase}$ and A_{phase} terms were obtained. In case of the MNP phase, the changeable viscosity μ_a is pre-calculated based on Equation (24), where the local strain rate of the MNP phase $\dot{\gamma}_a$ is determined from the velocity field of the previous time-step $\mathbf{U}_a^{old}$. In the momentum equation of the MNP phase the magnetic force of the magnet in Equation (20) is also included. In the original twoPhaseEulerFoam solver a face-based momentum equation formulation can be chosen, which was also used in our case. From the momentum predictor Equation (27) the $\mathbf{H}$ and A terms were interpolated to the cell faces as $\mathbf{H}_f$ and A_f, and then the guessed phase fluxes through the face are constructed as $\phi^*_{phase} = \frac{1}{A_{phase_f}} \mathbf{H}_{phase_f} \cdot \mathbf{S}_f$ for both phases, where the f symbol represents the face interpolated values. $\mathbf{S}_f$ is the cell face vector, whose magnitude is equal to the face area. According to the original solver the momentum transfer between the phases and the temporal derivative in the momentum equation are included at the calculation of the predicted fluxes. All of the equations after the momentum predictor are expressed with the fluxes.

Another important setting is the boundary condition for the pressure at the fixed velocity boundaries, like at the wall. At the wall both phase velocities are zero; i.e., the total flux $\phi = \alpha \phi_a + \beta \phi_b$ going through the wall should be also 0 at the new time step. In the original solver this is achieved by adjusting the pressure gradient boundary condition. This method's modified version was used in our case. The formula of the total flux with ∇p^{new} is represented in the pressure equation, where its divergence is set to zero. Based on it, and the known value of the total flux at the boundary the pressure gradient ∇p^{new} is updated before solving the pressure equation.

4. Experimental Setup

The aggregation procedure was also investigated experimentally. First an appropriate microfluidic device and the MNP suspension were prepared. Then the aggregation procedure was investigated at different flow rates. Finally, thermal measurements were performed both with MNP-filled and MNP-free devices.

4.1. The Microfluidic Chip

In order to verify our model with measurements, first we created a microfluidic structure. The requirement for the device was to be opaque, enabling us to observe the nanoparticle aggregate with a microscope during the aggregation. To reach this goal a unique microfluidic chip was manufactured, which basically consisted of a thin silicon layer between two opaque acrylic plates. The acrylic layers gave the mechanical stability while the silicon layer ensured the sealing. The sketch of the device is shown in Figure 2, while a photo of the device during a measurement is shown in Figure 3. The microfluidic channel was a long rectangular domain, which was cut in the silicon layer. The channel had a uniform width of 4 mm and its total length was 85 mm. The thickness of the silica layer, i.e., the height of the channel, was $h \approx 0.3$ mm. The three layers were held together by densely placed screws around the channel. The inlet and outlet ports of the channel were drilled through one of the acrylic plates. All the three layers, including the shape of the microchannel, were prepared with laser cutting. The final structure was found robust and stable even after a several days of usage.

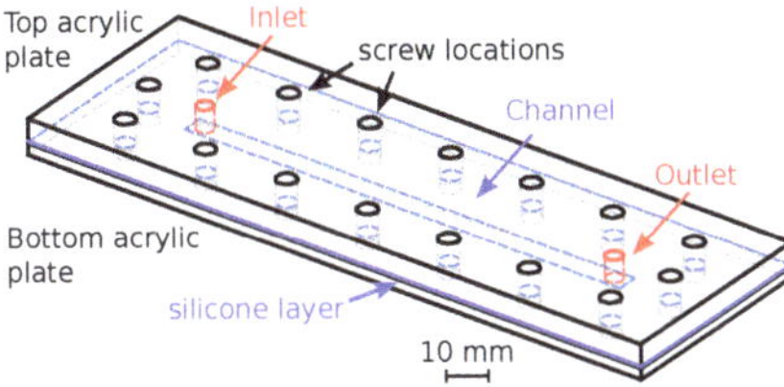

Figure 2. The schematic of the microfluidic chip. The rectangular channel was cut in a thin, approximately 0.3 mm thick silicone layer. The top and bottom surfaces were encapsulated by 4 and 2 mm thick acrylic plates, respectively. The layers were held together by screws. The channel had a width of 4 mm. The inlet and outlet ports were drilled through the top acrylic plate.

Figure 3. The photo shows the microfluidic device during a measurement. The MNP suspension continuously flowed thorough the channel from the inlet. A neodymium magnet was positioned under the center of the channel and the nanoparticles aggregated in its magnetic field. The aggregation procedure was monitored with a microscope.

4.2. Magnetic Nanoparticles

The nanoparticles were made of a magnetite core (Fe_3O_4) which had a diameter of 210 nm. The core was covered with a 20 nm thick silicon-dioxide shell, which resulted in a final particle diameter of $d_p = 250$ nm.

The magnetization curve of the MNPs was investigated in several papers [28–30]. Based on these works we concluded that the saturation magnetization of the nanoparticles can decrease as the size decreases with respect to the bulk magnetite saturation magnetization, which is $\mathbf{M}_{s\_bulk} = 92\,\mathrm{emu} \cdot \mathrm{g}^{-1} = 483\,\mathrm{kA} \cdot \mathrm{m}^{-1}$ [31]. In our case, however, we were far from the saturation, and therefore in the modeling parts the core was approximated to be linear with the bulk susceptibility of $\chi = 2.8$ [31].

The nanoparticle suspension was created by mixing the nanoparticles in distilled water. We experienced that the MNP self-aggregation can be significant in the suspension. Polyethylene glycol (PEG) and a small amount of detergent were added to avoid this phenomenon. Using a more diluted suspension, ultrasonic bath and pre-heating of the suspension up to $T = 60\,^{\circ}\mathrm{C}$ also helped to fragment most of these aggregates.

4.3. Measurement Setup

In the measurements the nanoparticle suspension was forced to flow through the microfluidic channel with different flow rates. The flow control was achieved by using a syringe pump (Fresenius Vial SA Pilote C). A cylindrical neodymium magnet was positioned under the center of the channel with the diameter of $d_m = 5\,\text{mm}$ and height of $h_m = 5\,\text{mm}$. The magnet's remanent magnetization was $B_r \approx 1.4\,\text{T}$ according to its datasheet.

To observe the aggregation in the magnetic field the device with the magnet was put under a microscope (Olympus BX51M); see Figure 3.

5. Experimental Results

5.1. Measurement of the MNP Aggregate in the Magnetic Field

The MNP aggregation in the microfluidic channel was studied at different flow rates; the measurement setup is shown in Figure 3. As it was observed, the aggregation started at the bottom wall close to the magnet. According to the expectations, the MNPs were accumulated in chain-like formulations, which were nearly parallel to the magnetic field. The aggregated chains became longer and denser over the time until a steady-state shape was achieved. Photos of the aggregation are shown in Figure 4. It was also noticed that shorter free chains were also developing in the fluid flow, which became longer, as the suspension passed thorough the magnet.

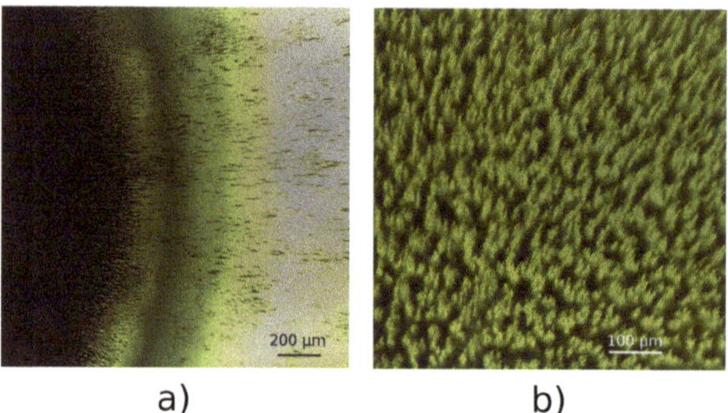

a) b)

Figure 4. (**a**) Microscopic image at the beginning of the aggregation procedure at the end of the aggregate, next to the bottom wall. The contour of the cylindrical magnet is also visible. Most of the MNPs aggregated into a chain-shaped forms, which were nearly parallel to the magnetic field. The fluid flowed from the left to the right. (**b**) A magnified dark-field image from the top of the steady-state aggregate close to its horizontal center. The ends of the individual aggregate groups are in focus. The fluid flowed from the left to right, which caused the bent shape of the aggregate ends.

The thickness of the nanoparticle aggregate over the magnet was seemingly non-homogeneous. It rather looked to have a spherical surface, deformed by the fluid flow. The maximum thickness of the aggregate was at the center, while at the side walls fewer nanoparticles were aggregated.

To have a comparison of the numerical model and the experiments, the gap between the thickest part of the steady aggregate and the top wall was measured at different flow rates. This was done by manually adjusting the focal point of the microscope. We measured the vertical movement of the microscope between the focus of the top of the aggregate and the top wall of the channel. The channel height was also determined with a similar method, and resulted in $h_{channel} \approx 320\,\text{µm}$. The depth of focus of the objective was $15\,\text{µm}$.

The aggregation procedure was measured at the different flow rates. Obviously at lower flow rates the quantity of the aggregate was larger because of the reduced drag. In these cases the horizontal shape was more evolved while the aggregate thickness was also higher. The gap sizes between the

MNP aggregate and the top wall are presented in Table 1. Note that at the smallest flow rate the top of the aggregate reached the top wall. It should be noted, however, that this was achieved only for a small part at the center of the aggregate, while a small gap still remained for the fluid next to the side walls.

Table 1. Measurement results of the gap sizes between the top of the MNP aggregate at the center and the top wall at different flow rates. At the lowest flow rate only a small part of the MNP aggregate reached the top wall at the center. In all cases the aggregate was thinner at the side walls than at the center. In the $Q = 300$ mL $\cdot$ h^{-1} case the aggregate thinning was more representative. The accuracy of the measurement was estimated to ± 15 µm.

Q (mL $\cdot$ h^{-1})	Gap Size (µm)
50	0–45 µm
100	60 µm
150	75 µm
200	90 µm
300	90–115 µm

Finally the volume fraction of the dense nanoparticle phase was measured. A flask was filled with nanoparticle suspension containing 20 mg of MNP. Then a big neodymium magnet was positioned under the flask, which resulted the MNPs to aggregate to the bottom. After this the MNP-free solvent was removed and the mass of the remaining aggregate with the encapsulated water was measured. Based on this measurement the estimated maximum volume fraction of the nanoparticles was $\alpha_{max} = 11\%$.

5.2. Thermal Measurements

Thermal measurements with MNP aggregate were also performed at different flow rates, as we were interested in the effect of the MNP aggregate on the thermal characteristics of the microfluidic channel. Pre-heated water was flown through the device first in a MNP-free channel (i.e., when there was no aggregate), and then in an MNP-filled channel. To measure the temperature loss in the chip, temperature sensors were fixed at the ends of the inlet and outlet tubes next to the device; see Figure 5. In the MNP-free channel case the temperature drop was measured at various flow rates. Then the channel was filled with the aggregate, and the temperature loss was measured again. In this case the fluid was the MNP suspension with a concentration of $c = 2$ mg $\cdot$ mL^{-1}. The thermal measurement at all flow rates started after the steady state shape of the aggregate was achieved. The temperature measurement procedure was started when the inlet-outlet temperature became steady and the given temperature data are the averages of 90 s of measurement.

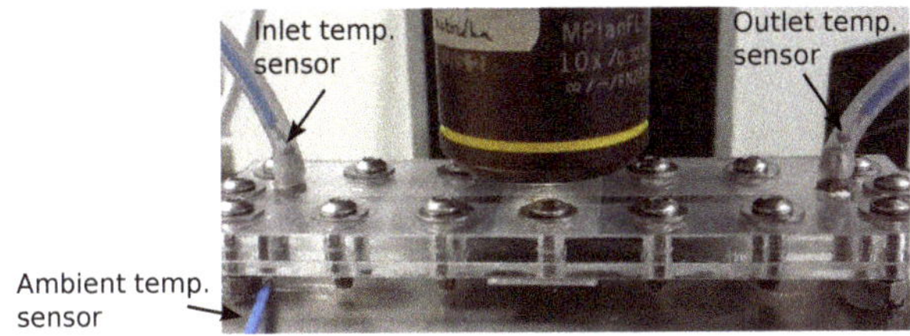

Figure 5. The figure shows the microfluidic chip prepared for the thermal measurement. Temperature sensors were installed into the inlet and outlet tubes next to the chip. An additional sensor was positioned under the channel to monitor the ambient temperature. The chip with the three sensors sank into icy water during the thermal measurement.

As the expected temperature differences between the two cases are not too high, it was important to perform the experiment in an environment with a fixed temperature. This was achieved by sinking the chip with the tubes in icy water, which provided a constant ambient temperature of $T_a = 0\,°C$. The measured temperatures and their ratios are shown in Table 2. The different inlet temperatures at the different flow rates were caused by the fact that the suspension was also cooled down in the inlet tube, whose end close to the chip was also in the icy water. As was expected, the temperature drop in the chip was higher in case of low flow rates. Moreover, the temperature ratio was consistently smaller in the case of the MNP aggregate's presence than in case of the MNP-free channel. This shows that the MNP aggregate can be used in the microchannel to enhance the heat transfer. At this stage of the research we cannot determine why the difference was evolved in the temperature. It could be the result of the elevated heat conductivity, the percolation of MNP in magnetic field or the changed flow field. Investigation of this issue can be one focus of future research.

Table 2. Thermal measurement results with the channel. The table shows the measured inlet and outlet temperatures and their ratio at the different flow rates in the MNP-free and the MNP-filled cases. The accuracy of the K-type thermocouple was $\pm 0.5\,°C$.

Flow Rate	MNP Free Channel			MNP Filled Channel		
Q (mL $\cdot$ h^{-1})	T_{in} (°C)	T_{out} (°C)	T_{out}/T_{in} (%)	T_{in} (°C)	T_{out} (°C)	T_{out}/T_{in} (%)
100	46.6	18.1	38.8	50.7	18.8	37.1
50	42.3	8	18.9	41.4	6.2	15.0
25	25.1	1.7	6.8	26.9	1.5	5.6
10	11.4	1.6	14.0	16.1	0	0

6. Numerical Results

6.1. Modeling the Chain Length Distribution at Different Concentrations

To calculate the magnetic phase's increased viscosity, the magnetic torque density needs to be identified at different nanoparticle concentrations. It is only an approximation, but by using Equation (22) with Equation (14) it is sufficient to know the chain length distribution to determine the torque density. To calculate it, several micro-domain simulations were accomplished. The numerical solver of the micro-domain model was implemented based on OpenFOAM's discrete particle method solver DPMFoam. The original solver was extended to calculate the magnetization of the particles and the magnetic interactions between them. In the simulation the particle mass and moment of inertia was calculated by setting the magnetite core density to $\rho_{core} = 5250\,\text{kg} \cdot \text{m}^{-3}$ and the silicon-dioxide shell density to $\rho_{shell} = 2196\,\text{kg} \cdot \text{m}^{-3}$. The simulation domain was a $20 \times 20 \times 20\,\mu\text{m}$ cube, where initially randomly placed particles were positioned. A homogeneous vertical magnetic field was set in the domain. Due to the magnetic field, the particles aggregated into chains, whose direction was parallel to the external magnetic field. The simulations were done at different nanoparticle concentrations: $1 \times 10^{-4}, 2 \times 10^{-4}, 5 \times 10^{-4}, 1 \times 10^{-3}, 2 \times 10^{-3}, 5 \times 10^{-3}$ and 1×10^{-2}. The formed chains for $\alpha_p = 1 \times 10^{-3}$ are shown in Figure 6.

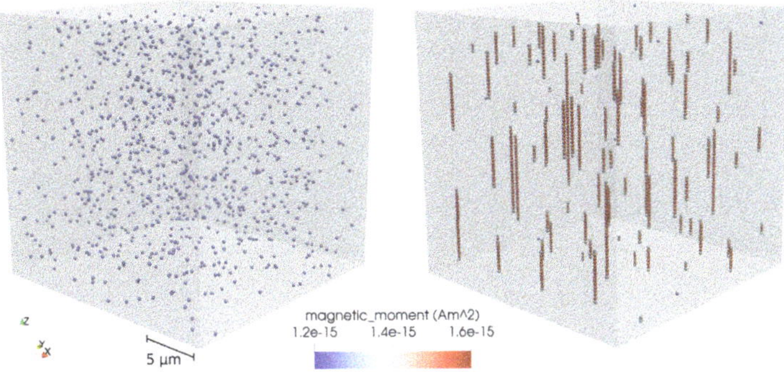

Figure 6. Aggregation procedure at $\alpha_a = 1 \times 10^{-3}$ concentration. The left picture shows the initial random particle positions, the right one shows the self-aggregated structure at $t = 0.14$ s. The magnetic field was $\mathbf{H} = 180\,\text{kA} \cdot \text{m}^{-1} \cdot \mathbf{e}_z \approx 0.23\,\text{T} \cdot \mathbf{e}_z$. The particles are colourised by their magnetic moment values.

To identify the chain length distribution a custom program was written in C++ to post-process the simulation data. In the program the close particles were collected into one group. The chain length was determined based on the difference between the highest and lowest particle positions. This way the chain length distribution and then the torque-density based on Equations (14) and (22) can be calculated algorithmically. By setting the fluid strain rate to $\dot{\gamma}_b = 500\,\text{s}^{-1}$ and its viscosity to $\mu_b = 8.9 \times 10^{-4}\,\text{Pa} \cdot \text{s}$ (water's viscosity) in Equation (14) the average torque densities at the different concentrations were calculated and are shown in Table 3.

Table 3. Calculated torque densities at different particle concentrations at the fluid strain rate $\dot{\gamma}_b = 500\,\text{s}^{-1}$. The torque densities were computed from Equation (22), where each particle chain's torque was calculated by Equation (14). The particle chain length distributions at the different concentrations were determined based on the micro-domain simulations.

α_a	τ_m (N $\cdot$ m^{-2})
1×10^{-4}	2.13×10^{-4}
2×10^{-4}	2.02×10^{-3}
5×10^{-4}	2.69×10^{-2}
1×10^{-3}	2.19×10^{-1}
2×10^{-3}	2.00
5×10^{-3}	7.23
1×10^{-2}	16.79

It can be seen that the torque density is rapidly increasing at low concentrations with the increasing concentration. Increasing α_a from 1×10^{-4} to 1×10^{-3} results in approximately three orders of magnitude of change in the torque-density. The reason for this is that at $\alpha_p = 1 \times 10^{-4}$ the particle cloud is sparse, and most of the particles are either single or only short chains are formed, containing 2 or 3 particles. The formed chains and particles are far from each other due to the low concentration. In turn at $\alpha_p = 1 \times 10^{-3}$ most of the particles are arranged into chains, whose average length is higher than at low concentrations; see Figure 6. Moreover, as it was shown previously the torque of the chains increases approximately with N^3, which results in a rapid change with the increasing chain lengths. The high values of the torque density appear as an increased viscosity of the nanoparticle aggregate, which will help the aggregation.

In the highest investigated concentration $\alpha_p = 0.01$ the length of the longest chains reached the size of the domain, which was 20 µm. Over that concentration we assume a linear change of the torque density with the concentration. It should be noted that the evolving chain formulation needs some time to arrange. At lower particle concentrations the time for the arrangement is higher, as initially the particles are further away from each other and the particle–particle force rapidly decreases with distance; see Equation (8). Moreover, the arrangement speed is obviously dependent on the magnetic field strength, as the particle–particle force is approximately proportional with H^2. To find the time when most of the chains are in a stable state, one way is to monitor the total kinetic energy of the particles [14]. After an initial peak this value decreases with time as the particles are arranging into chains. In our case we ran the lowest particle concentration case until $t = 0.65\,$s. Similarly to [14], by investigating the total kinetic energy of the system we found that at higher particle concentrations the aggregation procedure needed less time.

6.2. Simulation Setup for the Aggregation in the Microchannel

The aggregation in the microchannel was investigated experimentally with a neodymium magnet. In this section the measurement is modeled with the two-phase solver. The local magnetic torque density is determined from Equations (22) and (14), where the chain length distribution term $\sum(2i)^2$ is calculated with a piecewise-linear interpolation, fitted to the 7 results of the micro-domain simulations.

In Equation (14) the fluid's local strain rate $\dot{\gamma}_b$ is also needed, which can be identified from the velocity field of the fluid. Due to the magnetic field dependency of the chain length distribution, the torque density was set to $\tau_m \propto H^2$, as the magnetic force between the particles is proportional with the square of the magnetic field. It should be noted, however, that this dependency for the chain-length distribution is heuristic and needs further investigation. After the local torque density was calculated, the viscosity of the nanoparticle phase is set as the ratio of the torque density and local strain rate $\dot{\gamma}_a$; see Equation (24).

A two-dimensional simulation was prepared, assuming planar flow in the channel. Although treating the problem in 2D leads to some inaccuracies about the aggregate shape, the computational cost is significantly reduced compared to a 3D case. The simulation domain was a $16\,\text{mm} \times 0.32\,\text{mm}$ rectangular domain. The sketch of the simulation domain is shown in Figure 7. At the bottom wall, which was close to the magnet the local magnetic torque density, the MNP phase viscosity was increased to a four-times-higher value compared to Equation (14) according to the discussion of the linear chain model. The fluid phase was water with the density of $\rho_b = 997\,\text{kg} \cdot \text{m}^{-3}$ and the viscosity of $\mu_b = 8.9 \times 10^{-4}\,\text{Pa} \cdot \text{s}$.

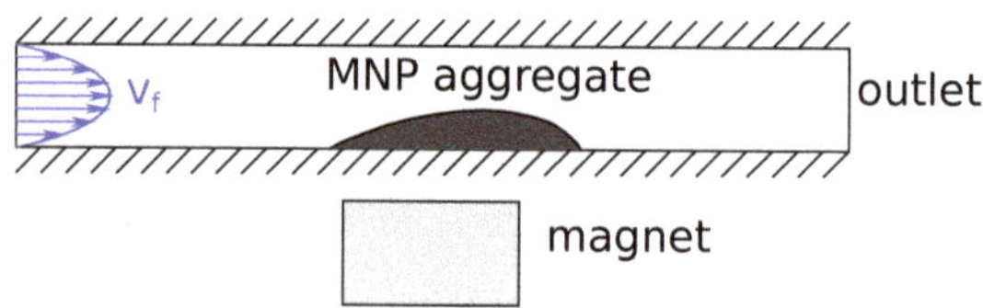

Figure 7. Sketch of the simulation of the microchannel. The fluid with the nanoparticles flows from the left to the right. The fluid has a parabolic velocity profile. The neodymium magnet is positioned under the center part of the magnet. As the particles arrive in the field of the magnet from the left, they start to aggregate at the bottom wall.

The magnetic field of the neodymium magnet was modeled with a modified version of OpenFOAM's `magneticFoam` solver. We extended the code to work on a 2D axisymmetric mesh, as the field of the cylindrical magnet is axisymmetric. This way a higher mesh resolution was achieved with the same computational cost compared to a 3D simulation. The axisymmetric field was mapped to the microchannel using custom code, which was implemented using OpenFOAM's interpolation classes. The interpolated field in the channel is shown in Figure 8. The magnet was positioned under the center of the simulation domain. The $\mathbf{H} \cdot \nabla \mathbf{H}$ term was also calculated, which was needed to identify the magnetic force on the nanoparticle phase in Equation (20).

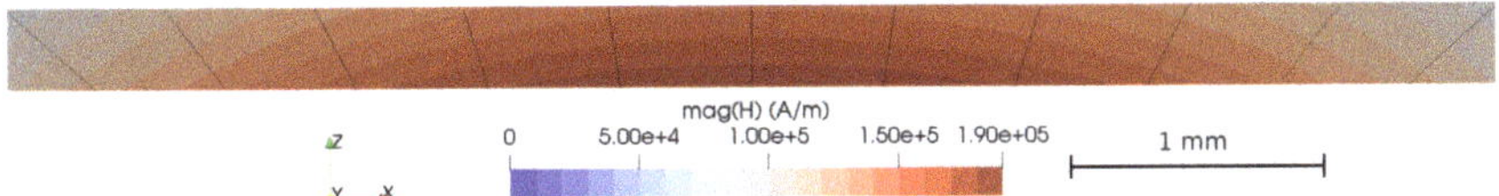

Figure 8. The calculated magnetic field in the center part of the microchannel. The peak values of the field reach the value of $H = 180\,\text{kA} \cdot \text{m}^{-1}$. The lines of induction are also shown.

In the first simulation the fluid and nanoparticle phase velocities at the inlet were set to be parabolic, with a flow rate of $Q = 150$ mL $\cdot$ h^{-1}, which was one of the used flow rates in the measurements. This flow rate represents an average velocity of $v_f = 3.26\,\text{cm} \cdot \text{s}^{-1}$ in the channel. Zero velocity boundary condition was set for the bottom and top walls for both phases, while at the outlet we used OpenFOAM's `inletOutlet` type boundary condition. The pressure was set to $p = 0$ at the outlet. At the other boundaries, the gradient of the pressure was set dynamically to fulfill an overall volumetric zero flux before solving the pressure equation, which was discussed in the theoretical section. The volume fraction of the MNP aggregate was set to $\alpha_{inlet} = 2 \times 10^{-4}$ in the simulation, which means a higher concentration compared to the measurements. This was done in order to speed up the aggregation and reduce the computation times. Based on Table 3, our assumption was that the magnetic torque density is so small at this volume fraction compared to the values at the higher concentrations that this artificial increase will not alter the results of the simulation. Based on [27,32] the convective terms in the momentum equations were discretized with the CUBISTA high-resolution scheme.

The numerical mesh was created with OpenFOAM's mesh generator `blockMesh` and it consisted of 86,400 hexahedral elements. It is important to note that the appropriate vertical mesh resolution was crucial to detecting the aggregation. In our case it was set to $\Delta z = 2.67\,\mu\text{m}$. Using a rough vertical mesh resolution, e.g., $\Delta z = 20\,\mu\text{m}$ caused the aggregation procedure to be limited to the fluid layers close to the bottom wall. The horizontal mesh resolution was $\Delta x \approx 14\,\mu\text{m}$ at the center part, while it was less dense at the inlet/outlet, as at these parts no aggregation was expected.

6.3. Results of Modeled Aggregation in the Microchannel

The simulation results showed the aggregation of the nanoparticle phase over the magnet in agreement with the measurements. The main cause of the aggregation was the increased viscosity of the nanoparticle phase. The phase fractions during the aggregation are shown in Figure 9. The aggregation started from the bottom wall over the magnet, similarly to the experiments. Then the aggregate started to grow in both vertical and horizontal directions. As the passage over the aggregate was narrowing over the time, the velocity was increasing here for both phases (see Figure 10), which made the aggregation for the incoming nanoparticles more difficult. Finally, a maximum quantity state was reached, where the gap size did not decrease any more. This means that at the aggregate border there was a balanced state between the drag force and the effect of the increased viscosity. The minimum gap between the top of the aggregate and the top wall was found to be $s \approx 69\,\mu\text{m}$, which is in agreement with the measurement result of $s_{meas} = 75\,\mu\text{m}$; see Table 1.

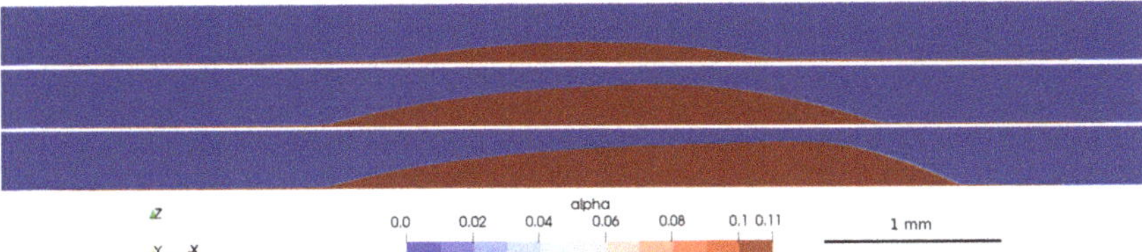

Figure 9. Nanoparticle phase evolution over the magnet during the aggregation procedure at the flow rate of $Q = 150$ mL $\cdot$ h^{-1}.

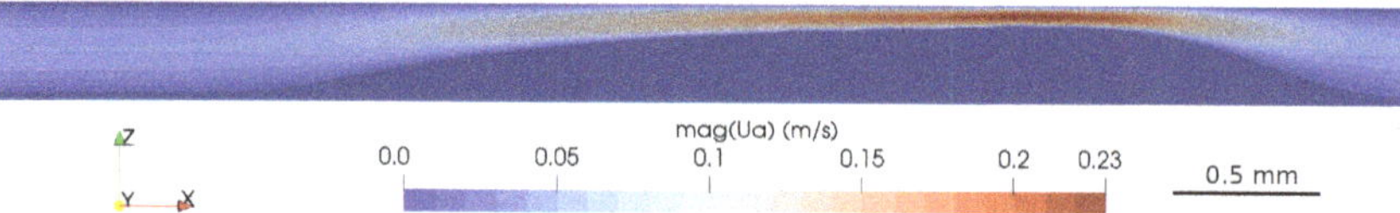

Figure 10. The velocity field of the nanoparticle phase at the third shown state of the aggregate in Figure 9. The aggregated part has nearly zero velocity, while the fluid with the non-aggregated particle phase bypasses it over the narrow passage.

Simulations were made done at the other flow rates which were used in the measurements. The shapes at the different flow rates are shown in Figure 11. The values of the gap thicknesses are shown in Table 4. At the higher flow rates the gap is larger, as the increased drag force results in less nanoparticles to aggregate.

In the simulations we found a dynamic balance rather than a steady shape of the final aggregate. This means that when the aggregate has reached its maximum quantity a small amount from its end was released, and then the filling procedure was repeated until the maximum quantity state was reached again. In Figure 11 the maximum quantity states are shown.

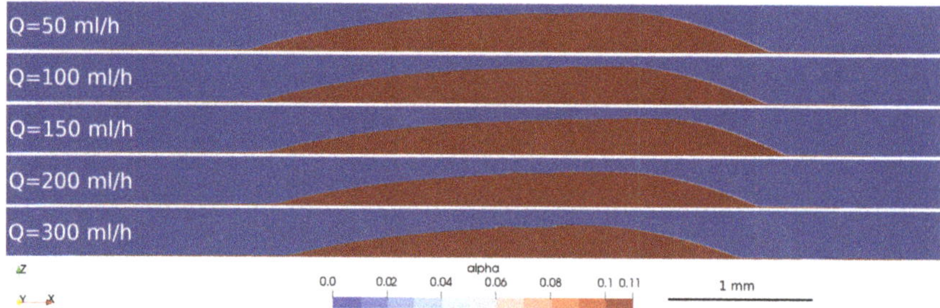

Figure 11. The simulated aggregate shapes at the different flow rates. As the flow rate is increasing, less MNPs are able to aggregate over the magnet.

Table 4. The simulated gap heights over the aggregate at different flow rates.

Flow rate (mL · h^{-1})	Simulated Gap (μm)
50	53
100	67
150	69
200	85
300	101

7. Comparison of the Experimental and Numerical Results

The measured and simulated gap heights and their differences at different flow rates are shown in Table 5. It can be seen that the simulated gap heights, and their change with the increasing flow rate corresponds to the measurement results, which shows that the proposed viscosity model relying on the micro-domain simulations can be used to solve such problems. It should be noted, however, that in the measurements at the side walls the aggregate thickness was smaller than at the center. This could be one reason for the difference between the experimental and simulated values.

Table 5. Gap heights over the aggregate at different flow rates in the simulation. The differences between the measured and simulated gap values based on Table 1 are also presented.

Flow Rate (mL · h^{-1})	Measured Gap (µm)	Simulated Gap (µm)	Difference (µm)
50	0–45 µm	53	8–53
100	60 µm	67	7
150	75 µm	69	6
200	90 µm	85	5
300	90–115 µm	101	0-14

Further Work

The presented two-phase solver is based on the results of the micro-domain simulations. We plan to improve the model by further investigating the magnetic field dependency of the chain distributions in the micro-domain. It should be noted, however, that to identify the exact chain distribution in the microchannel for a given part seems to be a complex problem, as it depends not only on the current MNP concentration and magnetic field, but also on other parameters that would be identified in a more detailed model. The difference between the theoretical micro-domain particles and the real MNPs may also lead to some deviations; i.e., in the model all particles have similar size, while in the reality the particles have a size distribution. The linear chain model itself is also not fully accurate, as is shown in Figure 1. A possible method to bypass the linear chain model is to insert the zero-flow chain distributions in a fluid flow with a constant strain rate, and then monitor the total magnetic torque represented by the magnetic forces between the particles. This way the magnetic torque density can be directly identified from the micro-domain simulation. Moreover, in this case the chain–chain interactions are also included. The viscosity model can be also improved, as was discussed previously. The model can be extended to include the heat transfer, which would be beneficial for the thermal-based applications.

8. Conclusions

In this work a specialized two-phase CFD model and a corresponding solver are presented that are capable of modeling the nanoparticle aggregation in a magnetic field. The model relies on the results of micro-domain simulations, where each particle is modeled separately. These simulations showed that the main reason for the aggregation in micro-scale is the magnetic particle–particle interaction which causes the particles to arrange into chains. The chains in a non-homogeneous fluid flow are bent, resulting in an additional torque-density in the fluid. To investigate the phenomenon a linear chain model is created. Using the model and the particle chain distributions at different MNP concentrations the local torque density of the chains in the fluid flow can be identified. The torque-density calculation was built later in the two-phase model as an increased viscosity. The results show that the torque-density increases rapidly with the increasing concentration of the nanoparticles.

After the micro-domain simulations the two-phase CFD model is presented, wherein the nanoparticles are treated as a second phase in the fluid. Using the two-phase method the simulation of the nanoparticle aggregate became possible in the macro scale. The effect of particle chains in the magnetic field is treated as an increased viscosity. A dedicated solver was created based on the OpenFOAM solver twoPhaseEulerFOAM. In our code the SIMPLEC algorithm was embedded into the two-phase solver, whose solution steps are discussed in detail.

Besides the numerical model, measurements were also performed to verify the simulation results. A microfluidic chip was prepared, containing a wide microchannel. A neodymium magnet was positioned under the channel and a nanoparticle suspension was flown through the channel. The aggregate was formed in the field of the magnet, which was observed with a microscope during the measurement. The aggregation procedure was observed at various flow rates. In all cases the aggregation started at the channel wall close to the magnet and later expanded in both the horizontal and vertical directions until a steady state shape was achieved. The maximum aggregate thicknesses at the different flow rates were measured. Lower flow rates resulted in higher aggregate quantity with a more extended horizontal shape and larger vertical thickness. Thermal measurements were also performed in order to investigate the effect of the aggregate on the heat transfer of the device. Hot water was flown through the channel, and the temperature drop was measured at several flow rates. Then the measurement was repeated using MNP suspension. In all flow rates, the aggregate filled case provided a higher temperature drop. This suggests that the MNP aggregate can be used to increase the heat transfer in the channel.

The two-phase solver was tested by simulating the observed aggregations in the experiment. Although the simulations were run only in 2D, the simulation results generally were in accordance with the measurements. The aggregation procedure in the simulations occurred similarly to the experiments, as the aggregate started to grow from the bottom wall close to the magnet in both vertical and radial directions until a maximum quantity shape. The gap size over the simulated aggregate corresponds to the measured values at the different flow rates.

The presented solver was able to give a prediction of the shape of the nanoparticle aggregate. This shows that the model can be used in the applications using magnetic nanoparticles and may be a useful novel tool in the design procedures of such devices.

Author Contributions: Conceptualization, P.P., M.N. and M.R.; methodology, P.P. and M.N.; software, P.P.; validation, P.P. and M.N.; writing—original draft preparation, P.P., M.N. and M.R.; writing—review and editing, P.P., M.N. and M.R.; supervision, M.R. All authors have read and agreed to the published version of the manuscript.

Funding: The research work was supported by the New National Excellence Program of the Ministry for Innovation and Technology of Hungary, financed from the National Research, Development and Innovation Fund of Hungary (ÚNKP-20-4-I). The research work was supported by the National Research, Development and Innovation Fund of Hungary (Budapest, Hungary; project SNN-125637). The research reported in this paper and carried out at the Budapest University of Technology and Economics was supported by the "TKP2020, Institutional Excellence Program" of the National Research Development and Innovation Office in the field of Artificial Intelligence (BME IE-MI-SC TKP2020).

Conflicts of Interest: The authors declare no conflict of interest.

Nomenclature

General notation a, $\mathbf{b}$	a : scalar, $\mathbf{b}$: vector $[b_x, b_y, b_z]$
$\mathbf{B}, \mathbf{H}$	magnetic field [T], [A · m^{-1}]
$\mathbf{M}$	magnetization [A · m^{-1}]
$\mathbf{m}_i$	magnetic moment of particle i [Am^2]
p	pressure [Pa]
Q	flow rate [m^3 · s^{-1}]
$\mathbf{U}$	velocity [m · s^{-1}]
α	volume fraction
$\dot{\gamma}$	strain rate [s^{-1}]
μ	dynamic viscosity [Pa · s]
ρ	density [kg · m^{-3}]
τ_m	magnetic torque density [N · m^{-2}]
χ	magnetic susceptibility
ϕ	volumetric flux on a face [m^3 · s^{-1}]
CFD	computational fluid dynamics
MNP	magnetic nanoparticle

References

1. Fan, J.; Gao, Y. Nanoparticle-supported catalysts and catalytic reactions–a mini-review. *J. Exp. Nanosci.* **2006**, *1*, 457–475. [CrossRef]
2. Ender, F.; Weiser, D.; Nagy, B.; Bencze, C.L.; Paizs, C.; Pálovics, P.; Poppe, L. Microfluidic multiple cell chip reactor filled with enzyme-coated magnetic nanoparticles—An efficient and flexible novel tool for enzyme catalyzed biotransformations. *J. Flow Chem.* **2016**, *6*, 43–52. [CrossRef]

3. Weiser, D.; Bencze, L.C.; Bánóczi, G.; Ender, F.; Kiss, R.; Kókai, E.; Szilágyi, A.; Vértessy, B.G.; Farkas, Ö.; Paizs, C.; et al. Phenylalanine Ammonia-Lyase-Catalyzed Deamination of an Acyclic Amino Acid: Enzyme Mechanistic Studies Aided by a Novel Microreactor Filled with Magnetic Nanoparticles. *ChemBioChem* **2015**, *16*, 2283–2288. [CrossRef] [PubMed]

4. Pálovics, P.; Ender, F.; Rencz, M. Towards the CFD model of flow rate dependent enzyme-substrate reactions in nanoparticle filled flow microreactors. *Microelectron. Reliab.* **2018**, *85*, 84–92. [CrossRef]

5. Pálovics, P.; Ender, F.; Rencz, M. Geometric optimization of microreactor chambers to increase the homogeneity of the velocity field. *J. Micromech. Microeng.* **2018**, *28*, 064002. [CrossRef]

6. Ahrberg, C.D.; Manz, A.; Chung, B.G. Polymerase chain reaction in microfluidic devices. *Lab Chip* **2016**, *16*, 3866–3884. [CrossRef] [PubMed]

7. Sosnovik, D.E.; Nahrendorf, M.; Weissleder, R. Magnetic nanoparticles for MR imaging: agents, techniques and cardiovascular applications. *Basic Res. Cardiol.* **2008**, *103*, 122–130. [CrossRef]

8. Reddy, L.H.; Arias, J.L.; Nicolas, J.; Couvreur, P. Magnetic nanoparticles: design and characterization, toxicity and biocompatibility, pharmaceutical and biomedical applications. *Chem. Rev.* **2012**, *112*, 5818–5878. [CrossRef]

9. Das, P.; Colombo, M.; Prosperi, D. Recent advances in magnetic fluid hyperthermia for cancer therapy. *Colloids Surf. B Biointerfaces* **2019**, *174*, 42–55. [CrossRef]

10. Gui, N.G.J.; Stanley, C.; Nguyen, N.T.; Rosengarten, G. Ferrofluids for heat transfer enhancement under an external magnetic field. *Int. J. Heat Mass Transf.* **2018**, *123*, 110–121.

11. Goharkhah, M.; Salarian, A.; Ashjaee, M.; Shahabadi, M. Convective heat transfer characteristics of magnetite nanofluid under the influence of constant and alternating magnetic field. *Powder Technol.* **2015**, *274*, 258–267. [CrossRef]

12. Lajvardi, M.; Moghimi-Rad, J.; Hadi, I.; Gavili, A.; Isfahani, T.D.; Zabihi, F.; Sabbaghzadeh, J. Experimental investigation for enhanced ferrofluid heat transfer under magnetic field effect. *J. Magn. Magn. Mater.* **2010**, *322*, 3508–3513. [CrossRef]

13. Li, Q.; Xuan, Y. Experimental investigation on heat transfer characteristics of magnetic fluid flow around a fine wire under the influence of an external magnetic field. *Exp. Therm. Fluid Sci.* **2009**, *33*, 591–596. [CrossRef]

14. Han, K.; Feng, Y.; Owen, D. Three-dimensional modelling and simulation of magnetorheological fluids. *Int. J. Numer. Methods Eng.* **2010**, *84*, 1273–1302. [CrossRef]

15. Karvelas, E.; Lampropoulos, N.; Sarris, I.E. A numerical model for aggregations formation and magnetic driving of spherical particles based on OpenFOAM®. *Comput. Methods Programs Biomed.* **2017**, *142*, 21–30.

16. Boutopoulos, I.D.; Lampropoulos, D.S.; Bourantas, G.C.; Miller, K.; Loukopoulos, V.C. Two-Phase Biofluid Flow Model for Magnetic Drug Targeting. *Symmetry* **2020**, *12*, 1083. [CrossRef]

17. Ghaffari, A.; Hashemabadi, S.H.; Bazmi, M. CFD simulation of equilibrium shape and coalescence of ferrofluid droplets subjected to uniform magnetic field. *Colloids Surfaces A Physicochem. Eng. Asp.* **2015**, *481*, 186–198. [CrossRef]

18. Jackson, J.D. *Classical Electrodynamics*; John Wiley & Sons: Hoboken, NJ, USA, 2012.

19. Guyon, E.; Hulin, J.P.; Petit, L.; Mitescu, C.D. *Physical Hydrodynamics*; Oxford University Press: Oxford, UK, 2001.

20. Saliba, A.E.; Saias, L.; Psychari, E.; Minc, N.; Simon, D.; Bidard, F.C.; Mathiot, C.; Pierga, J.Y.; Fraisier, V.; Salamero, J.; et al. Microfluidic sorting and multimodal typing of cancer cells in self-assembled magnetic arrays. *Proc. Natl. Acad. Sci. USA* **2010**, *107*, 14524–14529.

21. Bossis, G.; Volkova, O.; Lacis, S.; Meunier, A. Magnetorheology: fluids, structures and rheology. In *Ferrofluids*; Springer: Berlin, Germany, 2002; pp. 202–230.

22. Martin, J.E.; Anderson, R.A. Chain model of electrorheology. *J. Chem. Phys.* **1996**, *104*, 4814–4827. [CrossRef]

23. Weller, H.G.; Tabor, G.; Jasak, H.; Fureby, C. A tensorial approach to computational continuum mechanics using object-oriented techniques. *Comput. Phys.* **1998**, *12*, 620–631. [CrossRef]

24. Rusche, H. Computational Fluid Dynamics of Dispersed Two-Phase Flows at High Phase Fractions. Ph.D. Thesis, University of London, London, UK, 2002.

25. Wen, C.Y. Mechanics of fluidization. *Chem. Eng. Prog. Symp. Ser.* **1966**, *62*, 100–111.

26. Ferziger, J.H.; Perić, M.; Street, R.L. *Computational Methods for Fluid Dynamics*; Springer: Berlin, Germany, 2002; Volume 3.

27. Pimenta, F.; Alves, M. Stabilization of an open-source finite-volume solver for viscoelastic fluid flows. *J.Non-Newton. Fluid Mech.* **2017**, *239*, 85–104. [CrossRef]

28. Rajput, S.; Pittman, C.U., Jr.; Mohan, D. Magnetic magnetite (Fe_3O_4) nanoparticle synthesis and applications for lead (Pbr^{2+}) and chromium (Cr^{6+}) removal from water. *J. Colloid Interface Sci.* **2016**, *468*, 334–346. [CrossRef]

29. Goya, G.; Berquo, T.; Fonseca, F.; Morales, M. Static and dynamic magnetic properties of spherical magnetite nanoparticles. *J. Appl. Phys.* **2003**, *94*, 3520–3528. [CrossRef]

30. Daoush, W. Co-precipitation and magnetic properties of magnetite nanoparticles for potential biomedical applications. *J. Nanomed. Res.* **2017**, *5*, 1–6. [CrossRef]

31. Thompson, R.; Oldfield, F. Magnetic properties of natural materials. In *Environmental Magnetism*; Springer: Berlin, Germany, 1986; pp. 21–38.

32. Pimenta, F.; Alves, M. rheoTool. 2016. Available online: https://github.com/fppimenta/rheoTool (accessed on 21 May 2020).

Article

Digital Luminaire Design Using LED Digital Twins—Accuracy and Reduced Computation Time: A Delphi4LED Methodology

Marc van der Schans *, Joan Yu and Genevieve Martin

Signify, High Tech Campus 7, 5656AE Eindhoven, The Netherlands; joan.yu@signify.com (J.Y.); genevieve.martin@signify.com (G.M.)
* Correspondence: marc.van.der.schans@signify.com

Received: 16 July 2020; Accepted: 14 September 2020; Published: 22 September 2020

Abstract: Light-emitting diode (LED) digital twins enable the implementation of fast digital design flows for LED-based products as the lighting industry moves towards Industry 4.0. The LED digital twin developed in the European project Delphi4LED mimics the thermal-electrical-optical behavior of a physical LED. It consists of two parts: a package-level LED compact thermal model (CTM), coupled to a chip-level multi-domain model. In this paper, the accuracy and computation time reductions achieved by using LED CTMs, compared to LED detailed thermal models, in 3D system-level models with a large number of LEDs are investigated. This is done up to luminaire-level, where all heat transfer mechanisms are accounted for, and up to 60 LEDs. First, we characterize a physical phosphor-converted white high-power LED and apply LED-level modelling to produce an LED detailed model and an LED CTM following the Delphi4LED methodology. It is shown that the steady-state junction temperature errors of the LED CTM, compared to the detailed model, are smaller than 2% on LED-level. To assess the accuracy and the reduction of computation time that can be realized in a 3D system-level model with a large number of LEDs, two use cases are considered: (1) an LED module-level model, and (2) an LED luminaire-level model. In the LED module-level model, the LED CTMs predict junction temperatures within about 6% of the LED detailed models, and reduce the calculation time by up to nearly a factor 13. In the LED luminaire-level model, the LED CTMs predict junctions temperatures within about 1% of LED detailed models and reduce the calculation time by about a factor of 4. This shows that the achievable computation time reduction depends on the complexity of the 3D model environment. Nevertheless, the results demonstrate that using LED CTMs has the potential to significantly decrease computation times in 3D system-level models with large numbers of LEDs, while maintaining junction temperature accuracy.

Keywords: compact thermal model; LED; Delphi4LED; digital twin; digital luminaire design; computation time; Industry 4.0

---

1. Introduction

Rapid innovation and customization of light-emitting diode (LED)-based lighting products demand shorter design cycles, higher cost efficiency, and more reliable solutions from manufacturers. To meet these demands, digitalization of the design flow, also called an "Industry 4.0" approach, is required. Methods, processes, and tools that facilitate the usage of LED components in a digital design flow were developed and demonstrated in the European project Delphi4LED [1,2]. The proposed approach consists of the generation and implementation of multi-domain LED digital twins to enable fast and reliable computer simulations of LED-based lighting products. Multi-domain LED digital twins are models that accurately mimic the thermal-electrical-optical behavior of a physical LED, and can be integrated in larger system-level models, for example a luminaire-level model. Additionally, an LED digital twin should not carry proprietary information of the LED manufacturer, such as details related to the LED's construction, materials or production processes. This way, LED manufacturers or vendors can share them with end-users without disclosing their sensitive intellectual property.

An overview of the major steps involved in creating and implementing multi-domain LED digital twins is described in detail by Martin et al. [3]. The steps are summarized in Figure 1 and briefly outlined here. First, a thermal-electrical-optical characterization of the physical LED device is performed (step 1). The testing protocols and methods are discussed in [4–7] and take the established testing standards JEDEC JESD51-14, JESD51-51, JESD51-52, and CIE 225:2017 [8–11] into account. The results of the characterization are so-called iso-thermal current-voltage-flux (IVL) characteristics and thermal transient characteristics of the LED device. This data will be reported in future standard LED electronic datasheets [12] (step 2).

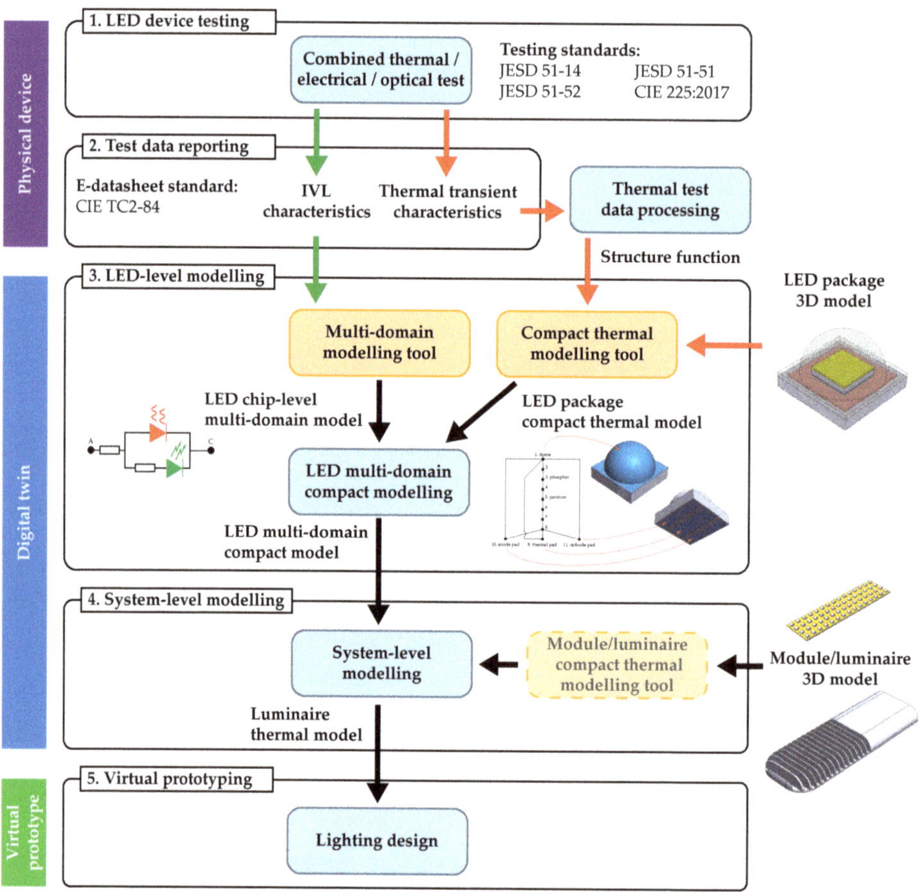

Figure 1. The Delphi4LED approach to creating and implementing LED digital twins (multi-domain compact models). The involved steps are indicated from top to bottom. Adapted from [3].

Next, the multi-domain LED digital twin, also referred to as the LED multi-domain compact model (MDCM), is extracted from the characterization data (step 3). The LED MDCM consists of two parts. The first part is a chip-level multi-domain model. It calculates the forward voltage, power dissipation, radiant flux, and luminous flux from the forward current and junction temperature. Poppe et al. [13,14]

discuss several sets of equations that can be used for this purpose. The extraction of the chip-level model is achieved by fitting the parameters of the equations to the IVL characteristics. The second part of the LED MDCM is a package-level compact thermal model (CTM). The LED CTM is a thermal RC-network attached to a simplified geometric representation of the LED. It calculates the relevant operating temperatures of the LED package, such as junction, phosphor and solder temperatures, from the power dissipation. The CTM extraction procedure is described by Bornoff et al. [15,16]. It involves the calibration of a detailed thermal model using the thermal transient characteristics, and the optimization of the RC-network to produce matching thermal dynamic behavior. In a fully realized LED digital twin, the two parts of the LED MDCM are coupled and solved self-consistently.

Finally, the LED digital twin is implemented in the larger system-level model of an LED-based lighting product, for example an LED module or an LED luminaire (step 4). There are different approaches to the system-level model. One option is to generate a compact thermal model of the LED module or LED luminaire. Poppe et al. [17] describe a method to create thermal network compact models for luminaires, which have subsequently been used in Spice-like luminaire simulations [3,18] and in an Excel spreadsheet application [3,19]. Alternatively, model order reduction could be used instead of thermal network compact models [20–23]. Another approach is to perform the LED module or LED luminaire simulations using the LED MDCM directly in a 3D computational fluid dynamics (CFD) model [3,24]. Ultimately, the LED module or LED luminaire model is used for virtual prototyping (step 5).

In this publication, we assess and compare the accuracy and computation time of 3D CFD system-level models equipped with LED CTMs and with LED detailed models. While this study does not include a multi-domain chip-level model, the LED thermal model is most demanding in terms of computation time in this case. First, an LED detailed model and an LED CTM are created according to the Delphi4LED methodology by performing a thermal characterization and LED-level modelling. The obtained LED thermal models are then implemented into 3D CFD software in the system-level model of two use cases: (1) an LED module-level model, and (2) an LED luminaire-level model. In previous research [24], we compared the computation time required to simulate an LED module-level model with up to 22 LEDs using LED detailed models and using LED CTMs. It showed that using the LED CTMs reduces the computation time by approximately a factor 10 for a steady-state, conduction only model. The novelty of this work is that the analysis is extended to the luminaire-level model. Typically, a LED luminaire contains several LED modules, and thus contains larger numbers of LEDs. Moreover, all methods of heat transport must be taken into account and their impact on the computation time is investigated.

2. Materials and Methods

The Delphi4LED approach as outlined Figure 1 is followed to create an LED detailed model and an LED CTM and to subsequently implement them in an LED module-level model and an LED luminaire-level model. The methods and processes of three of the involved steps are each described in a subsection. The first subsection briefly explains the test procedures used to obtain the required characterization data from physical LED samples (LED device testing). In the second subsection, the creation of the LED detailed model and the LED CTM are described (LED-level modelling). Finally, in the third subsection, the implementation of the LED detailed model and the LED CTM in the LED module-level model and the LED luminaire-level model is specified.

2.1. LED Device Testing

For the investigations, a phosphor-converted white high-power LED with a color rendering index (CRI) of 70 and correlated color temperature (CCT) of 4000 K is used. Four physical samples of the same type LED are each assembled on an insulated metal substrate (IMS) board for testing. Figure 2a shows the detailed 3D geometry of the LED sample placed on the test board. A cross-sectional view of the LED package geometry is provided in Figure 2b. The LED package and the test board together constitute the device under test (DUT).

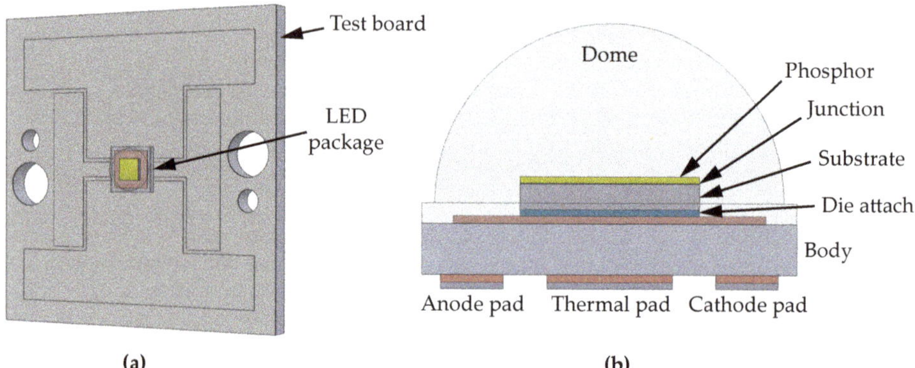

Figure 2. Detailed geometry of the LED under investigation: (**a**) The LED package assembled on the test board (DUT). (**b**) Cross-sectional view of the LED package geometry.

The device under test is placed on a cold plate. Thermal paste is applied between the test board and the cold plate to provide good thermal contact, and the board is fixed in place with two screws. To ensure reproducibility a torque screwdriver used. Simultaneous radiometric measurements and thermal transient measurements, i.e., the temperature response $T(t)$ to a power step, are performed using commercially available testing equipment (Simcenter T3ster and TeraLED) [13,25,26]. For these measurements a fixed cold plate temperature of $T_{\mathrm{ref}} = 50\,^{\circ}\mathrm{C}$ is used. In these tests, the DUT is first operated at a (total) forward current of $I_f = 1400\,\mathrm{mA}$ until steady-state is reached. Then a power step is applied by decreasing the current to a measurement current of $I_{\mathrm{meas}} = 10\,\mathrm{mA}$. The measured emitted radiant flux Φ_e is subtracted from the electrical power P_{el} to obtain the total thermal dissipation $P_{\mathrm{th}} = P_{\mathrm{el}} - \Phi_e$. The thermal dissipation is then used to normalize the transient temperature $\Delta T(t) = T(t) - T_{\mathrm{ref}}$ to obtain the transient thermal impedance $Z_{\mathrm{th}}(t) = \Delta T(t)/P_{\mathrm{th}}$, as well as the corresponding structure function (SF) and differential structure function (DSF). This is the thermal characterization data needed for the LED CTM in the LED-level modelling step.

2.2. LED-Level Modelling

Generating an LED CTM, in the form of a thermal RC-network, from the thermal transient characterization data involves two parts. In the first part, a detailed thermal model of the LED package is created and calibrated using the characterization data as described in [15]. Then, in the second part, the calibrated LED detailed model is subsequently used to generate training data for the LED CTM. This training data consist of thermal responses under several different boundary conditions. The training data is used to optimize the RC-values of the LED CTM, such that the errors in thermal behavior compared to LED detailed model are minimized [16]. Finally, after the LED CTM is optimized, the achieved accuracy is validated by subjecting both the detailed model and CTM to several additional boundary conditions, which were not used in the training, and determining the errors.

2.2.1. LED Detailed Model

For the LED detailed model, geometrical information is required. The outer dimensions of the LED are provided by the manufacturer. However, in order to have a sufficiently accurate model, additional information is extracted from microscope images, e.g., the chip size and phosphor layer size. For other internal geometrical characteristics, generally not provided by suppliers, an educated guess is made. This is for instance done for the die attach thickness. Minor mismatches in those thicknesses are later compensated during the calibration process by adjusting the thermal conductivity values. The geometric model of the LED package and test board shown in Figure 2 is used in the calibration process.

In the detailed model, thermal loads are applied to the junction as well as to the phosphor layer. They are considered the main contributors to the total heat dissipation. Indeed, there may be other package losses resulting from trapped light due to total internal reflections. Recently, Alexeev et al. discussed the effects of secondary heat sources on thermal transient analysis in detail [27]. However, since losses related to trapped light are difficult to quantify and localize without elaborate optical modelling, only the junction and phosphor losses are considered in this study. Since only the total dissipation P_{th} is known, the power split between the junction and phosphor is included in the calibration as an optimization parameter, together with the thermal conductivity values of the materials in the model.

The LED detailed model is calibrated by minimizing the errors between the modelled and measured $Z_{th}(t)$ responses and SFs. The model parameters are optimized to best match all four measured samples simultaneously. The calibration is performed using commercially available 3D CFD software (Simcenter Flotherm XT 2019.2).

To produce training data for the LED CTM, the calibrated LED detailed model is virtually taken off the test board and subjected to sets of different boundary conditions. This is done by applying uniform heat transfer coefficients (HTCs) to four selected peripheral faces: the bottom face of the anode solder pad, the bottom face of the cathode solder pad, the bottom face of the thermal solder pad, and the top face of the package/dome. The purpose of using multiple boundary conditions is to ensure that the extracted CTM will be boundary condition independent (BCI). The four sets of heat transfer coefficients that are used to generate the training data for the CTM are listed in Table 1. These HTCs training sets are chosen to represent practical operating environments of LEDs and are in the same range as the HTC sets used for the same purpose in [28]. The generated training data consists of the transient temperature profiles $\Delta T_{ih}^{\text{detailed}}(t)$ obtained for a power step P_{th}. Here, the index i indicates the junction layer, the phosphor layer, and each of the four faces to which HTCs are applied. The index h indicates the each of the four HTC training sets. This data is exported from the CFD software.

Table 1. HTC training sets used to generate training data for the LED CTM optimization.

Set	HTC (W/m^{2}K)		
	Anode/Cathode Pad	Thermal Pad	Dome
1	10,000	25,000	10
2	3000	75,000	20
3	1500	20,000	100
4	50,000	10,000	5

2.2.2. LED CTM

A CTM, in the form of a thermal RC-network, is optimized to produce matching thermal behavior under the same boundary conditions, i.e., the four imposed HTC training sets. The chosen network topology is illustrated in Figure 3. Each node i of the thermal network has a thermal capacitance $C_{th,i}$ to ground. A line between two nodes m and n indicates that the nodes are connected by a thermal resistor $R_{th,mn}$. Compared to the network topology used in earlier studies [16,28], our network topology has an additional node between the junction and phosphor nodes (node 4), and between the phosphor and dome nodes (node 2). It was found by trial and error that those nodes are necessary to better fit the dynamic behavior, particularly of the phosphor and dome nodes.

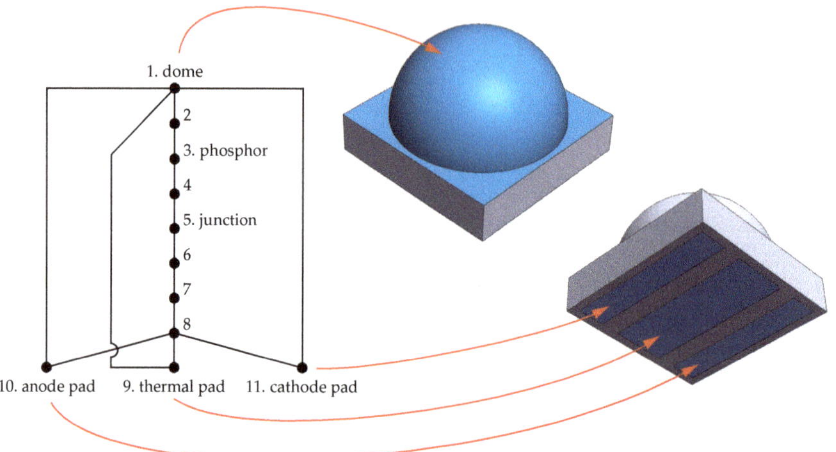

Figure 3. RC-network topology of the LED CTM (**left**), and two isometric views of the simplified 3D geometry of the LED CTM (**right**). The red arrows indicate to which surfaces (in blue) the temperatures of the peripheral nodes of the RC-network model are connected.

The RC-network optimization is performed by code developed in Python and works in a manner similar to that described by Schweitzer [29]. It uses the derivative-free BOBYQA algorithm [30] implementation from the NLopt library [31]. An advantage of performing the optimization separately outside the CFD software is that stand-alone RC-network evaluations are faster, which means that several 10,000 to 100,000 optimization iterations can be made in a few minutes.

First, the training data is imported and the corresponding thermal (transfer) impedances $Z_{\text{th},ih}^{\text{detailed}}(t) = \Delta T_{ih}^{\text{detailed}}(t)/P_{\text{th}}$ are calculated. Subsequently, the RC-network is numerically solved for the same power step P_{th} to obtain the CTM temperatures $\Delta T_{ih}^{\text{CTM}}(t)$ of nodes i under HTC training sets h. The corresponding thermal (transfer) impedances are again calculated as $Z_{\text{th},ih}^{\text{CTM}}(t) = \Delta T_{ih}^{\text{CTM}}(t)/P_{\text{th}}$. By varying the $C_{\text{th},i}$ and $R_{\text{th},mn}$ values, the difference in dynamic thermal behavior between the detailed model and the RC-network CTM is minimized using the following cost function:

$$f_{\text{cost}} = \sum_h \sum_i \sum_j \frac{\left(Z_{\text{th},ih}^{\text{detailed}}(t_j) - Z_{\text{th},ih}^{\text{CTM}}(t_j)\right)^2}{Z_{\text{th},ih}^{\text{detailed}}(t_j)} \tag{1}$$

where index h runs over all four HTC training sets, index i runs over the nodes included in the optimization, and index j runs over the time steps for which the simulation is performed. The included nodes are the junction, phosphor, and peripheral nodes, i.e., $i = \{1, 3, 5, 9, 10\}$. In this particular case node 11 is not explicitly included due to symmetry.

To assess the accuracy and boundary condition independence of the optimized LED CTM, both the LED CTM and the LED detailed model are tested under twenty additional HTC sets. The twenty HTC testing sets are listed in Table 2. These sets are combinations generated using the design of experiments functionality of the CFD software. For each of the peripheral faces, the lower and upper HTC bounds were set to the minimum and maximum values that occur in the training sets. Since the HTC testing sets were not used to train the model, they provide a better evaluation of the predictive temperature accuracy of the LED CTM compared to the detailed model. In the report on end-user specifications of the Delphi4LED project [32] the required junction temperature accuracy is stated as 2%.

Table 2. HTC testing sets used to generate test data for the LED CTM validation.

Set	HTC (W/m^{2}K)		
	Anode/Cathode Pad	Thermal Pad	Dome
1	42,725	13,250	66.75
2	1500	58,750	71.5
3	18,475	49,000	100
4	37,875	36,000	95.25
5	50,000	45,750	62
6	20,900	71,750	76.25
7	40,300	65,250	90.5
8	30,600	32,750	5
9	28,175	10,000	33.5
10	47,575	23,000	28.75
11	23,325	19,750	81
12	16,050	42,500	57.25
13	8775	16,500	47.75
14	35,450	68,500	52.5
15	6350	52,250	24
16	33,025	39,250	43
17	13,625	75,000	38.25
18	45,150	55,500	19.25
19	25,750	62,000	14.5
20	3925	29,500	85.75

Finally, to be able to interface with a larger 3D system-level model, the LED CTM is attached to simplified 3D geometric representation of the LED package, as indicated in Figure 3. While the outer contours of the simplified geometry are identical to those of the detailed model, it has no internal structure. The faces of the simplified geometry that are connected to the peripheral nodes of the RC-network have the same surface area as the corresponding faces of the detailed model.

2.3. System-Level Modelling

Two use cases are investigated to assess the predicted junction temperature accuracy and computation time performance of the LED CTM, compared to the LED detailed model, integrated in a 3D system-level model: (1) an LED module-level model, and (2) an LED luminaire-level model. The thermal environment of the LED CTMs and the LED detailed models is now explicitly simulated in these cases, instead of imposed by uniform HTCs. Please note that both the LED module-level model and the luminaire-level model presented here are solely intended for the purpose of numerically assessing the LED CTM accuracy and performance, compared to the LED detailed model. They are by no means optimized for thermal management or any other actual product requirements.

While the described methods and models can in principle be used in any 3D CFD software, we used Simcenter Flotherm XT 2019.2. All reported computation times are obtained on a workstation laptop with an Intel Core i7-6820HQ (2.7 GHz, 4 cores) processor. Unless stated otherwise, the default computational mesh settings ('Standard Resolution') of the software tool are used.

2.3.1. LED Module-Level Model

The LED module-level model consists of a simplified printed circuit board (PCB) populated with an array of LEDs, as illustrated in Figure 4. The board is 25 mm wide, 90 mm long and has a 1 mm thick dielectric layer (1 W/mK) and a 70 µm thick copper layer (386 W/mK). Several LEDs, N_{LED}, are uniformly distributed on top of the copper layer. The number of LEDs along the x-axis is N_x, and the number of LEDs along the y-axis is N_y. For N_{LED} up to 15, a single row of LEDs is used ($N_x = 1$), for N_{LED} between 16 and 30, two rows of LEDs are used ($N_x = 2$), for N_{LED} between 31 and 45, three rows of LEDs are used ($N_x = 3$), and for N_{LED} greater than 45, four rows of LEDs are used

($N_x = 4$). The bottom face of the dielectric layer is kept at a fixed uniform temperature T_{ref} and only solid conduction is considered in this model.

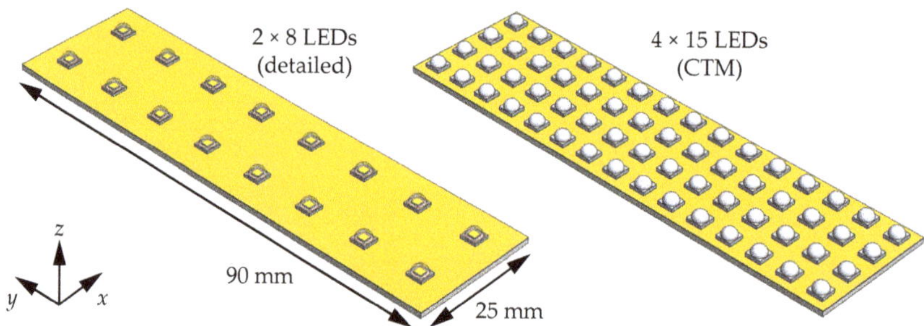

Figure 4. The LED module-level model. On the left an example is shown with $N_{LED} = 16$ LED detailed models ($N_x = 2$ and $N_y = 8$), and on the right an example is shown with $N_{LED} = 60$ LED CTMs ($N_x = 4$ and $N_y = 15$).

2.3.2. LED Luminaire-Level Model

The LED luminaire model consists of a simplified luminaire housing and five instances of the LED module, as illustrated in Figure 5. Each of the LED modules has $N_{LED} = 12$ LEDs ($N_x = 2$ and $N_y = 6$), resulting in total number of 60 LEDs. The luminaire housing is made of an aluminum alloy (140 W/mK) and has fins located above the LED modules for cooling to the surrounding air. In this luminaire-level model, conduction, convection and radiation are all taken into account, increasing the model complexity with flow simulation. All solid-fluid interfaces are assigned a surface emissivity of 0.8, including the anodized surface of the luminaire housing, and the ambient temperature is set at $T_{ref} = 25\,°C$.

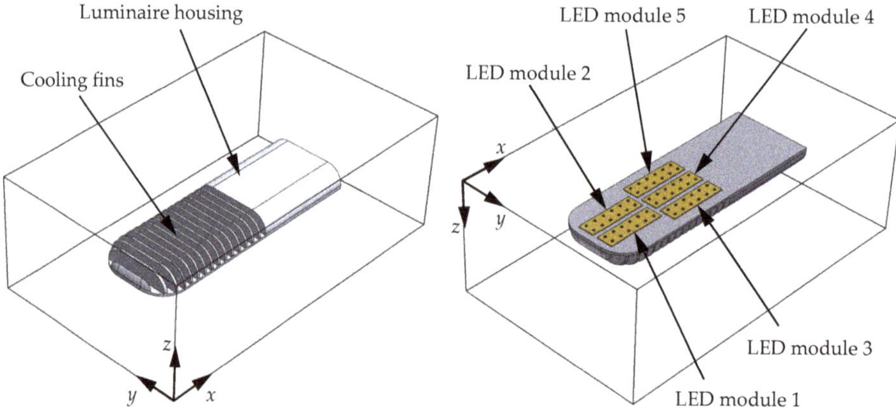

Figure 5. The LED luminaire-level model. On the left, the top part of the luminaire housing with cooling fins is visible, and on the right, the bottom part of the luminaire housing is visible, which supports five LED modules each with 12 LEDs ($N_x = 2$ and $N_y = 6$). The rectangular outlines indicate the computational domain.

3. Results

In this section, the created LED-level models (detailed model and CTM) and the results of using the LED-level models in a system-level model (module and luminaire) are presented. The first part presents the results of the physical LED device testing. Then, the second part presents the modelling results at LED-level. It includes the calibration of the LED detailed model, and the extraction and validation of the LED CTM. Next, the third part presents the accuracy and performance results of the LED module-level model. Finally, the fourth part presents the accuracy and performance results of the LED-luminaire model.

3.1. LED Device Testing Results

Subtracting the measured radiant flux from the supplied electrical power P_{el} results in the total dissipated thermal power $P_{th} = P_{el} - \Phi_e$ of the DUT. It is found that $P_{th} = (2.53 \pm 0.01)\,\mathrm{W}$. The reported uncertainty of 0.01 W indicates the standard deviation between the four measured samples.

Figure 6 shows the transient behavior of the DUT obtained from the measurements. The thermal impedance $Z_{th}(t)$ is presented in Figure 6a. The corresponding structure functions (SF) and differential structure functions (DSF) are shown in Figure 6b. The four measured samples show good reproducibility. The largest relative deviation in measured Z_{th} between the four samples is about 4%, and occurs in the early transient ($t < 150\,\mu s$), where initial correction has to be applied due to electrical transients present in the measurement signal. The measured steady-state Z_{th} has a relative deviation of around 0.1%. Using an additional 'dry' thermal transient additional measurement, i.e., without thermal paste applied between board and cold plate, the junction-to-board resistance was determined according to the standard JEDEC JESD51-14 [8]. The junction-to-board thermal resistance of the DUT is found as $R_{th,j-b} = (6.2 \pm 0.1)\,\mathrm{K/W}$.

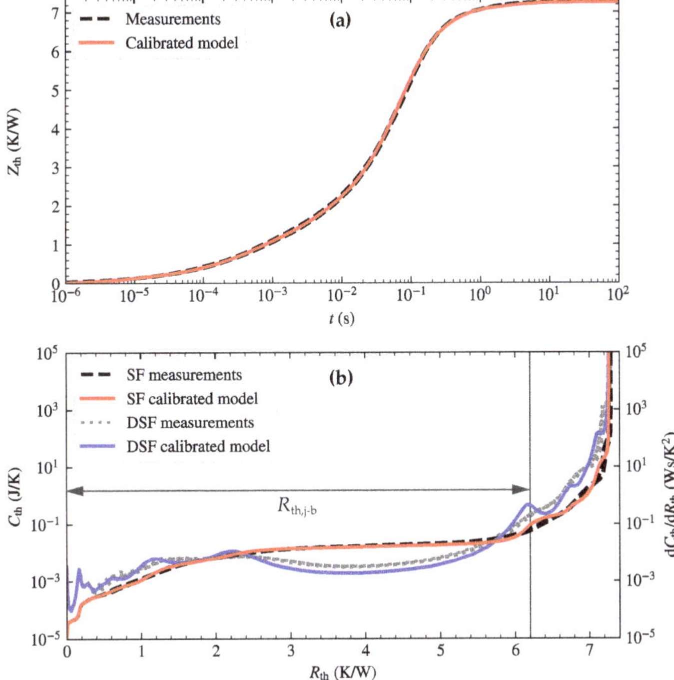

Figure 6. Thermal transient characteristics of the measured LED device and the calibrated LED detailed model: (**a**) Thermal transient impedance of the DUT and calibrated detailed model. (**b**) Structure function and differential structure function of the DUT and the calibrated detailed model.

3.2. LED-Level Modelling Results

First, the model parameters of the detailed model, i.e., the thermal conductivity values of the materials and junction-phosphor power split, were calibrated to match the $Z_{th}(t)$ and SF of the measured LED devices. The power split in the detailed model that best fits the measurements was found in the calibration as approximately 77% in the junction and 23% in the phosphor. The $Z_{th}(t)$, SF and DSF of the calibrated LED detailed model are shown together with those obtained from the measurements in Figure 6. Overall, the curves match well. Some deviations between the SFs and between the locations of the peaks and valleys of the DSFs are observed for approximately $R_{th} > 6\,\text{K/W}$. However, since we are ultimately only interested in the LED package, without the board, no further improvements to matching this part of the SF are required.

Next, training data was generated using the calibrated LED detailed model for four HTC training sets. This training data was subsequently used to optimize the RC-values of the LED CTM. The optimized RC-values are given in Table 3. The C_{th} column lists the thermal capacitance to ground for each of the nodes, and the R_{th} array lists the thermal resistances between connected nodes of the RC-network. Since $R_{th,mn} = R_{th,nm}$ only the lower triangular entries are displayed.

Table 3. Optimized thermal capacitance values and thermal resistance values of the LED CTM.

Node	C_{th} (J/K)	R_{th} (K/W)										
		1	2	3	4	5	6	7	8	9	10	11
1	8.533×10^{-4}											
2	9.906×10^{-3}	222.6										
3	1.157×10^{-4}		291.4									
4	2.793×10^{-4}			7.707								
5	5.306×10^{-5}				8.563							
6	2.365×10^{-4}					0.299						
7	2.567×10^{-3}						1.350					
8	9.209×10^{-3}							0.096				
9	1.113×10^{-4}	1268							1.822			
10	4.374×10^{-4}	1645							4.911			
11	4.374×10^{-4}	1645							4.911			

The thermal transient behavior of the calibrated LED detailed model and the optimized RC-network LED CTM are compared for the four HTC training sets in Figure 7. The largest absolute error $|Z_{th}^{detailed}(t) - Z_{th}^{CTM}(t)|$ after optimization is about 0.7 K/W and occurs between approximately 10^{-1} s and 10^{1} s for the phosphor node in HTC training set 3. For steady-state conditions, the relative errors $|Z_{th}^{detailed} - Z_{th}^{CTM}|/Z_{th}^{detailed}$ are all smaller than 2%, and smaller than 1% for the Z_{th} values corresponding to the junction. This results in junction temperatures that match within 0.2 K for the training data.

The relative errors in steady-state temperature rise, $|\Delta T^{detailed} - \Delta T^{CTM}|/\Delta T^{detailed}$, for the twenty HTC testing sets are plotted in Figure 8. The junction temperature errors range from about 0.6% to about 1.7%, remaining within the 2% requirement. All other temperature errors also remain within this limit, with the exception of the dome temperature for five of the twenty HTC testing sets. However, it should be noted that the temperature requirement is only specified for the junction temperature [32].

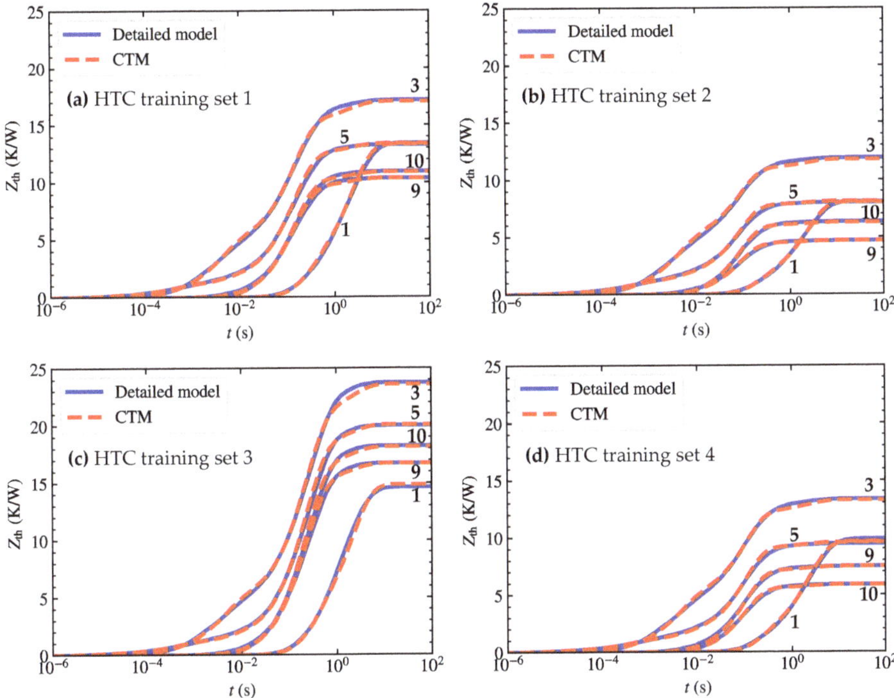

Figure 7. Transient behavior of the LED detailed model (blue) and optimized LED CTM (red) for (**a**) HTC training set 1, (**b**) HTC training set 2, (**c**) HTC training set 3, and (**d**) HTC training set 4. The numbers correspond to the nodes of the RC-network: 1-dome, 3-phosphor, 5-junction, 9-thermal pad, and 10-dome.

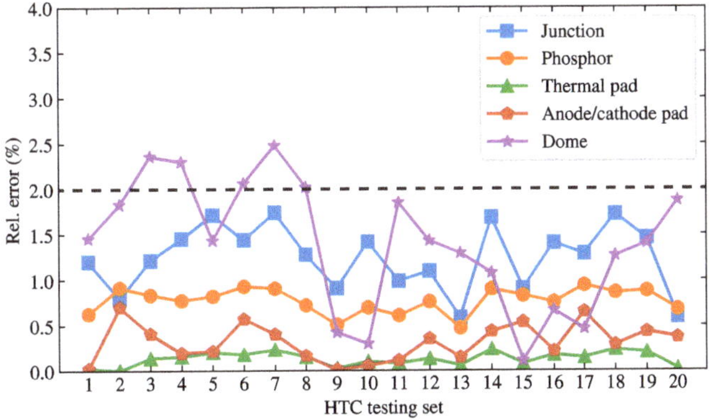

Figure 8. Relative errors in the steady-state ΔT of the LED CTM, compared to the LED detailed model, for each of the twenty HTC testing sets. The dashed line indicates the 2% junction temperature error requirement of the Delphi4LED end-user specifications [32].

3.3. LED Module-Level Model

To assess the accuracy and performance of the LED CTM, compared to the LED detailed model, implemented in the LED module-level model, it was simulated with multiple numbers of LEDs.

Since in each case the LEDs are uniformly distributed over the board, and the bottom face of the board is kept at a uniform temperature T_{ref}, the $\Delta T_{\text{j}} = T_{\text{j}} - T_{\text{ref}}$ varies less than 0.2 K between individual LEDs. For this reason only a single value, the average ΔT_{j} for all LEDs on the board, is reported for each case. The results are presented in Figure 9.

When there is only a small number of LEDs on the board, the distance between the individual LEDs is large enough that no mutual influence, or 'cross-talk', is experienced by the LEDs. This can be observed in the constant ΔT_{j} for N_{LED} up to 4 in Figure 9a. When the number of LEDs is further increased, ΔT_{j} gradually rises. As mentioned, between $N_{\text{LED}} = 15$ and $N_{\text{LED}} = 16$, we change from a single row ($N_x = 1$) to two rows ($N_x = 2$) of LEDs. This results in an effective increase in the distance between individual LEDs and causes the drop in ΔT_{j}. This occurs again, albeit less pronounced, when the number of rows is increased to three ($N_x = 3$) and to four ($N_x = 4$).

Comparing the average ΔT_{j} obtained using the LED CTM and the LED detailed model, the same behavior is observed. Nevertheless, the predictions of the LED CTM are systematically lower. The absolute difference in ΔT_{j} between the LED CTM and the LED detailed model decreases from about 1.6 K for $N_{\text{LED}} = 1$ to about 1.0 K for $N_{\text{LED}} = 60$. This corresponds to relative errors in ΔT_{j} between 6.0% for $N_{\text{LED}} = 1$ and 2.2% for $N_{\text{LED}} = 60$, as shown in Figure 9b.

Inspecting the peripheral faces of the thermal pad, anode pad and cathode pad, reveals discrepancies in the average surface temperatures and in the heat transfer distribution. For example, for $N_{\text{LED}} = 1$ the average surface temperature rise, $T - T_{\text{ref}}$, of the thermal pad and the total heat transfer through the thermal pad in the LED detailed model are 19.6 K and 1.02 W respectively, whereas in the LED CTM they are 17.4 K and 1.18 W respectively. For $N_{\text{LED}} = 60$ these differences are smaller, which results in a smaller junction temperature error. In this case the average surface temperature rise of the thermal pad and the total heat transfer through the thermal pad in the LED detailed models are 39.9 K and 1.07 W respectively, whereas in the LED CTMs they are 38.3 K and 1.18 W respectively. To ensure that the observed differences cannot be attributed to mesh convergence issues, the simulations for $N_{\text{LED}} = 1$, $N_{\text{LED}} = 16$, and $N_{\text{LED}} = 60$ were repeated with higher mesh density. The results were reproduced within 0.2 K.

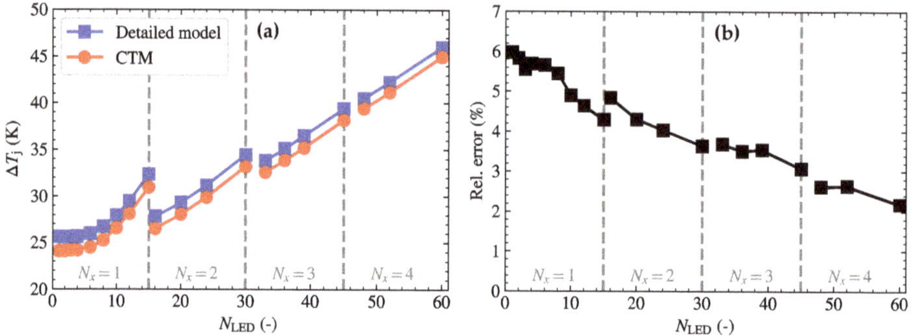

Figure 9. Average junction temperature rise of the LEDs in the LED module-level model: (**a**) comparison between the LED detailed model and the LED CTM, and (**b**) the relative error in the LED CTM values for different number of LEDs.

The time required by the central processor unit (CPU) to solve the model is shown as a function of the number of LEDs in Figure 10a. Up to 60 LEDs, the computation time increases approximately linearly for the LED CTMs, while the computation time using the LED detailed models increases super-linearly. The ratio between the CPU times using the LED detailed models and the LED CTMs is plotted in Figure 10b. It can be seen that for a single LED, using the CTM results in about a factor 2 reduction of computation time. For 60 LEDs the required computation time is reduced by nearly a factor 13 when the LED CTM is used instead of the LED detailed model. This is in line with our previous findings [24].

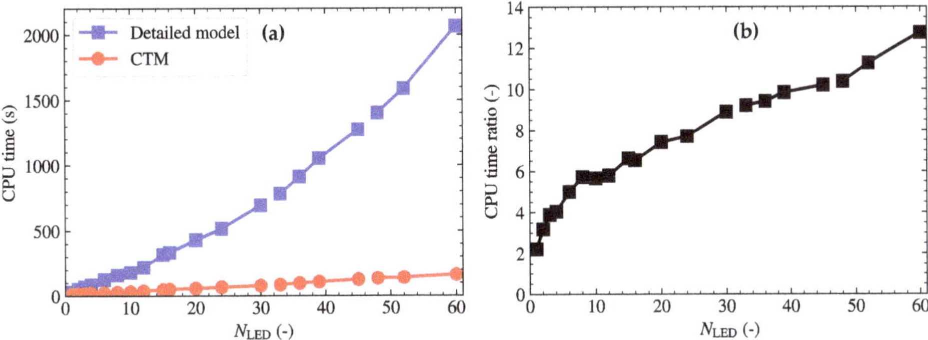

Figure 10. Time required to solve the LED module-level model: (**a**) comparison of CPU time between the LED detailed model and the LED CTM, and (**b**) the ratio in CPU time between the LED detailed model and the LED CTM.

3.4. LED Luminaire-Level Model

In the LED luminaire-level model, the bottom faces of the LED modules are in direct thermal contact with the luminaire housing, instead of being kept at a constant fixed temperature. This results larger gradients in the board temperature, in particular for LED modules 3, 4 and 5. This is illustrated in the temperature plot presented in Figure 11. As a consequence, there are also larger variations in ΔT_j among LEDs on the same board than in the previous case of the LED module-level model. For this reason not only the average ΔT_j for each of the modules is reported, but also the minimum and maximum ΔT_j are listed in Table 4. Since the geometry of the model has a plane of symmetry, the ΔT_j values should be the same for LED module 1 and 2, and for LED module 3 and 5. The values obtained from the model are indeed very similar for these boards, apart from some small variations of 0.1 K between LED module 1 and 2, which can be caused by asymmetries in the computational mesh.

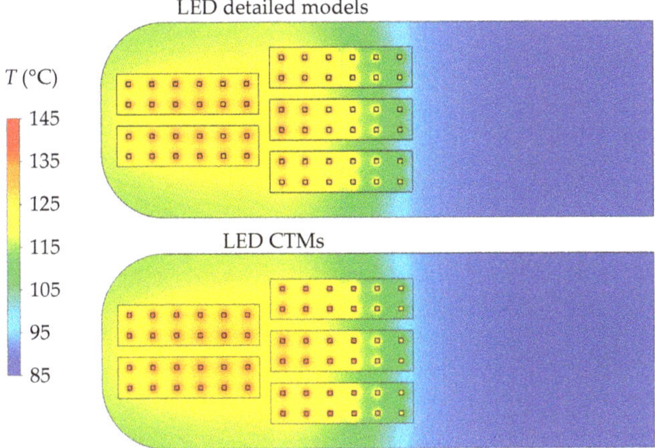

Figure 11. Surface temperature of the LED luminaire-level model with LED detailed models (**top**) and LED CTMs (**bottom**). For the LED CTMs, the temperature of the internal junction node is plotted on the LED geometry.

Table 4. Minimum, average, and maximum junction temperature rises and relative errors for each of LED modules in the LED luminaire-level model.

LED Module	ΔT_{j} (K)						Rel. Error (%)		
	Detailed Model			CTM					
	Min.	Avg.	Max.	Min.	Avg.	Max.	Min.	Avg.	Max.
1	118.8	121.1	122.2	118.2	120.3	121.7	0.4	0.7	0.9
2	118.9	121.1	122.2	118.3	120.3	121.6	0.5	0.7	0.9
3	100.7	113.2	120.7	99.8	112.2	119.4	0.5	0.9	1.1
4	101.5	114.3	121.8	100.8	113.6	120.7	0.3	0.7	0.9
5	100.7	113.2	120.7	99.8	112.2	119.4	0.5	0.9	1.1

As a general note, such high ΔT_{j} values indicate an improved thermal design would be necessary in practice. Comparing the ΔT_{j} values between the LED detailed models and the LED CTMs shows that the junction temperatures predicted by the LED CTMs are again systematically lower. The largest absolute error found in this case is 1.3 K, which is in the same range as the errors found previously in the LED module-level model, and occurs for the LEDs with the largest ΔT_{j} on LED modules 3 and 5. The relative errors in ΔT_{j} range from about 0.3% to 1.1% in this case. Inspecting the average surface temperature rise of the thermal pad and the total heat transfer through the thermal pad of those LEDs again reveals similar discrepancies as before. For the detailed model they are 114.6 K and 1.03 W respectively, and for the CTM they are 113.2 K and 1.12 W respectively.

The LED luminaire-level model containing the LED detailed models took 9016 s to solve, whereas the model containing the LED CTMs only needed 2153 s, resulting in a reduction of a factor 4.2. This is smaller than the computation time reduction that was found for 60 LEDs in the LED module-level model. Besides that the luminaire-level model comprises a larger geometry than the module-level model, convection and radiation are also simulated in this case. To assess the impact of convection and radiation on the computation time, the simulation of the luminaire-level model with LED CTMs is repeated with convection and radiation each turned off separately. Without radiation the model is solved in 1907 s, and without convection the model is solved in 706 s.

4. Discussion

An RC-network LED CTM was successfully generated. The steady-state junction temperature error of the LED CTM, compared to the LED detailed model, was evaluated between 0.6% and 1.7% on LED-level under imposed uniform HTCs. This meets the requirement of 2% stated in the Delphi4LED end-users' specifications [32].

A similar RC-network LED CTM, for a different LED package, is reported in [28]. The main differences are that a network topology with only nine instead of eleven nodes was used, the CTM was trained under three instead of four HTC training sets, using and a cost function based on the SF instead of Z_{th}. A slightly better relative error range of about 0.6% to 1.2% in the predicted steady-state junction temperature rise is reported there. Nevertheless, in both cases an acceptable accuracy for steady-state thermal behavior is achieved with an RC-network LED CTM. However, both in the present case and in [28] it appears more difficult to also achieve accurate dynamic behavior, in particular for the phosphor node. This indicates that further refinement of the extraction process or used network topology may still be necessary for cases in which the LED CTM is not operated in steady-state conditions. Another development in achieving accurate dynamic LED-level models that should be mentioned here is the BCI reduced order model (BCI-ROM) approach [20,21]. Recently, Bornoff and Gaal [28] compared this approach to the RC-network CTM and discussed its advantages related to extraction (no choices on a network topology have to be made) and accuracy (the required accuracy is prescribed by the user *a priori* for a wide range of HTCs). However, at the moment BCI-ROMs cannot be implemented yet in commercially available 3D CFD software.

When implemented in a 3D system-level, the LED CTM predicted slightly but systematically lower junction temperatures than the LED detailed model. The largest difference was 1.6 K among the two studied use cases. Inspecting the peripheral faces of the thermal pad, anode pad and cathode pad, revealed discrepancies between the LED CTMs and LED detailed models in the average surface temperatures and in the heat transfer distribution. This is likely caused by the fact that the LED CTM is extracted under uniform peripheral conditions, whereas gradients exist in the more realistic 3D thermal environment. This situation could be improved by splitting the anode, cathode and thermal pad surfaces in multiple surfaces and assigning each their own node in the RC-network. However, this is at the cost of making the LED CTM more complex. The error could also be partly related to the choice and number of HTC training sets. In the original DELPHI project for semiconductor CTMs, 38 HTC sets were proposed, and later even larger sets were tested [33]. However, it was shown that smaller subsets of five HTCs can still lead to accurate CTMs [33,34]. The range of thermal operating environments that is relevant for LEDs is much smaller, but to date no studies have been performed involving the number and type of HTC sets to apply for accurate LED CTM extraction.

In the LED module-level model, absolute errors in ΔT_j of up to 1.6 K and relative errors of up to 6.0% were found when comparing the model with LED CTMs and LED detailed models. In the LED luminaire-level model, absolute errors in ΔT_j of up to 1.3 K and relative errors of up to 1.1% were found when comparing the model with LED CTMs and LED detailed models. While the absolute errors are on the same scale in both cases, the relative errors are substantially smaller in the luminaire-level case. This is explained by the larger total thermal resistance to ambient of the luminaire system, resulting in a larger temperature rise. Although the CTM meets the 2% error requirement compared to the detailed model on luminaire-level, in the LED module-level model the relative errors are larger. Hence the aforementioned potential improvements may be necessary, depending on the end-user's needs. It should also be stressed however that we only considered the error of the extracted and implemented LED CTM compared to the LED detailed model. Compared to reality, for example if we were to measure a physical prototype, there may be various additional sources of errors. Some examples include: measurement uncertainties, uncertainties in the thermal dissipation of the components, and uncertainties related to the CFD simulation itself [35]. Additionally, the thermal resistance of the LED package may only represent a small fraction of the entire thermal resistance to ambient, especially at LED luminaire-level. When that is the case, the accuracy of the predicted T_j compared to a physical prototype will also largely depend on the accuracy of the part of the model besides the LEDs.

Regarding the computation time, using the LED CTM in the LED module-level model resulted in a reduction from about a factor 2 with one LED up to almost a factor 13 with 60 LEDs. In the LED luminaire-level model with 60 LEDs, a reduction of about a factor 4 was achieved using the LED CTMs. Of course, the exact computation time will be different in every case and depends on the complexity of the luminaire design. The difference in the achieved reduction between the studied cases can be explained by the fact that the luminaire-level model has a larger 3D environment, more complex shapes, and that convection and radiation are considered. As a result, the LEDs themselves constitute a relatively smaller part of the entire model than in the case of the LED module-level model, and hence their relative impact on the total calculation time decreases. In particular, the flow simulations were found responsible for a large part of the computation time in the studied case. When turned off, the computation time decreased by about a factor 3, from 2153 s to 706 s. Without radiation, the computation time decreased by about 10%. Nevertheless, the factor 4 reduction for the simplified LED luminaire-level model is still significant when a large number of scenarios needs to be simulated in a design parameter optimization. Furthermore, the gains will increase significantly for higher LED counts, as demonstrated by the LED module-level model. As an example, this approach could therefore be highly advantageous when modelling systems containing LED filaments, as each filament may contain several hundreds of LEDs.

5. Conclusions and Outlook

To summarize, we created an LED detailed model and an LED CTM following the Delphi4LED methodology and assessed the accuracy and computation time of 3D CFD system-level models equipped with these LED models. Compared to previous work [24], the analysis was extended to luminaire-level. This involved including higher number of LEDs, up to 60, and taking all heat transfer mechanisms into account. The cases discussed in this work demonstrate that using Delphi4LED LED CTMs in digital luminaire designs can provide a significant reduction in computation time while maintaining the required accuracy compared to a LED detailed model.

With this approach, any node could be added to the LED model in order to monitor a thermally critical part of the LED. In this case, the temperature of the node of interest does need to be monitored during the LED testing. This way it can be included in the detailed model calibration in order to obtain an accurate model and predictions for this temperature. It would, for example, be interesting to do this for the phosphor temperature. While our LED CTM has a phosphor node, no phosphor temperatures were measured and accounted for in the calibration of the LED detailed model. Therefore, the modeled phosphor temperatures cannot currently be validated. In many cases, it is also not straightforward to monitor the temperature phosphor temperature. In the present case, the silicone dome covering the phosphor precludes measurements using thermocouples or infrared (IR) thermography. However, methods based on the spectral distribution of the converted light [36,37], or specifically prepared phosphors with magnetic nano-particles [38] could in principle be used.

Finally, the present study focused on 3D thermal modeling of the LED-based lighting designs using an RC-network LED CTM. It is expected that a future implementation of LED BCI-ROMs in 3D system-level models will provide an alternative option for the RC-network CTM. Additionally, only the LED CTM was considered. The implemented model will be extended in future work with a chip-level multi-domain model to obtain a fully realized LED digital twin. Another useful addition to the luminaire-level model is to include lifetime prediction [39], which will be the subject of a follow up paper.

Author Contributions: Conceptualization, G.M.; methodology, M.v.d.S. and G.M.; software, M.v.d.S.; validation, J.Y. and G.M.; formal analysis, M.v.d.S.; investigation, M.v.d.S.; resources, J.Y.; data curation, M.v.d.S.; writing–original draft preparation, M.v.d.S.; writing–review and editing, G.M. and J.Y.; visualization, M.v.d.S.; supervision, G.M.; project administration, G.M.; funding acquisition, G.M. All authors have read and agreed to the published version of the manuscript.

Funding: This research received no external funding.

Conflicts of Interest: The authors declare no conflict of interest.

Abbreviations

The following abbreviations are used:

BCI	boundary condition independent
CCT	correlated color temperature
CFD	computational fluid dynamics
CPU	central processing unit
CRI	color rendering index
CTM	compact thermal model
DSF	differential structure function
DUT	device under test
HTC	heat transfer coefficient
IMS	insulated metal substrate
IR	infrared
LED	light-emitting diode
PCB	printed circuit board

MDCM multi-domain compact model
ROM   reduced order model
SF   structure function

References

1. Delphi4LED Project Website. Available online: https://delphi4led.org (accessed on 27 March 2019).
2. Bornoff, R.; Hildenbrand, V.; Lugten, S.; Martin, G.; Marty, C.; Poppe, A.; Rencz, M.; Schilders, W.H.; Yu, J. Delphi4LED—From measurements to standardized multi-domain compact models of LED: A new European R&D project for predictive and efficient multi-domain modeling and simulation of LEDs at all integration levels along the SSL supply chain. In Proceedings of the 2016 22nd International Workshop on Thermal Investigations of ICs and Systems (THERMINIC), Budapest, Hungary, 21–23 September 2016; pp. 174–180. [CrossRef]
3. Martin, G.; Marty, C.; Bornoff, R.; Poppe, A.; Onushkin, G.; Rencz, M.; Yu, J. Luminaire Digital Design Flow with Multi-Domain Digital Twins of LEDs. *Energies* **2019**, *12*, 2389. [CrossRef]
4. Bein, M.C.; Hegedus, J.; Hantos, G.; Gaal, L.; Farkas, G.; Rencz, M.; Poppe, A. Comparison of two alternative junction temperature setting methods aimed for thermal and optical testing of high power LEDs. In Proceedings of the 2017 23rd International Workshop on Thermal Investigations of ICs and Systems (THERMINIC), Amsterdam, The Netherlands, 27–29 September 2017; pp. 1–4. [CrossRef]
5. Hantos, G.; Hegedus, J.; Bein, M.C.; Gaal, L.; Farkas, G.; Sarkany, Z.; Ress, S.; Poppe, A.; Rencz, M. Measurement issues in LED characterization for Delphi4LED style combined electrical-optical-thermal LED modeling. In Proceedings of the 2017 IEEE 19th Electronics Packaging Technology Conference (EPTC), Singapore, 6–9 December 2017; pp. 1–7. [CrossRef]
6. Onushkin, G.A.; Bosschaart, K.J.; Yu, J.; van Aalderen, H.J.; Joly, J.; Martin, G.; Poppe, A. Assessment of isothermal electro-optical-thermal measurement procedures for LEDs. In Proceedings of the 2017 23rd International Workshop on Thermal Investigations of ICs and Systems (THERMINIC), Amsterdam, The Netherlands, 27–29 Sepember 2017; pp. 1–6. [CrossRef]
7. Farkas, G.; Gaal, L.; Bein, M.; Poppe, A.; Ress, S.; Rencz, M. LED Characterization Within the Delphi4LED Project. In Proceedings of the 2018 17th IEEE Intersociety Conference on Thermal and Thermomechanical Phenomena in Electronic Systems (ITherm), San Diego, CA, USA, 29 May–1 June 2018; pp. 262–270. [CrossRef]
8. *JEDEC JESD51-14: Transient Dual Interface Test Method for the Measurement of the Thermal Resistance Junction to Case of Semiconductor Devices with Heat Flow Trough a Single Path*; Technical Report; JEDEC Solid State Technology Association: Arlington, VA, USA, 2010.
9. *JEDEC JESD51-51: Implementation of the Electrical Test Method for the Measurement of Real Thermal Resistance and Impedance of Light-Emitting Diodes with Exposed Cooling*; Technical Report; JEDEC Solid State Technology Association: Arlington, VA, USA, 2012.
10. *JEDEC JESD51-52: Guidelines for Combining CIE 127-2007 Total Flux Measurements with Thermal Measurements of LEDs with Exposed Cooling Surface*; Technical Report; JEDEC Solid State Technology Association: Arlington, VA, USA, 2012.
11. Zong, Y.; Chou, P.; Dekker, P.; Distl, R.; Godo, K.; Hanselaer, P.; Heidel, G.; Hulett, J.; Oshima, K.; Poppe, A.; et al. *CIE 225:2017 Optical Measurement of High-Power LEDs*; Technical Report; International Commission on Illumination: Vienna, Austria, 2017. [CrossRef]
12. Martin, G.; Marty, C.; Bornoff, R.; Vaumorin, E.; Kleij, A.; Onushkin, G.; Poppe, A. From Measurements to Standardised Multi-Domain Compact Models of LEDs using LED E-Datasheets. In Proceedings of the 29th Quadrennial Session of the CIE. International Commission on Illumination, Washington, DC, USA, 14–22 June 2019; CIE: Vienna, Austria, 2019; pp. 379–386. [CrossRef]
13. Poppe, A. Multi-domain compact modeling of LEDs: An overview of models and experimental data. *Microelectron. J.* **2015**, *46*, 1138–1151. [CrossRef]
14. Poppe, A.; Farkas, G.; Gaál, L.; Hantos, G.; Hegedüs, J.; Rencz, M. Multi-Domain Modelling of LEDs for Supporting Virtual Prototyping of Luminaires. *Energies* **2019**, *12*, 1909. [CrossRef]

15. Bornoff, R.; Farkas, G.; Gaal, L.; Rencz, M.; Poppe, A. LED 3D thermal model calibration against measurement. In Proceedings of the 2018 19th International Conference on Thermal, Mechanical and Multi-Physics Simulation and Experiments in Microelectronics and Microsystems (EuroSimE), Toulouse, France, 15–18 April 2018; pp. 1–7. [CrossRef]

16. Bornoff, R. Extraction of Boundary Condition Independent Dynamic Compact Thermal Models of LEDs—A Delphi4LED Methodology. *Energies* **2019**, *12*, 1628. [CrossRef]

17. Poppe, A.; Hegedus, J.; Szalai, A.; Bornoff, R.; Dyson, J. Creating multi-port thermal network models of LED luminaires for application in system level multi-domain simulation using spice-like solvers. In Proceedings of the 2016 32nd Thermal Measurement, Modeling & Management Symposium (SEMI-THERM), San Jose, CA, USA, 14–17 March 2016; pp. 44–49. [CrossRef]

18. Poppe, A. Simulation of LED based luminaires by using multi-domain compact models of LEDs and compact thermal models of their thermal environment. *Microelectron. Reliab.* **2017**, *72*, 65–74. [CrossRef]

19. Marty, C.; Yu, J.; Martin, G.; Bornoff, R.; Poppe, A.; Fournier, D.; Fournier, D. Design flow for the development of optimized LED luminaires using multi-domain compact model simulations. In Proceedings of the 2018 24rd International Workshop on Thermal Investigations of ICs and Systems (THERMINIC), Stockholm, Sweden, 26–28 September 2018; pp. 1–7. [CrossRef]

20. Codecasa, L. A novel approach for generating boundary condition independent compact dynamic thermal networks of packages. *IEEE Trans. Components Packag. Technol.* **2005**, *28*, 593–604. [CrossRef]

21. Codecasa, L.; D'Alessandro, V.; Magnani, A.; Rinaldi, N.; Zampardi, P.J. Fast novel thermal analysis simulation tool for integrated circuits (FANTASTIC). In Proceedings of the THERMINIC 2014—20th International Workshop on Thermal Investigations of ICs and Systems, London, UK, 24–26 September 2014; Volume 2014, pp. 1–6. [CrossRef]

22. Codecasa, L.; Magnani, A.; D'Alessandro, V.; Rinaldi, N.; Metzger, A.G.; Bornoff, R.; Parry, J. Novel MOR approach for extracting dynamic compact thermal models with massive numbers of heat sources. In Proceedings of the 2016 32nd Thermal Measurement, Modeling & Management Symposium (SEMI-THERM), San Jose, CA, USA, 14–17 March 2016; pp. 218–223.. [CrossRef]

23. Lungten, S.; Bornoff, R.; Dyson, J.; Maubach, J.M.L.; Schilders, W.H.A.; Warner, M. Dynamic compact thermal model extraction for LED packages using model order reduction techniques. In Proceedings of the 2017 23rd International Workshop on Thermal Investigations of ICs and Systems (THERMINIC), Amsterdam, The Netherlands, 27–29 September 2017; pp. 1–6. [CrossRef]

24. Martin, G.; Yu, J.; Zuidema, P.; van der Schans, M. Luminaire Digital Design Flow with Delphi4LED LEDs Multi-Domain Compact Model. In Proceedings of the 2019 25th International Workshop on Thermal Investigations of ICs and Systems (THERMINIC), Lecco, Italy, 25–27 September 2019; Volume 2019; pp. 1–5. [CrossRef]

25. Farkas, G.; Vader, Q.; Poppe, A.; Bognar, G. Thermal investigation of high power Optical Devices by transient testing. *IEEE Trans. Components Packag. Technol.* **2005**, *28*, 45–50. [CrossRef]

26. Farkas, G.; Hara, T.; Rencz, M. Thermal transient testing. In *Wide Bandgap Power Semiconductor Packaging*; Elsevier: Amsterdam, The Netherlands, 2018; pp. 127–153. [CrossRef]

27. Alexeev, A.; Onushkin, G.; Linnartz, J.P.; Martin, G. Multiple Heat Source Thermal Modeling and Transient Analysis of LEDs. *Energies* **2019**, *12*, 1860. [CrossRef]

28. Bornoff, R.; Gaal, L. Comparison of Model Order Reduction and Thermal Network Approaches in the Extraction of Dynamic Compact Thermal Models of LEDs. In Proceedings of the 2019 25th International Workshop on Thermal Investigations of ICs and Systems (THERMINIC), Lecco, Italy, 25–27 September 2019; Volume 2019; pp. 1–7. [CrossRef]

29. Schweitzer, D. Generation of multisource dynamic compact thermal models by RC-network optimization. In Proceedings of the 29th IEEE Semiconductor Thermal Measurement and Management Symposium, San Jose, CA, USA, 17–21 March 2013; pp. 116–123. [CrossRef]

30. Powell, M.J.D. *The BOBYQA Algorithm for Bound Constrained Optimization without Derivatives (NA2009/06)*; Technical Report; Department of Applied Mathematics and Theoretical Physics, University of Cambridge: Cambridge, UK, 2009.

31. Johnson, S.G. The NLopt Nonlinear-Optimization Package. Available online: http://github.com/stevengj/nlopt (accessed on 28 January 2020).

32. Lungten, S.; Alexeev, A.; Onushkin, G. Delphi4LED D1.1—Report on End-User Specifications. Available online: https://delphi4led.org/pydio/public/b610a0 (accessed on 15 April 2019).
33. Lasance, C.J.M. Ten Years of Boundary-Condition- Independent Compact Thermal Modeling of Electronic Parts: A Review. *Heat Transf. Eng.* **2008**, *29*, 149–168. [CrossRef]
34. Lasance, C.; Den Hertog, D.; Stehouwer, P. Creation and evaluation of compact models for thermal characterisation using dedicated optimisation software. In Proceedings of the Fifteenth Annual IEEE Semiconductor Thermal Measurement and Management Symposium, San Diego, CA, USA, 9–11 March 1999; pp. 189–200. [CrossRef]
35. Lasance, C.J. The conceivable accuracy of experimental and numerical thermal analyses of electronic systems. *IEEE Trans. Components Packag. Technol.* **2002**, *25*, 366–382. [CrossRef]
36. Kusama, H.; Sovers, O.J.; Yoshioka, T. Line Shift Method for Phosphor Temperature Measurements. *Jpn. J. Appl. Phys.* **1976**, *15*, 2349–2358. [CrossRef]
37. Yang, T.H.; Huang, H.Y.; Sun, C.C.; Glorieux, B.; Lee, X.H.; Yu, Y.W.; Chung, T.Y. Noncontact and instant detection of phosphor temperature in phosphor-converted white LEDs. *Sci. Rep.* **2018**, *8*, 296. [CrossRef] [PubMed]
38. Du, Z.; Sun, Y.; Su, R.; Wei, K.; Gan, Y.; Ye, N.; Zou, C.; Liu, W. The phosphor temperature measurement of white light-emitting diodes based on magnetic nanoparticle thermometer. *Rev. Sci. Instruments* **2018**, *89*, 94901.. [CrossRef] [PubMed]
39. Hegedüs, J.; Hantos, G.; Poppe, A. Lifetime Modelling Issues of Power Light Emitting Diodes. *Energies* **2020**, *13*, 3370. [CrossRef]

Article

Mixed Detailed and Compact Multi-Domain Modeling to Describe CoB LEDs

László Pohl *, Gusztáv Hantos, János Hegedüs, Márton Németh, Zsolt Kohári and András Poppe *

Department of Electron Devices, Budapest University of Technology and Economics, Magyar tudósok körútja 2, bldg. Q, 1117 Budapest, Hungary; hantos@eet.bme.hu (G.H.); hegedus@eet.bme.hu (J.H.); nemeth@eet.bme.hu (M.N.); kohari@eet.bme.hu (Z.K.)
* Correspondence: pohl@eet.bme.hu (L.P.); poppe@eet.bme.hu (A.P.); Tel.: +36-1-463-2704 (L.P.); +36-1-463-2721 (A.P.)

Received: 6 July 2020; Accepted: 27 July 2020; Published: 5 August 2020

Abstract: Large area multi-chip LED devices, such as chip-on-board (CoB) LEDs, require the combined use of chip-level multi-domain compact LED models (Spice-like compact models) and the proper description of distributed nature of the thermal environment (the CoB substrate and phosphor) of the LED chips. In this paper, we describe such a new numerical solver that was specifically developed for this purpose. For chip-level, the multi-domain compact modeling approach of the Delphi4LED project is used. This chip-level model is coupled to a finite difference scheme based numerical solver that is used to simulate the thermal phenomena in the substrate and in the phosphor (heat transfer and heat generation). Besides solving the 3D heat-conduction problem, this new numerical simulator also tracks the propagation and absorption of the blue light emitted by the LED chips, as well as the propagation and absorption of the longer wavelength light that is converted by the phosphor from blue. Heat generation in the phosphor, due to conversion loss (Stokes shift), is also modeled. To validate our proposed multi-domain model of the phosphor, dedicated phosphor and LED package samples with known resin—phosphor powder ratios and known geometry were created. These samples were partly used to identify the nature of the temperature dependence of phosphor-conversion efficiency and were also used as simple test cases to "calibrate" and test the new numerical solver. With the models developed, combined simulation of the LED chip and the CoB substrate + phosphor for a known CoB LED device is shown, and the simulation results are compared to measurement results.

Keywords: Light-emitting diodes; power LEDs; CoB LEDs; multi-domain modeling; finite volume method; phosphor modeling

---

1. Introduction, Related Work

Commercial LED-based white lighting devices work in the following ways [1]:

a. Three individual monochromatic LED elements emitting red, green and blue colors are mixed, to produce light with the required chromaticity, including white;

b. Blue or near-ultraviolet LED chips are used to excite yellow phosphorous to provide white light (phosphor with emission peak in red is sometimes also added for the sake of improved color rendering).

The first way ensures the most versatile options for the user to tune the light, but arises a few serious problems: Very complex driving circuitry, bad long term stability, due to the different ageing of the three kinds of LEDs, high production cost and last, but not least, usually provide lower color rendering indexes than phosphor-converted white LEDs. Therefore, nowadays the mainstream lighting applications are based on the two latter ways, mostly on the last one. The second option enables easy tuning of the resulting light during the production technology. The last option provides the lowest production cost, but also limits the variability of the light the most; such LED devices are called phosphor-converted white LEDs (pc-WLEDs). In the subsequent parts of this paper, we shall also refer to such LEDs simply as white LEDs. This paper deals with multi-chip, large-area packages where the blue LED chips are directly attached to a common ceramics substrate, where this common chip carrier substrate also constitutes the LED package itself, and we focus on pc-WLED devices realized with single yellow phosphor-conversion, and chip-on-board (CoB) assemblies built of them.

Phosphor materials consist of a host compound and optical activator dopant ions. Appropriate phosphor materials used in pc-WLEDs should meet the following six basic criteria [2]:

1. An excitation spectrum showing good overlap with the pumping LED chips: High absorption of n-UV (360–420 nm) or blue light (420–480 nm).
2. An emission spectrum combination with the emission of LED, phosphors provide a pure white emission with a high color rendering index and allow to achieve low correlated color temperatures.
3. Efficient luminescence with a high quantum efficiency (QE).
4. Low thermal quenching of photoluminescence.
5. High stability against oxygen, carbon dioxide, chemicals, and moisture under application conditions.
6. Mild synthesis conditions, reasonable production costs.

Although many phosphor materials have been proposed in the literature in recent years, the number of phosphors effectively fulfilling all six requirements is relatively small [3]. Host materials include garnets, sulphides, (oxo-) nitrides, silicates, aluminates, borates, phosphates, and so on. The most frequently used activators are either broad-band emitting transitional metals Eu^{2+}, Ce^{3+}, Yb^{2+} ions, or line-emitting rare-earth ions Ln^{3+} and Mn^{4+}, etc. [3,4]. The first commercially available pc-WLEDs invented by Nichia Corporation was fabricated using blue InGaN LED chip and the yellow yttrium aluminium garnet $Y_3Al_5O_{12}:Ce^{3+}$ (YAG:Ce) phosphor.

For efficiency and long-term stability reasons, today, most commercial single-phosphor-converted white LED devices are still based on YAG:Ce [5]. A detailed discussion of the underlying physical effects (4f–5d transition, d-d transition) and of the structural design of phosphor materials can also be found there.

A much higher luminous emittance and conversion efficiency can be achieved by using nano-structured YAG:Ce ceramic phosphor plate and a high power blue laser diode for excitation [6]. The optimal Ce^{3+} dopant concentration, resulting in the highest luminous emittance and conversion efficiency was found at 0.5 mol%. Investigation of such solid-state light-sources, however, is beyond the scope of this paper.

The modeling of the phosphor layer of a white LED primarily means optical modeling; following the light-scattering, light absorption and light frequency (wavelength) conversion, which happen inside the phosphor layer. Simulating these processes calculated their thermal effects too, which results in a multi-domain model of the phosphor layer. There are many solutions for modeling these effects, from simple one-dimensional models through using the bidirectional scattering distribution functions to the detailed 3D models. Here we only summarize some examples that use these methods. 1D modeling of light is used in papers [7,8] where the model verification for thin phosphor layers is given. We also used a similar, modified model. This model [7,8] is improved to study the effect of non-homogenous phosphor concentration in Reference [9], although the simulation results, in that case, do not match the measurement results. The expected heat generation in the phosphor layer is calculated in article [10] without comparison to measurement.

The most commonly used method for establishing the optical model of a phosphor layer is to measure the bidirectional scattering distribution function, which gives the relationship between the radiance and emission of the phosphor layer by infinitesimal solid angle for both incoming and outgoing light. Once the bidirectional scattering function is recorded, the optical behavior can be modeled by simple integration. The method is used for phosphor layer modeling with experimental validation [11]. These measurements are made on phosphor plates only. In Reference [12], phosphor-coated LED optical modeling and measurements are reported. This modeling technique is quite accurate for optical modeling, but as the microscopic details are not known, only the macroscopic thermal model can be established.

Detailed models are also used for optical modeling of phosphor layers. There are several commercial software tools available for this purpose. [13] For example in paper [13] Tan et al report about the use of TracePro and ANSYS. For phosphor-converted CoB device modeling, FloTHERM from Mentor Graphics was also used [14]. In this work, an emphasis is put on the thermal aspects. LightTools from Synopsys allows detailed modeling of phosphor layers with user-defined properties [15], with a focus on simulating light properties, but without considering thermal effects. In their paper [16] Alexev et al. describe the combined use of ANSYS and LighTools for the study of single-chip, custom-made white LEDs; optical results of their simulations are checked by luminance measurements in a special test setup while the correctness of the thermal simulations is checked with the help of structure functions extracted from the simulation results and from thermal transient measurements performed by Mentor Graphics' T3Ster equipment [17]. In the special, custom-made mid-power LEDs investigated, they used their own custom-made phosphor composites. To help set up their combined thermal and optical simulation model, they measured the thermal conductivity, as well as the reflection, excitation and emission spectra of these phosphor composites. In their paper [18] Jeon et al. also report their measurements of phosphor properties aimed as input for optical modeling of white LEDs, though, this publication does not provide any information about the temperature dependence of these properties. The paper of Qian et al. [19] provides a detailed review of combined optical-thermal modeling of phosphor-converted white LEDs and presents an example for LED filament bulb. Unfortunately, in none of these publications is the interaction with the electric domain through the blue pump LED chip(s) included.

A few multi-domain models have already been created. For example, an optical-electrical- thermal compact model was published by Ye at al., where the phosphor layer is taken into account with temperature dependence [20]. In this paper, single and multi-chip white LEDs (with contact phosphor layers) and remote phosphor solutions are investigated. The multi-chip structure they studied is very close to the structures of the CoB LED devices. In their model, the electrical behavior of the LED chips is lumped into the energy conversion efficiency.

Compact model for multi-domain purposes with a remote phosphor layer presented in [21], where applying bidirectional resistances showed good agreement with the measurements. A similar solution can be found in Reference [22]. In Reference [21], a large are multi-chip white LED device is studied with a structure close to that of white CoB LEDs, through measured, so-called 'ensemble' characteristics (see later). For modeling heat transfer from the LED chips' junctions to the environment both in Reference [21,22] the so-called bidirectional thermal resistance model is used. The thermal resistance values needed for such a model are identified from thermal transient measurements with the help of the structure functions. In Reference [21], the authors provide a final single equation for the total luminous flux in which both the bidirectional resistance model and the Shockley type model for the IV characteristics are included. The heat dissipation coefficient introduced by the authors, in which the light propagation properties in the phosphor are embedded, is also part of this equation. In summary, the model presented in this paper can be well applied to represent the 'ensemble' characteristics of CoB LEDs, but is not able to provide detailed information on the lateral and vertical temperature distributions in the phosphor layer and cannot provide information on the individual junction temperatures of the blue pump chips of the LED array.

In our current paper, we summarize our multi-domain modeling solution for phosphor- converted LED devices, which is a mixed, compact-detailed model by using one of the "standard" chip-level multi-domain LED models for the description of the operation of the blue pump chips within CoB devices. Only the multi-domain nature of the operation of the blue LED chips can be represented by a compact model; the blue LED chips' thermal environment (substrate, phosphor) of a CoB device has to be considered by a distributed, detailed 3D model. This way detailed studying of thermal phenomena in the phosphor layer (both in the vertical and lateral direction), including the effect of local interaction of the phosphor and the blue pump LED chips within the CoB array is made possible. In Figure 1, we provide a summary of physical processes taking place in the different major structural elements of a phosphor-converted white CoB LED device that we aimed to cover with our simulation approach.

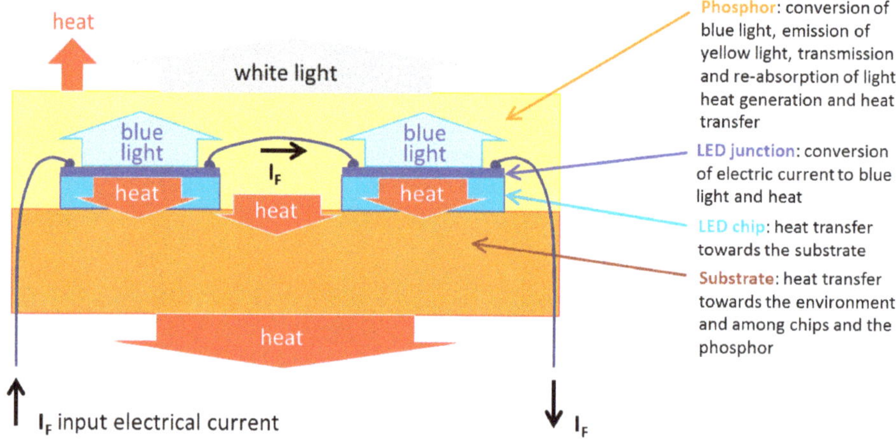

Figure 1. Overview of the physical processes in different, major structural elements of a phosphor-converted white chip-on-board (CoB) LED device to be captured by dedicated simulation models.

The organization of our paper is as follows: In Section 2, we describe the context and the goals of this work. Section 3 deals with the major bottleneck: How to identify the material properties of phosphor layers needed for multi-domain simulation. We were inspired by References [16,18] to also prepare stand-alone custom phosphor samples that measure the temperature dependency of the phosphor properties. Moreover, with the same phosphor mixtures, we prepared single-chip white LED samples with well-controlled properties in order to allow us to fine-tune and to validate our simulation methods and models. Details on this part of our work are also provided in Section 3. In Section 4, we introduce our coupled chip + phosphor multi-domain simulation model along with the description of the light and heat propagation models of different complexity. In Section 5, we apply the introduced models to a commercially available CoB LED device that has also been characterized by common thermal and optical measurements as well. The comparison of the simulation and measurement results obtained for this device is provided in Section 6. In Section 7, we provide a summary and conclusions. In the Abbreviations we provide a summary of abbreviations and symbols used in this paper.

This paper, as a significantly extended version of our THERMINIC 2019 conference paper [23] provides a comprehensive summary of our CoB LED multi-domain modeling related work (parts of which have already been published at other conferences as well [24,25]) completed with a few, recent measurement results.

2. Background, Related Own Work

The work described here has been carried out in the framework of the recently completed European H2020 ECSEL research project Delphi4LED [26]. The major focus of the project was placed on single-chip LED packages and luminaires made thereof, applying a modular approach in the overall luminaire design process [27]. In the Delphi4LED approach, multi-domain behavior is treated on

LED chip-level, by means of Spice-like compact (lumped) models [28]. The thermal effect of the LED package physical structure is also described by compact models [29]. To consider farther elements of the thermal environment, there are two options. On the one hand, the luminaire's thermal behavior and the effect of its thermal environment can be represented by yet another compact thermal model [30], and the entire model (including the LED chips models completed with the compact thermal models of their packages) forms a Spice netlist that can be simulated with any Spice compatible circuit simulator. On the other hand, another approach has also been developed within the Delphi4LED project: The 3D thermal environment of the LED chip is considered by a detailed 3D thermal model. In this approach, a 3D thermal simulator is modified in a way that it can iterate between the chip-level multi-domain LED model and the thermal solver, as illustrated in Figure 2. An implementation of such a scheme was used in the Delphi4LED project to demonstrate the use and benefits of the "industry 4.0" like design workflow suggested by the project [31].

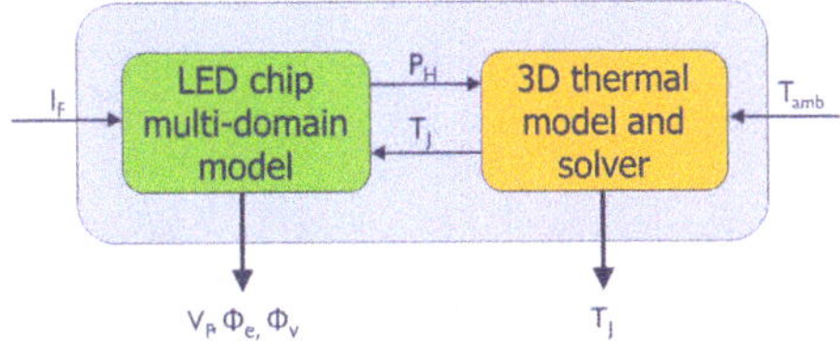

Figure 2. A chip-level multi-domain LED model embedded in a thermal simulator using a relaxation type iteration in order to realize an electro-thermal-optical solver used for the virtual prototyping of luminaires based on single-chip LED packages [27,29,31]. For the explanation of symbols used in this figure see the Abbreviations.

In the case of phosphor-converted white CoB LEDs, however, the first approach of using compact models only cannot be applied because of the distributed, multi-domain nature of the phosphor layer, involving:

- Considering the light propagation properties in 3D (absorption/transmission);
- Temperature dependence of phosphor properties (among those, that of the conversion efficiency);
- The resulting heat generation and temperature rise in the phosphor layer.

This is illustrated in Figure 1. The thermal effect of the phosphor layer (distributed heat source over the entire area of the CoB device) cannot be separated from the thermal behavior of the rest of the structural elements of a CoB LED, therefore:

- An appropriate optical-thermal model of the phosphor should be set up (with light absorption/emission, heat generation and temperature dependence described consistently);
- Integrated with the thermal model of the CoB device as it is illustrated in Figure 3.

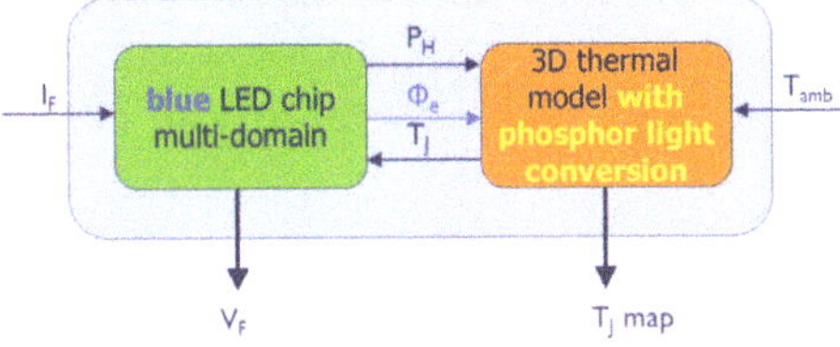

Figure 3. Application of a chip-level multi-domain LED model in a relaxation type implementation of an electro-thermal-optical solver, with a detailed 3D thermal model of the CoB LED chips' thermal environment completed with a thermo-optical model describing the light conversion and heat generation in the phosphor. For the explanation of symbols used in this figure see the Abbreviations.

Regarding the description of the multi-domain behavior of the blue LED chips, we relied on one of the Spice-like multi-domain LED models we developed earlier [28]. The heat transfer within the bulk of the LED chips, the ceramics substrate and within the phosphor layer is treated by BME's proprietary conduction-mode only thermal field solver [32,33] that has already been successfully adapted to the multi-domain modeling of large-area OLED devices [34,35].

The most important differences and similarities between the former multi-domain OLED simulator and the present approach for CoB LEDs are the following:

- The multi-domain behavior of OLED junctions is of distributed nature while in CoB LED devices, the junctions of the individual blue LED chips are represented by a compact (Spice-like) multi-domain model.
- Thermal modeling of the light-emitting polymer layer (LEP) of OLED and the phosphor layer of CoB LEDs is similar: In both cases heat, transfer in these layers need to be modeled along with the heat generated by conversion losses. The way how the dissipated heat in these layers is calculated, however, is different. In the OLED model, the LEP layer was considered two-dimensional; the heat was generated at the point of the light emission. However, the phosphor layer of CoB LEDs requires complex handling of a true three-dimensional model of light and conversion losses. For modeling the heat transfer, the layer thickness and thermal conductivity have to be known.
- Both in OLED LEP layers and in CoB phosphor layers conversion efficiency (electricity to light in the case of OLED LEPs, blue light to yellow light in CoB LED phosphors) depends on the local temperature.

The electrical interconnect network of blue LED chips of a CoB device is considered as zero-dimensional electrical nodes (with no voltage drop in the interconnects). The typical electrical configurations of LED chips inside a CoB device are shown in Figure 4.

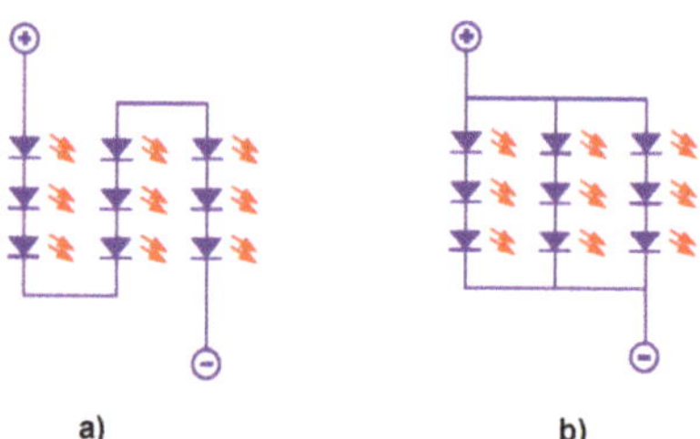

Figure 4. Typical electrical configurations of the arrays of blue LED chips within a CoB LED device package: (**a**) A single string of serially connected LED chips, (**b**) parallel connection of multiple serially connected strings of LED chips. (After [36]).

As seen in Figure 4, none of the electrical configurations of the arrays of blue LED chip inside a CoB device provides individual access to any of the chips within the array. This means that with the common I_F forward current we power the entire LED array and we can measure the total radiant/luminous flux of the entire array that is the sum of the fluxes emitted by the individual chips and converted by the phosphor. For a single LED string, the situation is the same for the overall forward voltage of the string. These measured characteristics are called 'ensemble' characteristics in the JEDEC JESD 51-51 standard [36].

The relationship between the overall, ensemble characteristics of an LED array and the individual chip characteristics (as illustrated in Figure 5 for the forward voltage) are

$$V_{F\_ensemble} = \sum_{i=1}^{N} V_{F\_i} \tag{1}$$

$$V_{F_{chip}}\left(I_F,\ T_{J\_ensemble}\right) = V_{F\_ensemble}\left(I_F,\ T_{J\_ensemble}\right)/N \tag{2}$$

$$\Phi_{X\_ensemble} = \sum_{i=1}^{N} \Phi_{X\_i} \tag{3}$$

$$\Phi_{x_{chip}}\left(I_F,\ T_{J\_ensemble}\right) = \Phi_{x\_ensemble}\left(I_F,\ T_{J\_ensemble}\right)/N \tag{4}$$

where Φ_X represents either the radiant flux, Φ_e or the luminous flux, Φ_V, N is the number of the LED chips in the LED string forming the LED array. $V_{F_{chip}}$ and $\Phi_{x_{chip}}$ represent the average forward voltage and flux data that can be related to an individual LED chip within the array.

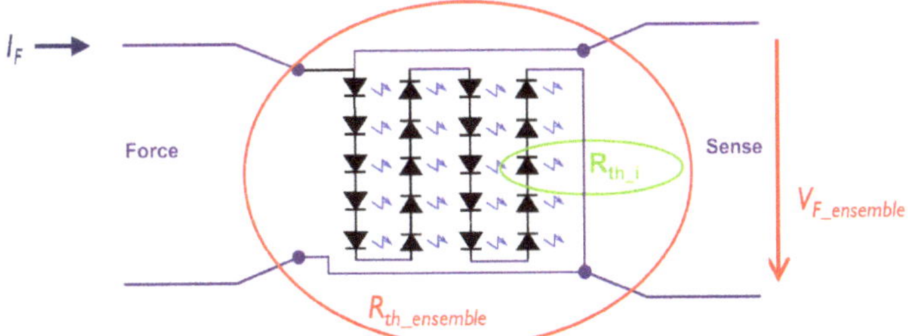

Figure 5. Illustration of the 'ensemble' thermal resistance and forward voltage of a serially connected LED array (as per Figure 4a) in the powering/measurement scheme recommended by the JEDEC JESD 51-51 standard [36] for thermal testing. For the explanation of symbols used in this figure see the main text and the Abbreviations.

This means, that the measured isothermal IVL characteristics of a CoB device need to be post-processed before applying the parameter extraction procedure to obtain the multi-domain chip-level model parameters to be used by the chosen Spice-like multi-domain LED model. Note, that in the case of using a combined thermal and radiometric/photometric test setup as suggested by the JEDEC JESD 51-51 and JESD 51-52 standards [36,37] for CoB LED measurements, there is no way to identify the T_{J_i} individual junction temperatures of the blue LED chips within the entire array. As the best approximation, one has to calculate with the $T_{J\_ensemble}$ temperature as if it was a uniform temperature for each LED chip within the CoB device. Thus, the set of isothermal IVL characteristics is given by the $V_{F_{chip}}\left(I_F,\ T_{J\_ensemble}\right)$ and $\Phi_{x_{chip}}\left(I_F,\ T_{J\_ensemble}\right)$ data as given by Equation (2) and Equation (4), respectively.

To test the validity of the Equations (1)–(4) we created an array from individual LED packages on a cold plate with a diameter of 12 cm that was attached to a 50 cm integrating sphere. In this setup, the LEDs could be powered and characterized both individually and together as an array of LEDs. It was found that the chip-level characteristics derived from the measurement results of the entire array were very close to the averages of the individually measured voltages and fluxes, suggesting that the above-mentioned approximation for the individual characteristics of the individual LED chips is acceptable.

3. Characterization of the Phosphor

3.1. Methodology to Set up and to Validate the Multi-Domain Model of the Phosphor

The major bottleneck in the modeling of phosphor-converted white LEDs like the CoB devices is to get access to the properties of the phosphor layers as manufacturers do not share such information. Regarding the thermal properties, one can find multiple approaches in the literature. Papers [38,39] describe numerical simulation methods for the calculation of the effective thermal conductivity of different phosphor-resin mixtures with different phosphor particle concentrations. In both papers the simulation results are compared to measurements: In Reference [38], the laser flash method, while in Reference [39] the transient hot-wire method is used to measure the effective thermal conductivity of phosphor layers. A. Alexeev et al. also investigated the effect of phosphor particle concentration on the overall thermal resistance of white LED packages [40]. Wenzl et al. in their paper [41] describe how heat generation in the phosphor layers depends on the extinction coefficient (i.e., light absorption). In this work, besides the thermal conductivity, further material properties of the phosphor layers, such as quantum efficiency are also considered in the simulations. The authors used a simple LED reference structure that inspired our present work (see later). Data available in these papers, unfortunately, did not help us set up our own models, especially regarding temperature dependence of the light conversion.

Though Bachmann in his PhD dissertation [42] provides detailed measurement data of different kinds of phosphors, including a few graphs showing the temperature dependence of luminescence intensity, there is no data on efficiency and thermal properties of phosphor powder-resin composites; the limited data on the temperature dependence of luminescence intensity could not be used for our modeling purposes. As a workaround to the problem of lack of sufficient data on temperature-dependent behavior of phosphors, we had the following approach [25]:

- We used commercially available phosphor powders for the preparation of a large-area (cca. 5 cm in diameter) phosphor sample using spin-coated PDMS (polydimethylsiloxane) as a host matrix. PDMS-phosphor composites with different mass fractions of the two constituents were created. The photon conversion properties and their temperature dependency, as well as the thermal conductivity of the samples, were measured.
- Using the same phosphor-PDMS composites the original phosphor-lens structure of flip-chip assembled LED packages were replaced by our own, custom-made phosphor-lens structures. Before attaching our own phosphor-lens structures, the bare blue LEDs were fully characterized by isothermal IVL measurements.
- All custom-made white LEDs have also been fully characterized by the measurement of their isothermal IVL characteristics. After these measurements, the custom-made lenses were removed from the blue LED chips, and the phosphor layer thicknesses were measured by cross-sectioning. With each composite and phosphor layer thickness, multiple white LED samples were prepared.
- These custom-made white LEDs were used as simple reference structures (see later in Section 3.3) to set up, test and validate our multi-domain phosphor model [24].
- The phosphor multi-domain model validated this way was built into our proprietary finite volume based thermal simulation code (SUNRED). The multi-domain compact model of LED chips was also included in this solver, following the scheme sketched in Figure 3.
- The CoB LED device (Lumileds 1202s CoB) fully characterized during the round-robin test of the Delphi4LED project [43] was modeled and simulated with multi-domain simulation engine. The properties of the phosphor layer of these CoB LEDs were measured (see later), and some layer thicknesses of these CoB LEDs were also identified by cross-sectioning.

The subsequent (sub)sections of this paper describe the details of the steps listed above.

3.2. Study of the Relevant Properties of Phosphor Layers

In order to establish a thermo-optical model for the phosphor layer, first, we investigated the features of some phosphor materials. Our initial idea was to derive phosphor properties from measured spectra of blue and phosphor-converted white LEDs of the same LED family (XP-E LEDs of Cree) in the hope that in the white LEDs the same blue chips were used. In the measured spectra the blue peak wavelengths differed; therefore, the assumption, that the only difference between the two kinds of LEDs was the phosphor, was questioned.

As a next step in setting up a proper multi-domain simulation model for CoB LED devices, standard, commercial CoB LED devices in their unpowered state were considered as stand-alone phosphor samples (Figure 6). We prepared different PDMS/phosphor powder composites to be characterized as stand-alone samples (see Figure 7b) to obtain certain parameters of the phosphors.

(**a**) (**b**)

Figure 6. Commercially available CoB LED devices in their unpowered state were also measured as stand-alone phosphor samples: (**a**) A Lumileds 1202 s CoB LED device; (**b**) multiple such CoB LED devices attached to a temperature-controlled stage to be measured as stand-alone phosphor samples.

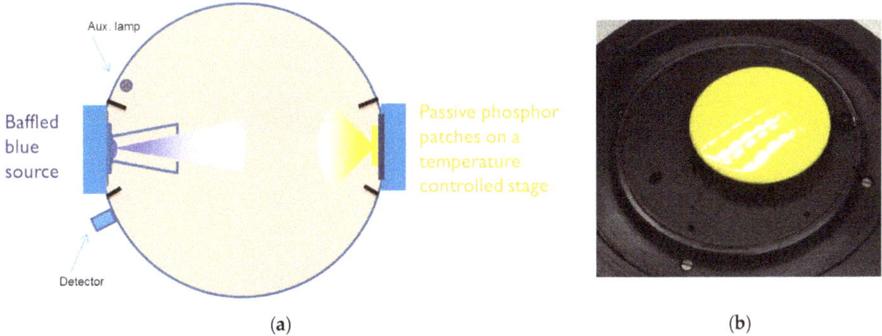

(**a**) (**b**)

Figure 7. Test setup for phosphor sample measurements: (**a**) Schematic of the integrating sphere arrangement with excitation blue light source and passive phosphor layers attached to a temperature-controlled stage; (**b**) photograph of a custom-made phosphor sample attached to a temperature-controlled stage.

A 50 cm integrating sphere with dual DUT (device under test) ports was used in an arrangement, as seen in Figure 7, to capture the spectral power distribution (SPD) of the secondary emission of the phosphor samples with a CAS-140CT spectroradiometer. All phosphor samples were attached to a temperature-controlled stage. By sweeping the temperature of that stage spectra, were captured at phosphor temperatures between 15 °C and 150 °C.

The integrating sphere that we used had two DUT ports facing each other along the equator of the sphere (Figure 7a). This arrangement allowed us to install a blue excitation light source at one port, to focus the excitation blue light on the phosphor sample mounted on a temperature-controlled stage on the other port of the sphere. A cone with a black outer surface with a small aperture was used to decrease the amount of blue light inclining not the sample, but the sphere. This ensured to capture reasonable levels of converted light without saturating the spectroradiometer with the blue excitation.

The samples were exposed to variable intensities of blue light. The base material of the phosphor samples was PDMS. 1 mm thick PDMS medals with different mass fractions of phosphor powders were prepared on a black-painted aluminum plate that was attached to the above-mentioned temperature-controlled stage (as shown in Figure 7b). Also, the whole thermostat was painted black to minimize the backscattering of blue light into the sphere. Figure 8 presents a set of spectra of the blue excitation and the secondary emission of a phosphor sample measured at different phosphor temperatures.

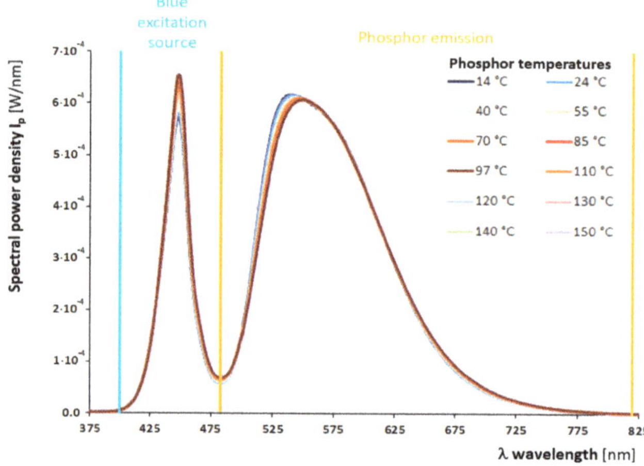

Figure 8. Temperature-dependent spectral power distribution (SPD) of a custom-made remote phosphor sample.

As our integrating sphere is not calibrated in the given geometrical arrangement, only an estimated blue reference SPD could be used for temperature-dependent efficiency calculations. For the definition of the different efficiency parameters of the phosphor refer to F. Schubert's widely known book on LEDs [44]. Figure 9 shows the calculated temperature dependence of different properties, such as conversion efficiencies for one of the characterized custom-made PDMS samples.

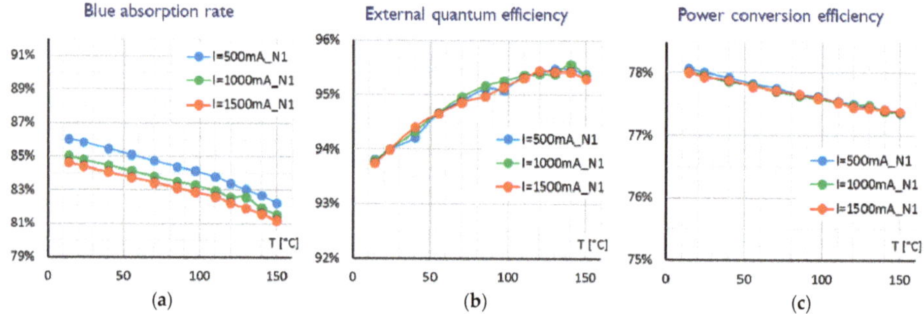

Figure 9. Measured temperature dependence of (**a**) the blue absorption rate, (**b**) the external quantum efficiency, and (**c**) the power conversion efficiency for a phosphor sample prepared from one of the commercially available phosphor powders.

As seen in the diagrams presented in Figure 9, linear or second order relationships can well be used as good approximations for the temperature dependence of the measured properties. The measurement results show that not only the efficiency of the phosphor layer depends on the temperature, but slightly the wavelength of the converted photons (thus, the emission spectra) as well. Because of the experienced linearity in conversion efficiency, this effect can be built in the multi-domain model. Based on these experimental data, we could set up the thermo-optical phosphor model that we built into our thermal solver; the parameters regarding the temperature dependence can be well fitted to data of other phosphor materials.

Besides the temperature dependence of the light conversion properties, the thermal conductivity of the manufactured PDMS and phosphor-powder mixtures was measured with the DynTIM equipment of Mentor Graphics [45], see Figure 10. The thermal conductivity measurement results for one of the phosphor powder types are shown in Table 1.

Figure 10. One of our custom-made phosphor samples on the measurement stage of a Mentor DynTIM (dynamic thermal interface material thermal conductivity measurement) equipment [45].

Table 1. Measured thermal conductivity of phosphor layer with different phosphor powder concentration.

Mass Fraction [m/m %]	Thermal Conductivity [W/mK]	Variance in Thermal Conductivity Measurement
0	0.22	0.004
25	0.29	0.009
50	0.4	0.002
66	0.61	0.006
75	0.66	0.001

3.3. Characterization of Custom-Made Phosphor-Converted White LEDs

To test the validity of the phosphor model we created custom-made phosphor-converted white LEDs with precisely known structure, both in terms of the bare, packaged blue LED chips and the added phosphor layers. These devices were used as reference structures for fine-tuning the phosphor model attached to our thermal solver. Also, with these single-chip LEDs we avoided all uncertainties associated with the 'ensemble' characteristics of the actual CoB structures, as well as obtained blue LED spectra and white LED spectra where the blue peaks precisely matched.

Figure 11 provides some details of the 'fabrication process' of our own custom-made white LEDs. XPG3 flip-chip power LEDs from Cree were used such, that their original lenses were removed (Figure 11a). At this stage, each blue LED was characterized by isothermal IVL measurements. Phosphor-converted white LEDs were then fabricated by proximate conformal phosphor deposition before forming the clear lens (Figure 11b,c). We used PDMS + phosphor powder mixtures as for the characterization of stand-alone phosphor samples discussed in the previous section. The phosphor powder was mixed with PDMS in 50–50 m/m %, and light conversion layers of four different thicknesses were deposited on the already characterized bare blue LEDs. This way, white LEDs with four different spectral power distributions (thus, four different correlated color temperatures) were achieved. With varying the phosphor thickness, our aim was to convert a different number of photons using the same blue excitation every time. With this technique, we assured that the only differences between the measurement results for the blue and the white LEDs were caused by the phosphor layers themselves.

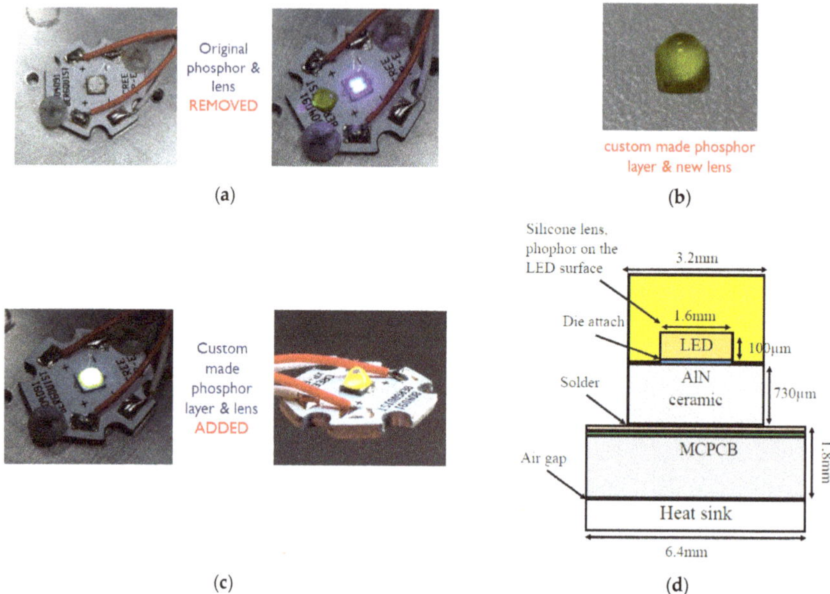

Figure 11. Creating our own white LEDs from Cree XPG3 LEDs: (**a**) Bare blue LED packages with original phosphor and lens removed, (**b**) custom-made phosphor layer with known composition and new lens, (**c**) blue LED packages with the new phosphor + lens structure attached, (**d**) simplified cross-sectional view of the custom-made LED structures used for simulations [24] to validate the simulator. For the explanation of abbreviations used in this figure see the Abbreviations.

The flip-chip assembly of the base LED device was chosen to make sure that while the original dome with phosphor is removed and our own custom-made phosphor layers are added, all electrical connections of the LED chips remain safely untouched.

For each phosphor layer setup, complete isothermal IVL characterization was performed for six forward current values at five junction temperatures.

After measurement of the isothermal IVL characteristics, the custom lenses were dismounted, and cross-sectioned to measure the thickness of the phosphor layers. As we experienced, the temperature of the phosphor layer has a significant impact on the external quantum efficiency, and on blue absorption, which not only affects the efficiency of the LED, but the color of the resulting light too. That is one reason why it is necessary to establish joint compact and detailed multi-domain (thermal, electrical and optical) models in the case of white LEDs, especially for CoB devices.

The thickness dependence of the total dissipated power and the temperature rise of the phosphor layers is shown in Figure 12. This set of data was used to validate the multi-domain phosphor model when applied to the custom-made single-chip reference LED devices.

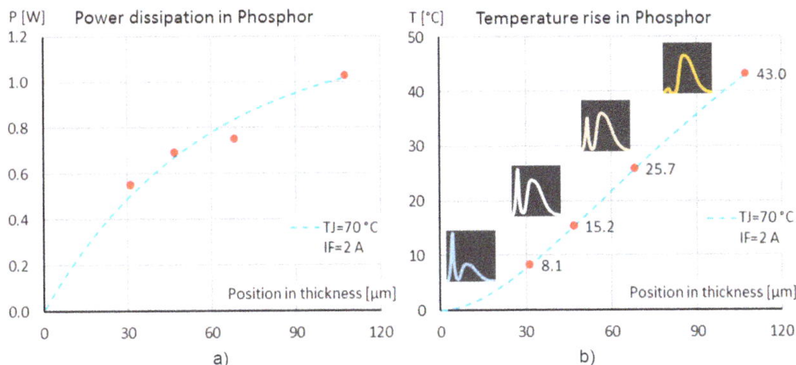

Figure 12. The layer thickness dependence of the major thermal properties of the custom-made phosphor layers identified from the measurements: (**a**) Thickness dependence of the total dissipation in the phosphor layer; (**b**) temperature rise of the phosphor layer.

4. Multi-Domain Modeling: Chip-Phosphor Interaction, Light and Heat Propagation

From now on, we call the color of the absorbed light of the primary emitter LED chips as blue and the converted and re-emitted light color as yellow. For modeling phosphor layers different effects need to be considered, such as blue absorption, blue scattering, blue-to-yellow conversion (Stokes shift), yellow absorption, yellow scattering and the temperature dependence thereof, if applicable.

We distinguish our phosphor models according to the way the light path is followed in 1D (distance from the source), or in 3D. In certain cases, the absorption and reflection on the LED chip/substrate surface are taken into account. In the following section, we discuss these approaches with their possible limitations.

4.1. A 1D Phosphor Model

In our 1D model, we consider the blue and yellow absorption and the wavelength conversion. We use simple formulae to approximate the optical (radiant) power of the blue and yellow light, and for the heat generated due to conversion and absorption losses.

When the blue light propagates through the phosphor layer, it may be absorbed or converted by the phosphor particles. Assuming that the particle concentration in the phosphor layer is homogeneous, the blue photon number follows the Lambert-Beer law:

$$N_B(x) = N_e \cdot e^{-\mu_b x} \tag{5}$$

where N_B is the blue photon number at distance x (measured from the source), N_e is the originally emitted photon number (calculated from the optical power), μ_b is the sum of the attenuation and conversion coefficients $\mu_b = \mu_{ba} + \mu_{bac}$, and is proportional to the probability of hitting a phosphor particle. As there is no yellow light emitted from the LED chips, the source of yellow photons is the conversion by a phosphor particle, so the number of yellow photons can be written as:

$$N_Y(x) = N_e \cdot (1 - e^{-\mu_c x}), \tag{6}$$

where μ_c is the conversion coefficient.

Considering the yellow attenuation from the distribution of yellow source:

$$N_Y(x) = N_e \frac{\mu_c (1 - e^{-\mu_b x})}{\mu_y + \mu_b} \tag{7}$$

229

Note that the converted yellow photons do not follow the direction of blue photons. The direction of propagation of the converted yellow photons can be assumed to be isotropic. Therefore, only half of the converted yellow photons will propagate away from the LED.

The other half starts to move in the direction of the surface of the LED chip. Let the thickness of the phosphor layer be d and let the yellow source (x') be above point x of the blue source, where the photon number sought can be described as:

$$N_{ys}(x') = N_e \cdot \mu_c e^{-\mu_b x'} \tag{8}$$

The attenuation of that source from x' also given by the Lambert-Beer law:

$$N_y(x) = N_e \cdot \mu_c e^{-\mu_b x'} \cdot e^{\mu_y (x-x')}. \tag{9}$$

With the help of integration, we can get the formula for the photons arriving above point x:

$$N_{yB}(x) = \int_x^d N_e \cdot \mu_c e^{-\mu_b x'} \cdot e^{\mu_y (x-x')} dx' = N_e \frac{\mu_c \left(e^{-\mu_b x} - e^{-\mu_b d}\right)}{\mu_y + \mu_b}. \tag{10}$$

The number of yellow photons reflected from the LED chip surface is calculated from the power of yellow light incident at the surface. The latter is

$$N_y(0) = N_e \cdot \frac{\mu_c \left(1 - e^{\mu_b d}\right)}{\mu_y + \mu_b}. \tag{11}$$

Considering also the r reflection coefficient (the ratio of reflected and absorbed photons), and the attenuation from the LEDs surface we get:

$$N_{yR}(x) = r \cdot N_e \frac{\mu_c \left(1 - e^{\mu_b d}\right)}{\mu_y + \mu_b} \cdot e^{-\mu_y x}. \tag{12}$$

Considering that in each direction half of the number of photons is emitted, the final formula for the yellow photon number is:

$$N_y(x) = \frac{1}{2}\left(N_e \frac{\mu_c \left(1 - e^{-\mu_b x}\right)}{\mu_y + \mu_b} + N_e \frac{\mu_c \left(e^{-\mu_b x} - e^{-\mu_b d}\right)}{\mu_y + \mu_b} + r \cdot N_e \frac{\mu_c \left(1 - e^{\mu_b d}\right)}{\mu_y + \mu_b} \cdot e^{-\mu_y x}\right). \tag{13}$$

From the known wavelengths of the photons, the energies of the blue and yellow photons and the energy difference, due to wavelength conversion can be calculated as follows:

$$E_{ba} = h\frac{c}{\lambda_b}, \quad E_{ya} = h\frac{c}{\lambda_y}, \quad E_c = h\frac{c}{\lambda_b} - h\frac{c}{\lambda_y}. \tag{14}$$

With the above energy values, the dissipation density at point x can be expressed with the original blue photon number, N_e as follows:

$$N_e\left(E_{ba}\mu_{ba} \cdot e^{-\mu_b x} + E_{ba}\mu_c \cdot e^{-\mu_b x} + E_{ya}\mu_y \cdot \frac{1}{2}\left(\frac{\mu_c\left(1 - e^{-\mu_b x}\right)}{\mu_y + \mu_b} + \frac{\mu_c\left(e^{-\mu_b x} - e^{-\mu_b d}\right)}{\mu_y + \mu_b} + r\frac{\mu_c\left(1 - e^{\mu_b d}\right)}{\mu_y + \mu_b} \cdot e^{-\mu_y x}\right)\right) \tag{15}$$

With this analytical formula, we can determine the parameters of the phosphor layer by measurement, although,

- We neglected that the light emitted from the chip has an angle distribution;
- There is no scattering considered in the model;

- Furthermore, in Equation (14), we assumed a single blue and yellow wavelength and did not deal with the actual spectral power distributions of the original blue and the converted longer wavelength light.

The neglected effects do not result in a significant error in the thermal model, if the phosphor layer is much thinner than the dimension of the light emitting surface of the blue LED chips.

4.2. Simplified 3D Phosphor Model

The simplified 3D model is an extension of the 1D model to 3D. The scattering is still neglected, but the spatial light distribution is taken into consideration now. To keep this model simple, we do not calculate with the attenuation of the yellow light. The main question to answer with the use of this model is how the overall heat generation distribution will be affected by the losses in the phosphor.

In this approximation the Lambert-Beer law is used again, which gives us the relation between a point on the surface of the LED chip ($\mathbf{r}'$) and one in the phosphor layer ($\mathbf{r}$):

$$N_B\left(\mathbf{r}-\mathbf{r}'\right) = N_{en} \cdot e^{-\mu_b(\mathbf{r}-\mathbf{r}')}, \tag{16}$$

where N_{en} is the number of nodal (r') emitted photons. Since every location of the lighting surface is considered as a point-like source we have to correct the formula as follows:

$$N_B\left(\mathbf{r}-\mathbf{r}'\right) = \frac{N_{en} \cdot e^{-\mu_b(\mathbf{r}-\mathbf{r}')}}{2\pi\left|\mathbf{r}-\mathbf{r}'\right|^2} \tag{17}$$

To get the number of blue photons at location $\mathbf{r}$, an integration over the lighting surface is needed:

$$N_B(\mathbf{r}) = \iint \frac{N_{en} \cdot e^{-\mu_b(\mathbf{r}-\mathbf{r}')}}{2\pi\left|\mathbf{r}-\mathbf{r}'\right|^2} dA' \tag{18}$$

Considering now the $\Theta(\theta)$ spatial distribution of the blue light we get

$$N_B(\mathbf{r}) = \iint \Theta(\theta) \frac{N_{en} \cdot e^{-\mu_b(\mathbf{r}-\mathbf{r}')}}{2\pi\left|\mathbf{r}-\mathbf{r}'\right|^2} dA' \tag{19}$$

The computation for this surface integral for a 100 by 100 mesh lasts a few minutes for each plane that is modeled in the phosphor layer. In order to calculate with the yellow light propagation, we can use the calculated blue photon number, as it is proportional to the source of yellow light:

$$N_Y(\mathbf{r}) = \int\int\int \frac{\mu_c N_B\left(\mathbf{r}'\right)}{4\pi\left|\mathbf{r}-\mathbf{r}'\right|^2} \cdot e^{-\mu_y(\mathbf{r}-\mathbf{r}')} dV' \tag{20}$$

which increases the computation time to hours for every point. The reflection can be handled by a mirrored blue light source (situated on the backside). This computational cost implies that the analytical formulas should not be used even with a simplified 3D model. The practical workaround to this computational cost issue is to use numerical methods, such as the Finite Volumes Method (FVM).

4.3. Detailed 3D Phosphor Model with a Numerical Approach

The approach of overall modeling of a CoB device described in the present study was inspired by our previous work that targeted multi-domain simulation of large-area OLEDs [34,35]. For OLED simulations the FVM was applied to get a 3D network of multi-domain elementary cells. This network was solved by the Successive Network Reduction (SUNRED) method [32,33] and the solution provided voltage, temperature, radiance and luminance maps as a response to a given, assumed driving current of the investigated device.

For the simulation of phosphor covered (inorganic) LEDs, we created two special elementary cell models: One for the pn-junction of the LED (describing the multi-domain behavior of the blue LED chips by a Spice-like model) and another multi-domain simulation grid cell type for the phosphor, as illustrated in Figure 13 (for the sake of simplicity for a 2D case). These models will be detailed in the next subsections.

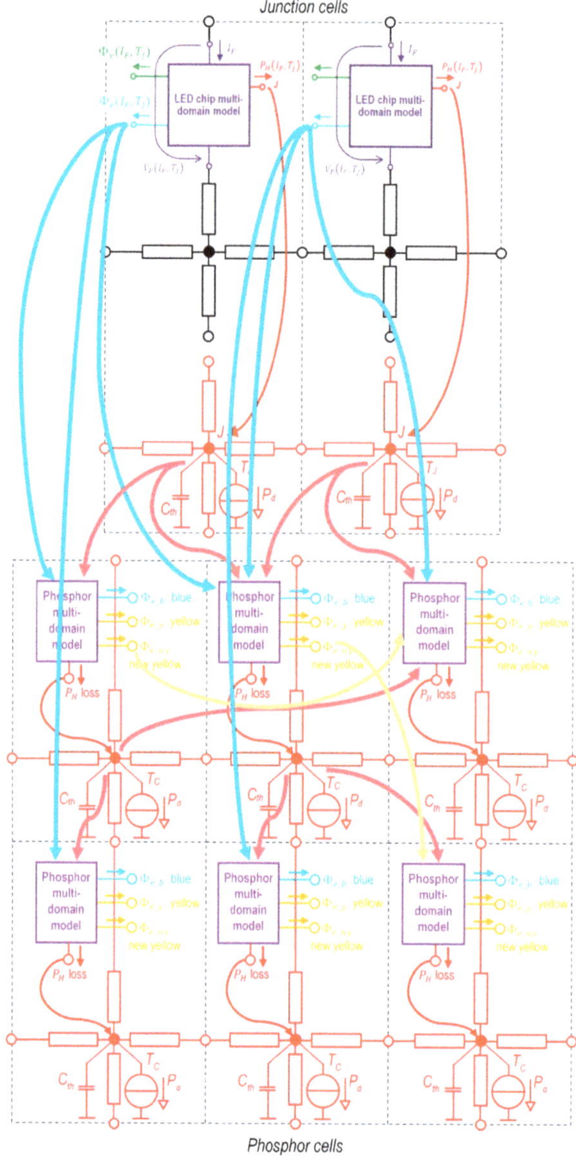

Figure 13. Part of the realization of the overall scheme of Figure 3 within our Finite Volumes Method (FVM) numerical solver: The multi-domain FVM cell models depicted in 2D—black circuit elements

represent electrical connections, red circuit elements represent mesh grid cells of the thermal subsystem. The mesh grid level multi-domain phosphor model takes local temperature and material parameters, plus the incident blue and yellow light fluxes as input and provides the transmitted blue/yellow light and converted yellow light fluxes as output. (For a detailed explanation of the symbols used in the drawing refer to the main text). For the explanation of symbols used in this figure see the main text and the Abbreviations.

4.3.1. Junction and Phosphor Cells

A 'junction cell' is basically a special cell that represents the bulk material of the LED chip from thermal perspective, but it also incorporates the new, constant forward current-driven formulation of the Shockley-model based multi-domain LED model that we developed previously within the Delphi4LED project [28]. For every single blue LED chip in the CoB device a local instance of this model is applied, considering the local junction temperature (T_J) in the FVM simulation grid cell where this model is attached to, see Figure 13. Such a grid cell is called a 'junction cell'.

This LED chip model is connected to the electrical model of the cell at its interface to other regions of the device, see the black network elements in Figure 13. The two input quantities of the junction model are the constant forward current flowing through it (I_F) and the temperature of the junction (T_J.).

Its four output quantities are the forward voltage (V_F), the generated heat flux (P_H), the emitted radiant flux (Φ_e) and the emitted luminous flux (Φ_V).

The P_H power provided by the LED chip multi-domain model instances is the local heat-source of the FVM grid cell, represented by the P_d generators in the generic thermal grid cell. The calculated Φ_e radiant flux of a junction cell is split among the blue light rays that are propagated towards the 'phosphor type' simulation grid cells of the FVM solver (see the light blue arrows in Figure 13).

The 'phosphor cells' include the same thermal part as an ordinary structural material or the 'junction cells' (red thermal network elements), and they also include a phosphor multi-domain model, describing the light conversion and propagation and the corresponding thermal losses in the phosphor material.

The calculated thermal loss is the local P_d heat-source of the given FVM grid cell. Note, that in Figure 13, symbol P_H is used both for the multi-domain model of the LED chips and for phosphor regions to denote the local heating power.

4.3.2. Modeling the Light Conversion in the Phosphor

The chip-level multi-domain LED model that we use calculates the total emitted radiant/luminous flux only; it does not provide actual spectral power distribution of the emitted light Therefore, we considered only the blue and converted yellow fluxes, without any details about their actual spectral power distributions. Note, however, that the following discussion is also valid for all colors in the entire spectral range of the emitted converted light, ranging up to red as well and with an additional model for the absorption and emission spectra of phosphors, temperature-induced slight spectral changes like the ones seen in Figure 8 could also be included.

The wavelength conversion results in a loss of energy, which heats up the phosphor. Because the phosphor particles are mixed with transparent but poorly heat-conductive material (e.g., silicone) as a host matrix, their temperature can be significantly, even tens of degrees higher than in other parts of the LED package. The conversion is temperature-dependent, therefore, a multi- domain simulation of the phosphor is required.

Figure 14 shows some light paths in an LED package. Blue light is emitted in different directions from the LED chip, which may reflect from the surfaces, even crossing the phosphor layer several times. Meanwhile, the blue light is partially converted to yellow. The emitted yellow follows different paths than the blue light absorbed by a phosphor particle; it may propagate in any direction. The yellow light can be reflected several times, and it can be absorbed and re-emitted in different directions. The energy loss due to the absorption of yellow light also contributes to the temperature rise of the phosphor layer.

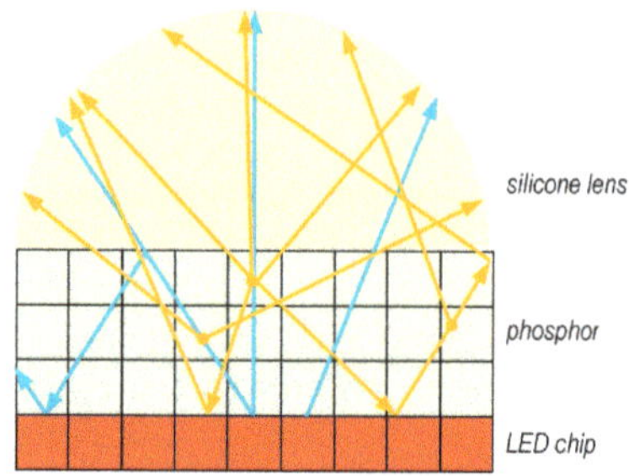

Figure 14. Some light paths in an LED: Blue light comes from the chip, yellow light is generated in the phosphor.

We assume again, that the attenuation of the blue and yellow fluxes follows the Lambert-Beer law:

$$\Phi_{e\_blue\_out} = \Phi_{e\_blue\_in} e^{-\alpha_{blue}(T)\cdot d} \tag{21}$$

$$\Phi_{e\_yellow\_trans} = \Phi_{e\_yellow\_in} e^{-\alpha_{yellow}(T)\cdot d} \tag{22}$$

where α_{blue} and α_{yellow} are the blue and yellow attenuation coefficients, d is the effective thickness of the phosphor layer. Equation (22) refers to the case where the yellow light is not produced in the layer but enters from the outside (transmitted yellow). The absorbed blue and yellow fluxes can be calculated as

$$\Phi_{e\_blue\_absorb} = \Phi_{e\_blue\_in} - \Phi_{e\_blue\_out} \text{ and} \tag{23}$$

$$\Phi_{e\_yellow\_absorb} = \Phi_{e\_yellow\_in} - \Phi_{e\_yellow\_trans}. \tag{24}$$

The yellow flux converted from the absorbed blue flux can be calculated as

$$\Phi_{e\_yellow\_conv} = \Phi_{e\_blue\_absorb} \cdot \eta_{conv}(T) \tag{25}$$

where η_{conv} is the conversion efficiency. A part of the absorbed yellow may be re-emitted:

$$\Phi_{e\_yellow\_re} = \Phi_{e\_yellow\_absorb} \cdot \eta_{yellow\_re}(T) \tag{26}$$

where η_{yellow_re} is the "yellow-to-yellow conversion efficiency", characterizing the re-emission. The yellow output of a phosphor layer is the sum of the transmitted, converted and re-emitted yellow:

$$\Phi_{e\_yellow\_out} = \Phi_{e\_yellow\_trans} + \Phi_{e\_yellow\_conv} + \Phi_{e\_yellow\_re} \tag{27}$$

The heating power due to the conversion loss is the difference between the absorbed blue and the converted yellow radiant fluxes:

$$P_{loss\_conv} = \Phi_{e\_blue\_absorb} - \Phi_{e\_yellow\_conv}. \tag{28}$$

The heating power due to the yellow transmission loss is the difference between the input and the transmitted and re-emitted yellow fluxes:

$$P_{loss\_yellow\_trans} = \Phi_{e\_yellow\_in} - \Phi_{e_{yellow_{trans}}} - \Phi_{e\_yellow\_re}. \tag{29}$$

The full heating power of the phosphor layer is the sum of the absorption and transmission losses:

$$P_{H\_phosphor} = P_{loss\_conv} + P_{loss\_yellow\_trans} \tag{30}$$

4.3.3. Rays split over the FVM Simulation Grid Cells

The FVM simulation calculates the quantities per elementary cell—therefore, the absorption and transmission losses, as well as the blue and yellow output fluxes, are determined for the elementary cells of the phosphor layer. The idea behind the method comes from the ray tracing [46] and path tracing [47] algorithms used in global illumination.

The goal is to create a generic cell model that can handle all-optical and thermal phenomena presented in Section 4.3.2 (as illustrated in Figure 13), and leaves the definition of the light paths to the user. The quantities are determined for each elementary cell individually.

In a cell-split model, light propagates in a beam from the internal volume or from a surface of a cell to the outer surface of another cell. For example, blue light typically propagates from the surface of junction cells to the outer surfaces of a phosphor layer, as illustrated in Figure 15a Handling 3D beams would require integral calculation. To avoid this complicated and time-consuming method, we use 1D rays as in ray tracing methods, see Figure 15b. A ray is defined by the coordinates of its start and end points. In this model, based on Equation (21), the blue output flux for Ray 1 is:

$$\Phi_{e\_blue\_out} = \Phi_{e\_blue\_in} e^{-\alpha_{blue\_1}(T_1)\cdot d_1} e^{-\alpha_{blue\_4}(T_4)\cdot d_4} \tag{31}$$

where α_{blue_n}, T_n and d_n are the attenuation coefficient, temperature and distance traveled by the light in cell n, respectively. $\Phi_{e\_blue\_in}$ is the flux of the junction cell in the direction of the output surface. As d is the distance between the entry and exit points of the ray in the cell, and the temperature and the material are considered homogeneous in a cell, this is a simple calculation. The other equations presented in Section 4.3.2 are similarly simple to re-write for this discretized view.

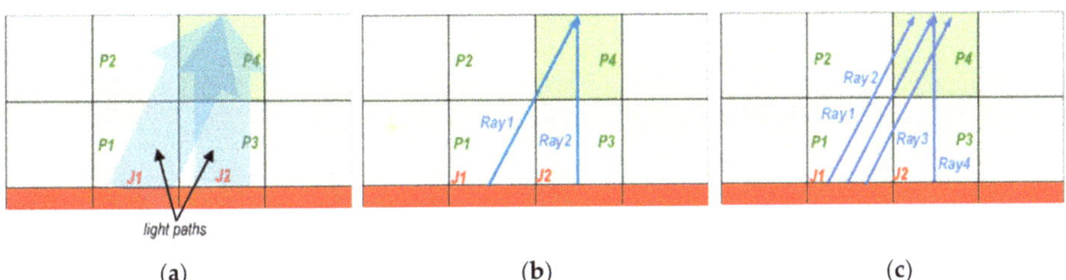

Figure 15. Illustrations for modeling the propagation of the blue light; (a) 3D paths of the blue light; (b) Single 1D rays of the blue light; (c) Multiple 1D rays of the blue light.

If the 3D beam is replaced with a 1D ray, an error is introduced in the calculation that can be reduced by considering multiple rays from the starting surface to the target surface, as illustrated in Figure 15c. (This method is well known in computer graphics for antialiasing purposes).

Different strategies are possible to handle yellow rays. If the 1D light propagation, shown in Section 4.1 is used, the blue ray and the converted yellow will further propagate together. The method is also applicable for 3D blue propagation (Figure 16a), although the result will obviously be inaccurate. A more accurate result can be obtained by determining the flux of the yellow light emitted from a blue ray in a cell and how much of it is radiated towards an outer surface. This will be a yellow ray (Figure 16b). The re-emitted yellow rays can come from the yellow rays in a similar way.

How many rays will be in total if the phosphor region is a rectangular cuboid of $X \times Y \times Z$ cells with one XY surface in contact with the blue chip's junction cell? The number of junction surfaces, in this case, is $N_J = X \times Y$. If 1D light propagation is used, a single ray will leave every elementary junction surface, resulting in a total of N_J rays. Using our 3D light propagation model, blue rays start from every junction, one for each outer surface. The number of outer surfaces of the phosphor is $N_P = X \times Y + 2 \times X \times Z + 2 \times X \times Z$, so the number of blue rays $N_B = N_J \times N_P$. Each blue ray triggers a yellow ray from each intersected cell to each outer surface. If the number of intersected cells is estimated to be $N_C \approx 0.7X + 0.7Y + 0.6Z$, then for the number of yellow rays we obtain $N_Y = N_C \times N_B \times N_P$. If, for example, $X = Y = 10$, $Z = 5$, then $N_J = 100$, $N_P = 100$, $N_B = 30,000$, $N_C = 17$, $N_J = 153,000,000$.

The number of yellow rays is by several orders of magnitude greater than the number of blue rays, and the re-emitted yellow rays have not yet been addressed. A realistic model of a real LED device requires a higher spatial resolution of the FVM grid—thus, the number of yellow rays would increase to such a high value that one cannot manage. Therefore, we need a modeling approach where the number of yellow rays to follow is significantly reduced: This is the "indirect yellow ray model" illustrated in Figure 16c. In this model, for each cell, the amount of yellow flux generated by the blue rays passing through and the re-emitted flux generated by the passing yellow rays are cumulated. In a subsequent iteration step, this cumulated flux is considered as the total yellow flux emitted by the cell, and one yellow ray per cell will be started for every outer surface. (Since this uses the flux calculated in the previous iteration, it is not derived from the flux generated by the junction cell in the current iteration, so the calculation is indirect). The number of yellow rays, thus, depends on the number of phosphor cells and the number of the outer surfaces. Using numbers of the previous example the number of the yellow rays is $N_{Y_i} = X \times Y \times Z \times N_P = 150,000$—by three orders of magnitude less, therefore, manageable.

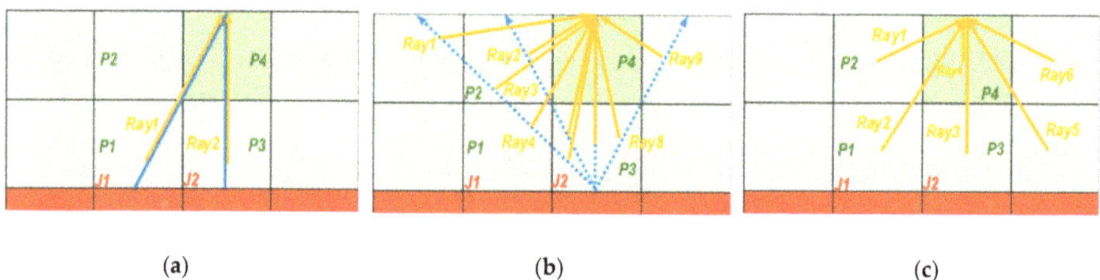

Figure 16. Strategies to count the number of yellow light rays: (**a**) Common yellow and blue rays; (**b**) yellow rays triggered by blue rays arriving directly from an elementary junction surface; (**c**) indirect yellow rays.

4.3.4. Phosphor Cell Model

The phosphor multi-domain model is built in the thermal FVM model of the phosphor cells, as shown in Figure 13. The model delivers four output quantities:

1. Local heating power, P_H: The power to heat the simulation grid cell resulting from conversion losses. This is the value of the heat-flux forced by generator P_d into the thermal network model associated with every finite material volume represented by a simulation grid cell;
2. The new yellow radiant flux, Φ_{e_ny}, the sum of the yellow radiation produced by conversion from blue light passing through the cell and the re-emitted radiation from the yellow light passing through the cell;
3. The blue radiant flux leaving the cell, Φ_{e_b}, the remainder of the input blue flux after conversion;
4. The yellow radiant flux leaving the cell, Φ_{e_y}, the sum of the input yellow radiant flux remaining after the attenuation in the cell and the calculated new yellow flux, Φ_{e_ny}.

Two of these quantities, P_H and Φ_{e_ny} directly influence the operation of the actual simulation grid cell where they were calculated, the other two values (Φ_{e_b}, Φ_{e_y}) are stored as simulation results.

The model calculates the total output quantities for rays starting from the junction cells, taking into account the temperature and material parameters of each cell crossed by them.

The structure of the phosphor cell model is shown in Figure 17. The output quantities are calculated individually per ray, and then they are summed. There are two types of ray models.

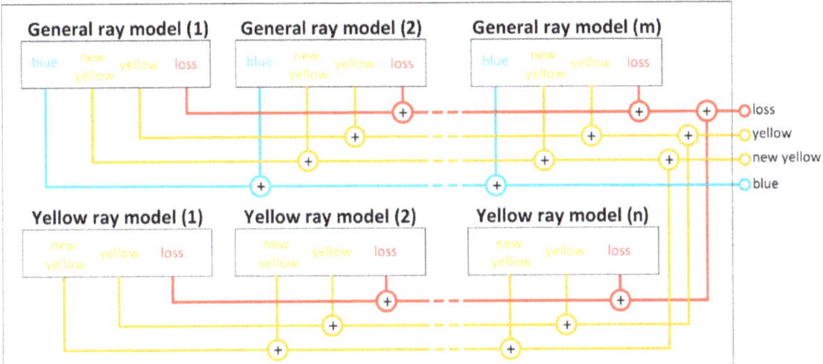

Figure 17. Structure of the phosphor cell model. The number of general and yellow ray models can be different.

The general model describes one ray that travels from a junction cell to a particular phosphor cell. It can be a blue, a yellow or a mixed ray (such as in Figure 16a). Blue rays were shown in Figure 15b, yellow rays are shown in Figure 18: the ray starts as blue at the junction, then it is converted in a cell and continues its path to a given destination as a yellow ray.

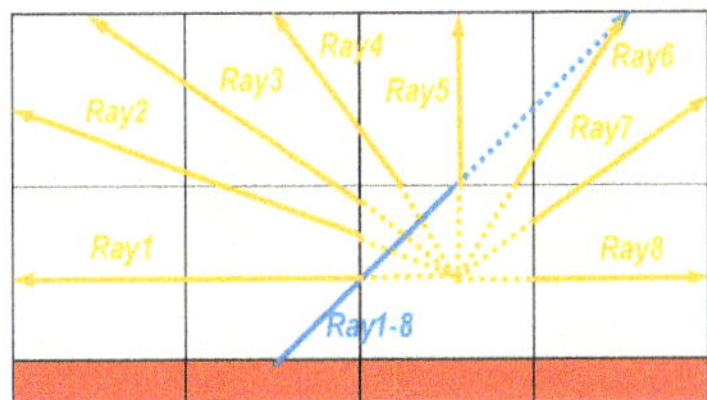

Figure 18. Yellow rays in the general ray model: The eight yellow rays start with the same blue part.

The yellow ray model describes a ray starting from a phosphor cell containing only yellow component, see Figure 16c. The yellow ray model uses less memory than the general ray model. The general model must be used at junction cells, and the yellow model can be used at phosphor cells.

The ray models represent a ray by sections corresponding to individual cells it crosses, see Figure 19. Generally, multiple rays leave the starting cell, the amount of flux in a given ray is controlled by the proper adjustment of the K_0 multiplier factor. The internal ray sections calculate the input blue and/or yellow flux of the current cell, the ray model of the cell cascade in the path calculates the output quantities of the current cell.

Each internal ray section corresponds to a cell in the space between the starting cell and the current cell, using the temperature and material parameters of that cell. The structure of the general and yellow ray sections is shown in Figure 20.

If a ray suffers an imperfect reflection at the i-th section of its path, we can model it by adjusting the K-values of section i. In the case of reflection, sections i and $i + 1$ usually belong to the same phosphor cell.

The fluxes of the input rays are transferred to the α_{blue} and α_{yellow} blocks of the model. These blocks represent attenuation according to Equations (21) and (22), as well as absorbed fluxes according to Equations (23) and (24). The α_{blue} and α_{yellow} attenuation factors are temperature-dependent material parameters of the cell. The η_{conv} and η_{yellow_re} blocks represent the blue-to-yellow and yellow-to-yellow conversion efficiencies, which are temperature-dependent material parameters, as seen in Equations (25) and (26).

The K blocks control the amount of the resulting flux propagated to the next ray section. These blocks correspond to a constant multiplication, most often zero or one.

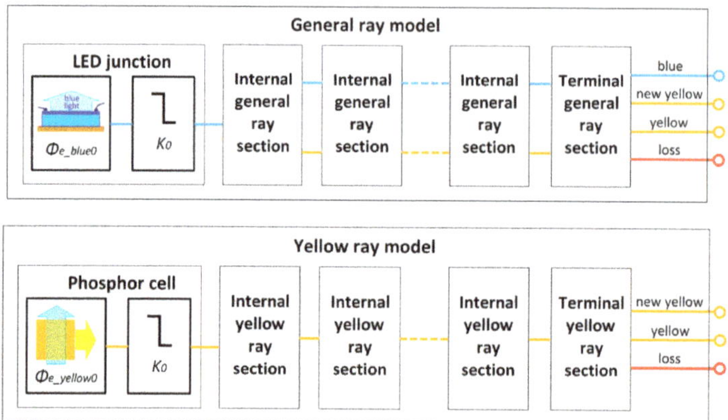

Figure 19. Structure of the ray models. $\Phi_{e\_yellow0}$ is the sum new yellow of the starting phosphor cell. K_0 is the ratio of the flux treated by the ray to that of the starting cell.

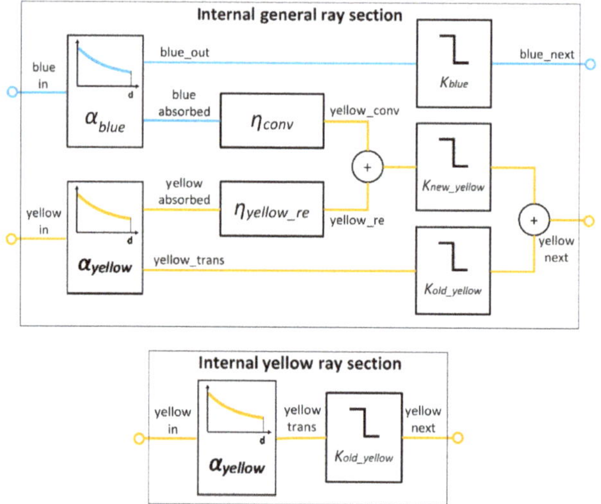

Figure 20. Structure of the internal ray sections.

The structures of the general and yellow ray sections are shown in Figure 21. The terminal sections differ from internal ones in that they calculate both new yellow flux and P_{loss_conv} power loss of the conversion as described by Equations (28) and (29).

4.4. The Use of the Simple 1D and of the Complex 3D Models, Extraction of Phosphor Model Parameters

The 1D model provides fairly good accuracy for thin phosphor layers (much thinner than the lateral dimensions of the layer). Therefore, with an appropriate set of phosphor samples, it can be used for the extraction of phosphor material properties (model parameters), such as absorption rate or conversion efficiency. With the set of model parameters identified, the 3D phosphor model is used for accurate simulations.

As described in Section 3.3, phosphor-converted white LEDs have been created using fully pre-characterized bare blue ones, with five different phosphor layer thicknesses, realizing five different spectral power distributions, thus, realizing light output with different CCTs. (The mass fraction of the phosphor powder has also been varied; see the applied mass fractions in Table 1).

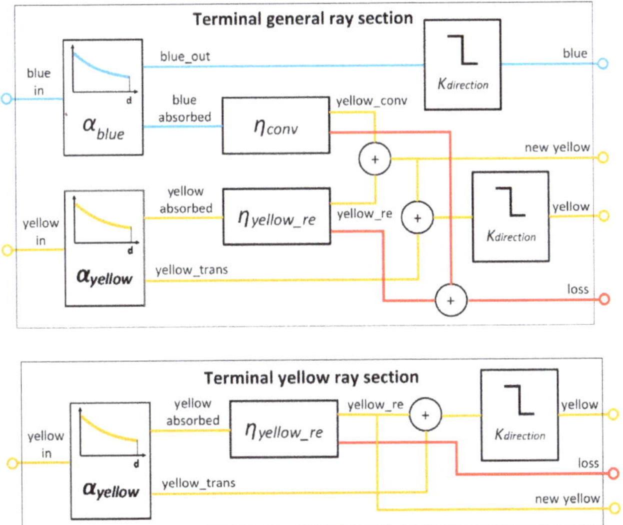

Figure 21. Structure of the terminal ray sections.

On the one hand, these LEDs with known properties have been used for extracting material properties for the multi-domain model as mentioned above [25], and on the other hand, for validating the overall simulation approach (with the blue chip multi-domain compact model embedded), simulating them characterized single-chip white LEDs well. The major steps of extracting the phosphor layer parameters were the following:

(a) Spectral power distributions for all manufactured custom-made white LEDs have been measured (along with the spectra of the used bare blue LEDs before attaching the phosphor + lens structure).

(b) We separated the spectra into 'blue' and 'yellow' parts (as illustrated in Figure 8 for large, stand-alone phosphor samples), in order to allow us to determine the number of 'blue' and 'yellow' photons. (For the calculation of the number of yellow photons a rough spectral power distribution was assumed).

(c) From the blue photon number distribution, the sum of the attenuation coefficient for blue light and the conversion coefficient can be determined.

(d) From the yellow photon number distribution (four points), the remaining coefficients (blue attenuation, yellow attenuation and reflection) can be determined.

Note, that the spectral power distribution of the 'yellow' light can be better approximated by assuming multiple wavelength bands of the 'yellow' light (in an extreme case bands correspond to the wavelength resolution to the measured spectra). This would assume, however, as many yellow ray models as many yellow wavelength bands are assumed. (The execution time of the ray model would linearly scale with the number of the assumed yellow wavelength bands).

We measured the SPDs of all custom-made white LEDs. The spectrum measurement was done on five samples with different phosphor layer thicknesses (0 µm (blue LED), 32 µm (cool white LED), 47 µm (neutral white LED), 68 µm (warm white LED), and 99 µm (amber LED)), at five different blue LED junction temperatures (30 °C, 50 °C, 70 °C, 90 °C and 110 °C) and each with six different forward currents (100 mA, 350 mA, 700 mA, 1000 mA 1500 mA, 2000 mA). From the total of 150 spectrum measurements, we present data for five measurements only, for demonstration purposes (the applied measurement conditions were: Fifty degree Celsius ambient temperature and 1000 mA forward current). The calculated photon numbers are summarized in Table 2.

Table 2. Calculated photon numbers for different LED phosphor layer thicknesses.

Phosphor Layer Thickness (µm)	Blue Photon Number ($\times 10^{16}$)	Yellow Photon Number ($\times 10^{16}$)
0	373	0
32	109	241
47	62	272
68	44	284
99	8	281

For the blue photon numbers, we can fit an exponential, with a coefficient of $\mu_b = 0.039/\mu\text{m}$, from the yellow data, we can determine the missing coefficients. As Equation (12) is transcendent there are more solutions, but physically only one is found to be correct: $(r = 1 \pm 0.08; \mu_c = 0.038 \pm 0.04; \mu_y = 0.037 \pm 0.006; \mu_b = 0.039 \pm 0.001)$, where the length is given in µm, so the dimension of the coefficients is $1/\mu\text{m}$.

The measured blue and yellow photon number versus the 1D prediction is shown in Figure 22. As the diameter of the LED chip is 1.6 mm, theoretically, the accuracy range of the 1D model is 0–160 µm in phosphor layer thickness. Comparing the measurement data to simulation results suggests that the method provides sufficient accuracy. The measurements at other LED operating points (junction temperature, forward current) show less difference in the extracted parameters than the inaccuracy of the calculated parameters. For different concentrations of phosphor powder the coefficients can be scaled, as the original relationship between the coefficients and the physical effect was the Lambert-Beet law, where the attenuation coefficient is proportional to the attenuation cross-section, which is proportional to the phosphor powder concentration: $\mu \sim \sigma \sim C_{phos}$. The phosphor temperature rise inside the phosphor layer with respect to the junction temperature can be determined through the integration of the heat distribution.

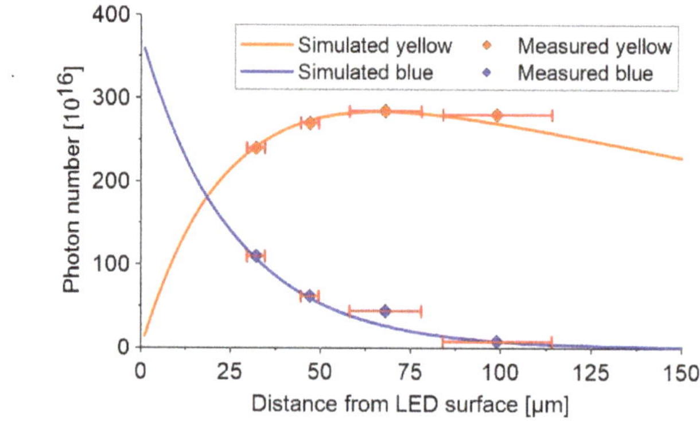

Figure 22. The photon numbers of blue and yellow light emitted at the top surface the phosphor layer, for different phosphor layer thicknesses.

5. Simulation of a Commercially Available CoB Device

We measured and modeled a Lumileds 1202s CoB LED device which consists of 24 LED chips (two strings of 12 LEDs, connected in parallel) with a lateral dimension of 600 µm $\times$ 700 µm each, placed on an aluminium pad with solder layer between them, covered by a phosphor layer, see Figure 23a. The parameters of its phosphor coating were measured as outlined in Section 3.2, see further details in Reference [25]. A 3D model was created for simulation (Figure 23b). In the numerical simulation model the 650 µm thick phosphor was divided into nine layers of equal thickness. The parameters of the multi-domain LED chip model were extracted from the measured 'ensemble' characteristics. The coefficients of the phosphor model were extracted from the measured spectral power distributions.

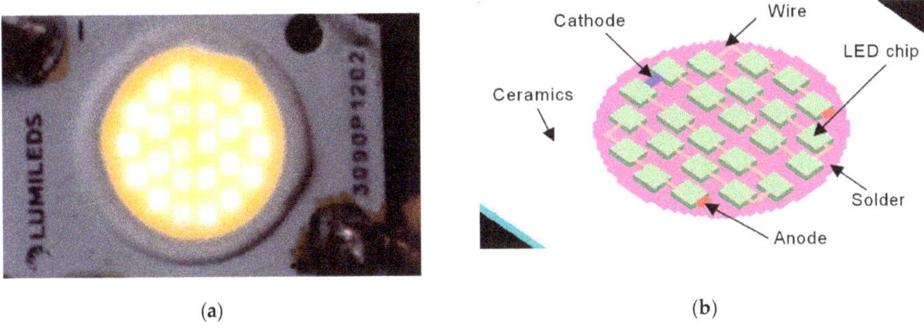

(a)   (b)

Figure 23. A Lumileds 1202s CoB LED: (**a**) Photograph of a physical device; (**b**) axonometric view of its 3D simulation model (internal structure, without the visualization of the phosphor layer).

5.1. Different Simulation Setups

We studied seven strategies that describe the propagation of light in the phosphor with different level of approximations, as shown in Figure 24: 1D propagation only (Figure 24a) or propagation in multiple directions with a uniform spatial distribution (Figure 24b–f), as follows:

- (a) One general ray goes from the blue chips' junction in a direction perpendicular to the outer edge of the phosphor. This ray includes decreasing blue flux and the increasing yellow flux.
- (b,c) A single general ray or multiple ones may propagate from the junction to each outer surface of the phosphor. The rays include the blue and yellow fluxes. Case (c) covers the full hemisphere, while case (b) covers only the spatial area of the top of the phosphor. This assumption means that the converted light follows the path of the blue light, but in reality, the converted light propagates in all directions.
- (d–f) The blue flux corresponds to general rays as assumed in cases (b) and (c), but the yellow light starts from the center of each phosphor cell, taking into account the blue and yellow light absorbed in that cell. The propagation of the yellow light is modeled with uniform spatial distribution. In case (d), the light can propagate only in the direction of the top surface. In case (e), the light goes in the direction of the top surface and the sides, and in case (f), it goes in all directions. When the light reaches the edge of the phosphor, it "disappears".
- (g) Same as (f), but light may be reflected in the selected structures (LED chip, solder, etc.). We always used full reflection in the simulations.

As it can be seen in Figure 24, even with a small number of cells, the number of possible rays and ray sections is already very high. The full model, shown in Figure 23, contains 720 blue chip junction cells and 20,628 phosphor cells. Due to the symmetry, we can use half of the model with 360 and 10,314 junction and phosphor cells, respectively. If all rays were taken into account in the calculation, we would have an unmanageable number of rays, especially for yellow rays; therefore, we weight the rays by the estimated flux they carry (i.e., we assign 'importance' to them) and sort them by this. The rays with the lowest flux are discarded until the total flux of the remaining rays reaches the desired level. In the simulations presented, we use two *levels of importance*: Ninety percent and ninety-nine percent, that is, rays representing 10% and 1% of the flux are discarded. Of course, the discarded flux is distributed proportionally among the remaining rays so the total flux will be finally propagated by the rays considered. The estimated flux transported by a ray is the product of the input flux of the ray and the solid angle, $\Omega_{output\_surface}$ of the target surface seen from the starting point:

$$\Phi_{e\_estimated\_transported} = \Phi_{e\_estimated\_in} \cdot \Omega_{output\_surface} \tag{32}$$

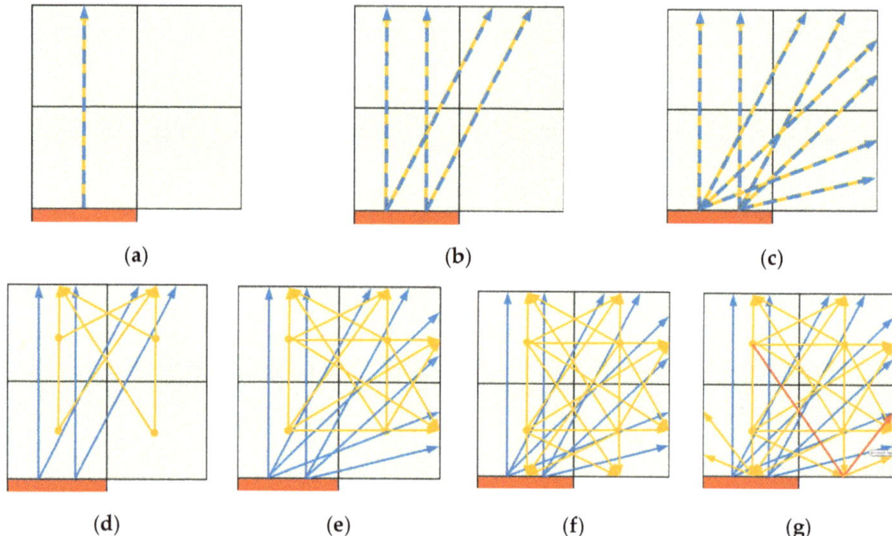

Figure 24. Different strategies of modeling light propagation, tested in our CoB simulations: (**a**) 1D; (**b**) general rays to top; (**c**) general rays to top and sides; (**d**) general + yellow rays to top; (**e**) general + yellow rays to top and sides; (**f**) general + yellow rays to all phosphor sides; (**g**) general + yellow rays to all sides with bottom side reflection. (One of the reflected yellow rays was highlighted in red).

In the case of the half model and five rays per junction, for cases (c), (e), (f), and (g) shown in Figure 24, the total number of blue rays was 3,663,900. The number of yellow rays was 3,075,078 at a 99% importance level, while at a 90% level it was 1,648,216 only. Table 3 shows the number of the general (or blue) rays and yellow rays for the different importance levels, the total number of sections of the rays, and the average length of the rays weighted by the estimated flux that they carry.

Table 3. General and yellow ray numbers of the half CoB LED model in the case of five blue rays/junction.

Ray Strategy	General Ray, 90% №/Section №/Weighted Average Length [μm]	General Ray, 99% №/Section №/Weighted Average Length [μm]	Yellow ray, 90% №/Section №/Weighted Average Length [μm]	Yellow Ray, 99% №/Section №/Weighted Average Length [μm]
(a) [1]	360/3.6k/650	360/3.6 k/650	-	-
(b)	824 k/17 M/988	1.7 M/46 M/1106	-	-
(c)	1.6 M/37 M/1099	3.1 M/93 M/1263	-	-
(d)	824 k/17 M/988	1.7 M/46 M/1106	1.9 M/26 M/388	6.7 M/144 M/470
(e)	1.6 M/37 M/1099	3.1 M/93 M/1263	5.3 M/89 M/540	14 M/384 M/724
(f)	1.6 M/37 M/1099	3.1 M/93 M/1263	7.4 M/117 M/543	21 M/539 M/703
(g)	1.6 M/37 M/1099	3.1 M/93 M/1263	7.4 M/159 M/955	21 M/642 M/1106

[1] In the case of the ray strategy (a), always one blue ray/junction is used, and the importance level is 100%.

For the simulations, we used a workstation with an AMD Threadripper 2920x processor having 12 cores and 32 GB of RAM. With this machine, we achieved the following execution times and memory need.

For the half model the simulation of strategy (a), which contains 360 rays, takes 6.9 s and requires 1.0 GB of memory, while strategy (g), with a 90% importance level and nine million rays, takes 435 s and consumes 5.1 GB of RAM. For the full model, strategy (g) with 90% level and 32 million rays, the execution time is 2215 s and memory need is 20.0 GB. Simulating the full model would have required about 80 GB of RAM at a 99% importance level.

In the simulations, the bottom surface of the models was set to a fixed 25 °C, the other sides were modeled with constant convection boundary condition with a heat transfer coefficient of 10 W/m^{2}K, at an ambient temperature of 25 °C.

Based on the findings of A. Alexeev and his co-workers [48–50], from the point of view steady-state behavior, the effect of the heat transfer from the top of the phosphor (silicone dome in the case of LED packages with lenses) can be neglected. Therefore, in all simulation scenarios, we neglected radiation—but due to the large, open phosphor surfaces, we assumed cooling by natural convection through the large top surface area of a CoB device. (Alexeev pointed out that in the phosphor, as a secondary heat-path towards the ambient, the heat storage has a significant effect though. Through the thermal capacitance associated with every FVM simulating grid cell, this is inherently accounted for in our thermal simulation model, as it was shown in Figure 13).

The model parameters used in Equations (21)–(30) contain four phosphor material parameters that were extracted from the measurement results, as outlined earlier. We examined two models:

- In one case, we considered that the phosphor absorbs blue light only, allowing yellow to pass completely, i.e., the loss arises exclusively from blue light.
- The other model follows the real behavior of the phosphor. It can be seen in Section 4.4 that the blue and yellow attenuation are approximately equal ($\mu_y = 0.037 \pm 0.006$; $\mu_b = 0.039 \pm 0.001$), so the conversion efficiency is also considered to be the same.

Temperature dependency of the parameters is under 0.02%/°C, therefore, it is considered constant in the simulation.

5.2. Influence of the Chosen Light Propagation Model on the Phosphor Temperature

With our modeling approach, the main target was to accurately describe the thermal behavior of white CoB LEDs; the accurate calculation of the distribution of emitted light (i.e., radiance/luminance maps of the CoB surface) was a secondary target for us. Since we cannot measure the temperature distribution inside the phosphor, the question is how accurately the simulated temperature distributions obtained with the simpler models match the results obtained with the most accurate model. To reduce the need for computational resources, we took advantage of the symmetry of the CoB LED device, and only half of the detailed model, shown in Figure 23, was used. In these simulations, the driving current of the CoB LED was 200 mA, half of the 400 mA of the full model, which resulted in 7.3 W of electrical input power, and the blue flux radiated by the LEDs was 2.5 W.

Tables 4 and 5 show the simulation results. We obtained roughly 28% higher phosphor temperature rise with the simplest 1D model (ray strategy (a)) than with the most detailed, most accurate model (ray strategy (g)) while the obtained junction temperatures were practically not affected by the chosen light propagation model.

Table 4. Simulation results of the half CoB LED model in the case of 400 mA driving current, no yellow absorption, 90% and 99% importance levels, five blue rays/junction.

Ray Strategy	90% Importance Level Proportion of Output Blue/Yellow/Loss/ T_{max} Junction/T_{max} Phosphor	99% Importance Level Proportion of Output Blue/Yellow/Loss/ T_{max} Junction/T_{max} Phosphor
(a) [1]	9.3%/66.4%/24.3%/65.2 °C/108.8 °C	9.3%/66.4%/24.3%/65.2 °C/108.8 °C
(b)	9.3%/66.4%/24.3%/65.2 °C/104.1 °C	9.3%/66.4%/24.3%/65.2 °C/103.5 °C
(c)	9.3%/66.4%/24.3%/65.0 °C/100.2 °C	9.3%/66.4%/24.3%/65.0 °C/98.4 °C
(d)	9.3%/66.4%/24.3%/65.2 °C/104.1 °C	9.3%/66.4%/24.3%/65.2 °C/103.5 °C
(e)	9.3%/66.4%/24.3%/65.0 °C/100.2 °C	9.3%/66.4%/24.3%/65.0 °C/98.4 °C
(f)	9.3%/66.4%/24.3%/65.0 °C/100.2 °C	9.3%/66.4%/24.3%/65.0 °C/98.4 °C
(g)	9.3%/66.4%/24.3%/65.0 °C/100.2 °C	9.3%/66.4%/24.3%/65.0 °C/98.4 °C

[1] In the case of the (a) ray strategy, always one blue ray/junction is used and the importance level is 100%.

Table 5. Simulation results of the half CoB LED model in the case of 400 mA driving, equal blue and yellow absorption, 90% and 99% importance levels, five blue rays/junction.

Ray Strategy	90% Importance Level Proportion of Output Blue/Yellow/Loss/ T_{max} Junction/T_{max} Phosphor	99% Importance Level Proportion of Output Blue/Yellow/Loss/ T_{max} Junction/T_{max} Phosphor
(a) [1]	9.3%/66.4%/24.3%/65.2 °C/131.9 °C	9.3%/66.4%/24.3%/65.2 °C/131.9 °C
(b)	9.3%/64.2%/26.5%/65.5 °C/131.4 °C	9.3%/62.9%/27.8%/65.7 °C/137.9 °C
(c)	9.3%/63.1%/27.6%/65.2 °C/122.9 °C	9.3%/61.4%/29.3%/65.5 °C/125.5 °C
(d)	9.3%/66.4%/24.3%/65.1 °C/124.2 °C	9.3%/66.2%/24.5%/65.1 °C/126.8 °C
(e)	9.3%/66.4%/24.3%/64.7 °C/113.0 °C	9.3%/66.4%/24.3%/64.7 °C/114.5 °C
(f)	9.3%/66.5%/24.2%/64.8 °C/107.5 °C	9.3%/66.5%/24.2%/64.8 °C/108.7 °C
(g)	9.3%/66.3%/24.4%/64.9 °C/107.8 °C	**9.3%/66.4%/24.3%/64.9 °C/108.5 °C**

[1] In the case of the (a) ray strategy, always one blue ray/junction is used and the importance level is 100%.

5.3. Simulated Temperature, Radiance and Luminance Distributions at 400 mA Driving Current, Using Different Phosphor Models

With the models described in the previous sections, steady-state simulations have been carried out for the real CoB device. In all simulations, 400 mA forward current was applied, and we used the phosphor models (a–g) and compared them. Figure 25 shows the temperature distribution at the top of the CoB LED. Several degrees of differences develop in the temperature distribution obtained by simulating the full and the half structure, see Figure 25b. At the center of the CoB device, the difference between the results obtained by the full and half models reaches 7.1 °C. The difference obtained for the two structures is due to the different light propagation caused by the symmetry plane as an artificial boundary. The difference in the light propagation can be best visualized by the different distributions of the radiance at the phosphor surface (and also at the internal surfaces of the layered phosphor model) as illustrated in Figure 26. Thus, this problem can be mitigated to some extent by a modified optical model at the symmetry plane: The cell surfaces in contact with the symmetry plane are also the target surfaces of the rays, and the rays hitting them are reflected on the surface as if coming from the symmetrical other side of the CoB structure such that the total flux of the reflected rays is the same as if the rays would have originated from the missing other half of the structure. This still would result in a somewhat different ray distribution and flux compared to the full model because the rays 'reflected' from symmetry plane carry more flux than the rays passing the same plane in the full structure.

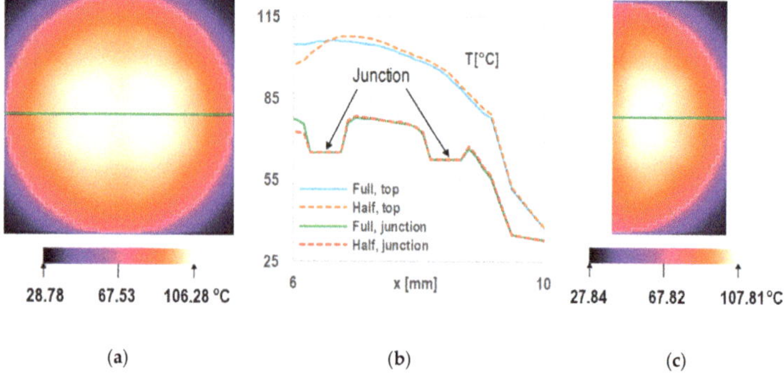

Figure 25. Simulated distributions of the temperature rise (model setup: 400 mA driving current, 25 °C ambient temperature, light propagation described by model g) presented in Section 5.1, with equal blue and yellow absorption, 90% importance level) (**a**) surface temperatures: full model; (**b**) cross-sectional plots of temperature distributions at the blue chip junctions and at the top surface along the green lines; (**c**) surface temperatures: half model.

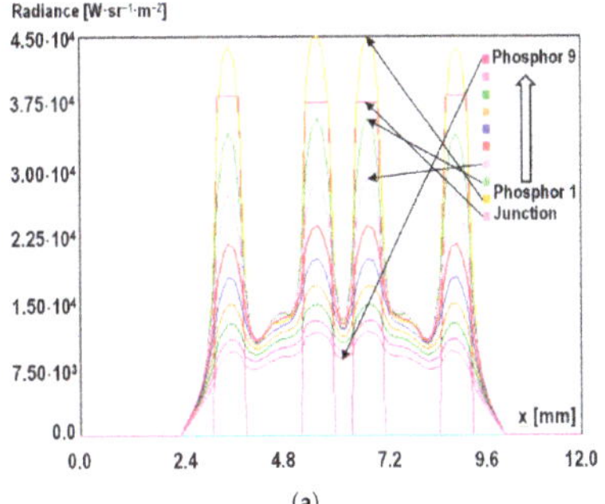
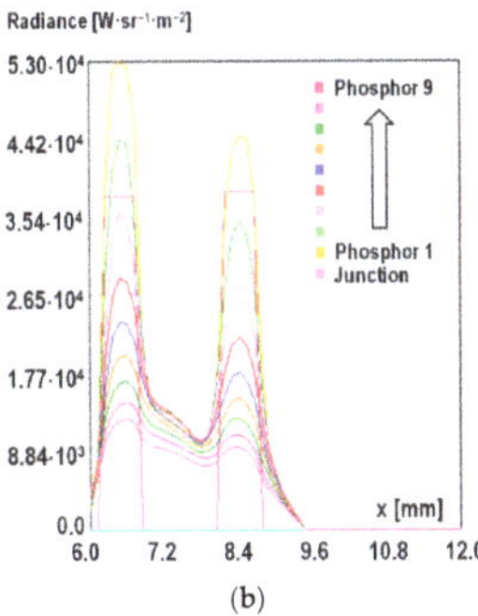

Figure 26. Cross-sectional plots of the radiance distributions along the green lines of (**a**) Figure 25a for the full model and (**b**) Figure 25c for the half model. (Simulation model setup—400 mA driving current, 25 °C ambient temperature, light propagation described by model g) presented in Section 5.1, with equal blue and yellow absorption and 90% importance level).

Further analysis showed, that even though with the optical model modified at the symmetry plane the radiance distribution of the half model became more similar to the radiance distribution of the full model, but in terms of temperature distribution the deviation of phosphor temperature at the top in the half model was found to be 5.5 °C higher than in the case of simulating the full structure. Overall, taking advantage of the symmetry of the structure to reduce model complexity (which is a common practice in numerical thermal simulations) is not recommended in this multi-domain simulation problem that involves modeling of light propagation as well.

Figure 26 shows that as we move away from the chip, the radiance decreases. The junction layer breaks this pattern, where the radiance is less than in the phosphor layer immediately above it. The reason for this phenomenon is that in the model, the yellow light is reflected from the surface of the chip and does not reach the junction. In Figure 27, we present simulated radiance and luminance maps at the top of the entire phosphor layer of the CoB device by using the model based on the full geometry of the device structure. (Note, that since in the presented light propagation model the spectral power distribution of the converted light was not resolved, the calculated luminance maps are based on approximated luminous flux values associated with the radiant fluxes carried by the rays, therefore all simulated luminance maps presented here are approximate ones only.)

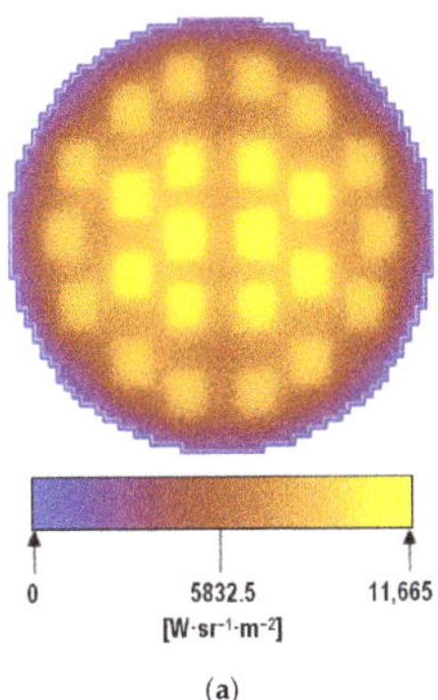
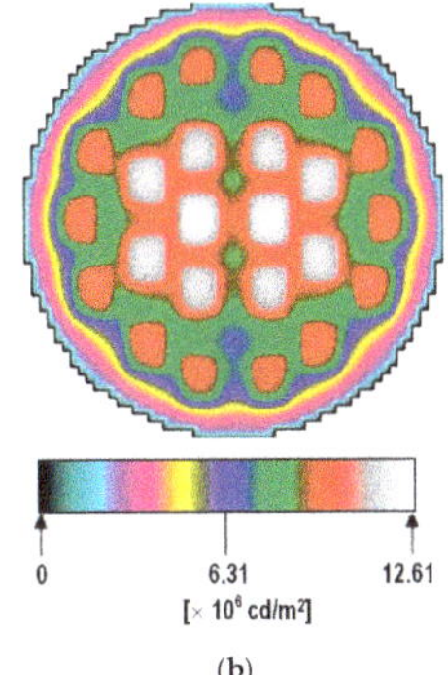

Figure 27. Simulated radiance and luminance distributions on the top of the phosphor (model setup: 400 mA driving current, 25 °C ambient temperature, light propagation described by model g) presented in Section 5.1, with equal blue and yellow absorption, 90% importance level): (**a**), radiance map; (**b**) luminance map.

6. Simulation Results, Comparison with Measurements

The means of comparing the detailed multi-domain simulation results of a CoB device to measured data are very limited. As mentioned in Section 2, with usual LED package level testing tools, only the 'ensemble' characteristics, such as the overall forward voltage and the emitted total radiant or luminous flux can be measured. The junction temperature identified with the help of the JEDEC JESD 51-51 electrical test method for LEDs will also be an average value, without any information about the differences of the individual chip temperatures within a CoB device.

With imaging methods, however, we have some hope to measure properties that we can also obtain by simulations, such as the temperature distribution or the luminance distribution at the top surface of a CoB device, using an infrared camera or an imaging luminance meter (luminance measuring camera). Both measurements are problematic. In the case of infrared thermography, one has to make sure that the emitted light does not introduce false information in the IR image. In the case of luminance measurement cameras, the problem is that such cameras are not designed to characterize high-intensity light sources, when a CoB LED is driven by its nominal forward current, it is so bright that a usual luminance measuring camera gets saturated. Therefore, with such a camera we could measure the luminance distribution of our CoB LED device only at very small forward currents (e.g., 80 mA) where the luminance did not cause the camera to saturate yet. To have a considerable temperature rise, for measurements by an IR camera, the investigated CoB LED was driven by 100 mA forward current. During the measurements the CoB device was attached to a temperature-controlled stage, providing a targeted ambient temperature of 25 °C (in practice achieving an actual temperature of 24.4 °C).

During the simulations an ambient temperature of 25 °C was assumed, the thermal boundary conditions were the following: At the bottom of the ceramics substrate of the CoB structure, we assumed a 25 °C constant temperature; while at the other outer surfaces of the device (at the sides and at the top), a heat transfer coefficient of 10 W/m^{2}K was applied (representing heat transfer by natural convection roughly). In Figure 28, we present the transient of the average temperature of the CoB device together with two temperature maps grabbed during the heating up process. The image on the top corresponding to 43.7 °C average temperature represents already the thermal quasi-steady-state of the device. This is compared to the simulated surface temperature distribution in Figure 29: The peak temperature and the shape of the temperature distribution are well estimated by the simulation, the difference in the measured and simulated peak temperature is 0.68 °C only.

Comparing the measured and simulated luminance maps (Figure 30) we can see that the difference between the measured and simulated maximal luminance is 19% while in the average luminance the difference is 7% (measured—2.721×10^6 cd/m^2, simulated—2.528×10^6 cd/m^2). One reason for the higher maximum simulated luminance value is the 90% importance level (meaning that 10% of the radiated power was distributed among the major light paths). We could not apply a higher importance level on the available computers. Another probable reason for the discrepancy is that there may be differences in the actual internal structure of the CoB LED compared to the modeled structure, and the actual reflections may differ from the ideal case used in the model.

Note, that at this stage of the development of our FVM simulation model, it is very hard to judge the accuracy of the simulations from the differences between the measured and simulated luminance maps. On the one hand, even at the low driving currents the CoB LED device was too bright for accurate direct imaging with the luminance measuring camera; on the other hand, due to the lack of resolving the spectral power distribution of the converted light in the optical part of our multi-domain model, the luminance estimated from the radiance is only a very rough approximation. Nevertheless, the properly matching orders of magnitude of the simulated and measured luminance values obtained for the brightest spots and the 19% relative difference between maximal values and 7% difference between average values are promising.

A further result from the multi-domain simulations of the CoB device is the voltage distribution on the chip interconnect metallization layers, see Figure 31. The differences between the discrete values corresponding to the chip locations represent the actual forward voltages of the individual chips, calculated by the instances of the chip-level multi-domain LED model, embedded into our FVM solver. The actual voltage drops are determined by the local temperatures of the chips. Unlike in the case of a real, physical CoB device, in the simulation model, we have access to the individual forward voltage values of the chips, present in the LED array of a CoB device.

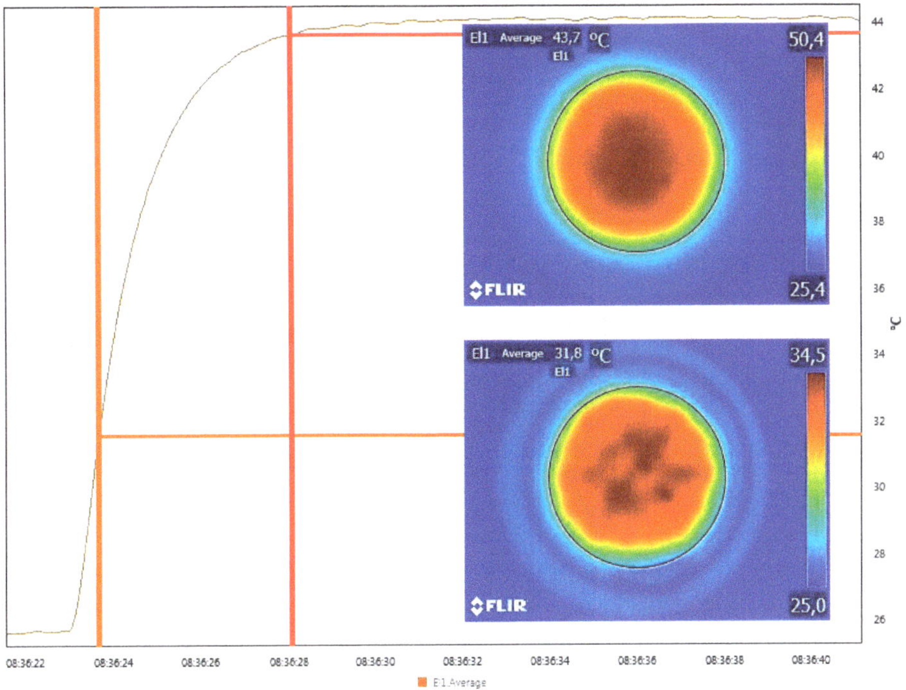

Figure 28. The measured transient of the average temperature of the CoB device and two thermal images grabbed during the heating up process of a physical sample of the investigated CoB device.

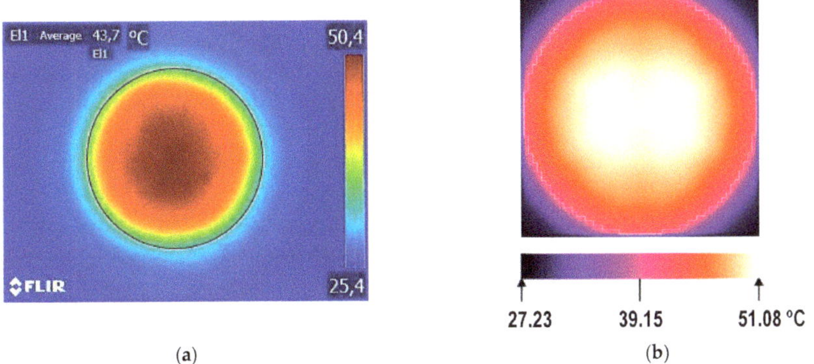

(a) (b)

Figure 29. Steady-state temperature distribution at the top surface of the investigated CoB device at 100 mA driving current: (**a**) measured; (**b**) simulated.

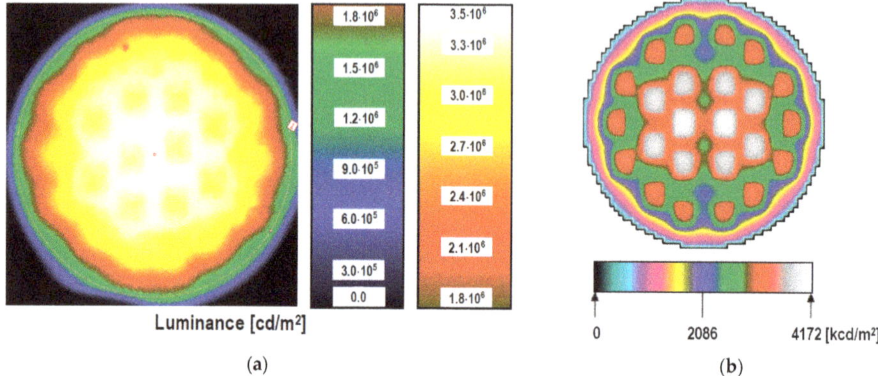

Figure 30. Luminance distribution at 80 mA of driving current and at an ambient temperature of 25 °C: (**a**) Results as presented by the software of the luminance measurement camera, (**b**) luminance distribution estimated from the simulated radiance map.

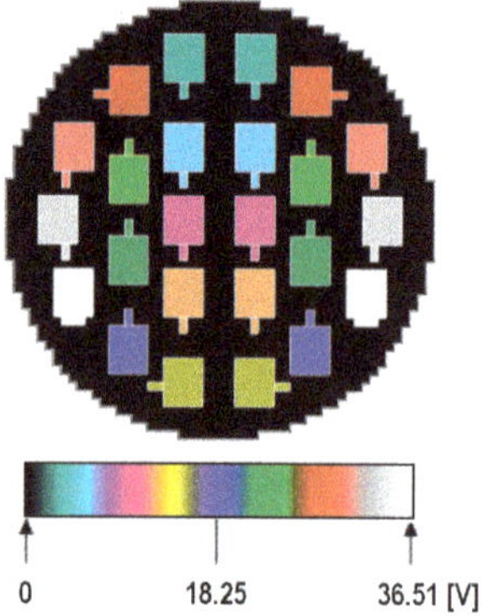

Figure 31. Simulated voltage distribution in the metallization layer of the anodes of the blue LED chips (400 mA driving current, 25 °C ambient temperature).

The simulation also provides insight into the lateral and vertical distributions of other properties, such as the actual junction temperatures of the individual LED chips, radiance at a given plane within the phosphor parallel to the substrate, phosphor temperature as a function of distance from the blue chip surfaces, as illustrated in Figure 32.

Note that, a similar vertical characteristic was published in [23] where the yellow light continuously increases away from the junction, while in Figure 32c it decreases continuously. The phenomenon is caused by the difference between the two light propagation models: In Reference [23] propagation model (a) was used where blue and yellow light travel together, perpendicular to the chip surface, while Figure 32c corresponds to the light propagation model (g) where the yellow light exits the cells of the phosphor model with equal probability in all directions, so a significant portion of it starts down, then it is reflected from the surface of the chip and travels towards the surface of the phosphor. This means that the yellow light passes through many surfaces in both directions. The consequence of this phenomenon is that there is greater radiance near the chip and dissipation occurs closer to the chip than in the case of applying model (a), resulting in a lower average phosphor temperature.

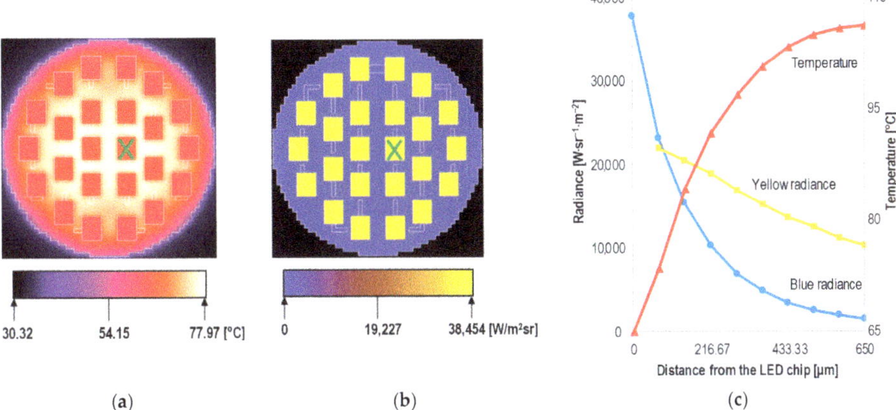

Figure 32. Simulated distributions within the CoB device: (**a**) Junction temperature of the blue LED chips; (**b**) blue radiance at the top of the LED chips; (**c**) vertical distribution of the phosphor temperature, blue radiance and yellow radiance at the location indicated by the green X marker (400 mA driving current, 25 °C ambient temperature).

As mentioned before, one cannot compare the simulated internal distributions of the forward voltages, chip junction temperatures and vertical phosphor temperature distributions to measurements, the 'ensemble' characteristics though, such as the overall forward voltage or the emitted total radiant flux can be both measured and calculated from the simulation results. In Table 6, we provide a comparison of these quantities. Table 6 also provides temperature data: The 'ensemble' junction temperature, $T_{J\_ensemble}$ as measured in compliance with the JEDEC JESD51-51/51-52 standards and an average junction temperature, $T_{J\_average}$, calculated from the distinct junction temperatures used as input to the instances of the multi-domain chip-level LED compact model. Since these temperature values are obtained in different ways, it does not make sense to calculate any relative temperature error from them.

Table 6. Measured and simulated 'ensemble' properties of the investigated CoB LED device (400 mA driving current, 25 °C ambient temperature).

Quantity	Measured	Simulated	Relative Error [%]
$V_{F\_ensemble}$ [V]	36.33	36.51	0.495
$F_{e\_ensemble}$ [W]	3.560	3.759	5.59
$T_{J\_ensemble}$ [°C]	62.7	n.a.	n.a.
$T_{J\_average}$ [°C]	n.a.	63.17	n.a.

7. Summary and Conclusions

In this paper, we proposed and described a methodology for electrical-thermal-radiometric multi-domain modeling and simulation of white CoB LEDs by the combination of compact and distributed modeling methods.

We have also presented a method for multi-domain light modeling in the phosphor, also considering the local temperature dependence of some phosphor properties, such as the conversion efficiency. We implemented different physics/analytic formulae based light propagation models of various complexities and with a few, added heuristics, offering different trade-offs between the need for computational resources and expected accuracy.

A major bottleneck in every numerical simulation is to provide accurate/realistic input data, especially in terms of material properties. Phosphor layers in LEDs pose special problems in this regard, since the composition of the applied materials are usually not disclosed to the public, not even general data, such as thermal conductivity or exact absorption/emission spectra and their possible temperature dependence. Therefore, in order to set up realistic models, we created our own phosphor samples and our own phosphor-converted white LEDs that were characterized in details, in order to extract the phosphor properties as input for modeling and also, to serve as simple reference structures with controlled properties for validating our simulation models.

With the developed opto-thermal model for phosphor layers we carried out trial simulations for assumed CoB LED structures with different phosphor thicknesses, and with these models we also compared our different light propagation models. We found that for single-chip white LEDs with thin phosphor layers, even the simplest 1D light propagation model may provide sufficiently accurate results. For a general case of any complex CoB device, however, we found that the complex 3D light propagation model is better suited. We found that certain heuristics can be used to speed up the simulations without compromising the accuracy of the results. This 3D light propagation model was implemented in our FVM based numerical simulation code. With our previously proposed Spice-like chip-level multi-domain LED model included, the electrical behavior of the LED chips is also included in the CoB model, allowing to study different chip-level, package level and phosphor level problems (such as the effect of increased local junction heating, due to thermally degraded die attach layers on the luminous flux output) simultaneously, spatially resolved to chip-level or even to smaller scales.

The use of this detailed, distributed thermo-optical model, combined with the LED chips' compact multi-domain model, was demonstrated through the example of the Lumileds 1202s CoB LED devices. This was among the test samples of the round-robin test of the Delphi4LED project and such was already characterized in great detail by multiple independent LED testing laboratories [43]. Using this example, we found that our present modeling approach provides satisfactory accuracy during multi-domain simulation of CoB devices.

The work reported here is far from complete. An obvious approximation is that the spectral power distribution of the converted light is not yet considered. This issue, though, does not impose any theoretical problem; only the 'yellow ray model', indicated in Figure 17, needs to be multiplied according to the number of spectral ranges to which the detailed emission spectrum of the converted light is resolved. This would be important to model the visual performance (such as the total luminous flux, the surface luminance map, color point/correlated color temperature) of the CoB devices accurately enough. Note, however, that in terms of calculations of the radiant properties, our present model already provides accurate results since in the calculation of the yellow photon number the wavelength dependence of the radiant power of the photon flux is considered.

Author Contributions: Conceptualization, A.P., L.P., M.N., and Z.K.; methodology, L.P., M.N., Z.K., G.H.; software, L.P., M.N.; validation, G.H. and J.H.; formal analysis, L.P. and M.N.; data curation, G.H. and J.H.; writing—original draft preparation, L.P., M.N., G.H., J.H.; writing—review and editing, A.P., L.P., Z.K., G.H.; supervision, A.P.; project administration, A.P.; funding acquisition, A.P. All authors have read and agreed to the published version of the manuscript.

Funding: This research received funding from the European Union's Horizon 2020 research and innovation program through the H2020 ECSEL project Delphi4LED (grant agreement 692465). Co-financing of the Delphi4LED project by the Hungarian government through the NEMZ_16-1-2017-0002 grant of the National Research, Development and Innovation Fund is also acknowledged. The work related to phosphor characterization was co-funded by the K 128315 grant of the National Research, Development and Innovation Fund. Final validation tests and writing this paper were supported by the Higher Education Excellence Program of the Ministry of Human Capacities in the frame of Artificial Intelligence research area (BME FIKP-MI/SC) and the Nanotechnology research area (BME FIKP-NAT) of the Budapest University of Technology and Economics. The support of the Science Excellence Programs at BME under the grant agreement 341 NKFIH-849-8/2019 and BME NC TKP2020 of the Hungarian National Research, Development and Innovation Office is also acknowledged.

Acknowledgments: The help of Z. Sárkány (from Mentor, a Siemens business, Budapest, Hungary) in measuring the thermal conductivity of phosphor samples is acknowledged.

Conflicts of Interest: The authors declare no conflict of interest.

Abbreviations

Abbreviation	Meaning
AlN	aluminium nitride
ANSYS	multiphysics engineering simulation software of ANSYS Inc.
BME	Budapest University of Technology and Economics
CFD	computational fluid dynamics
CoB	chip-on-board
Delphi4LED	European H2020 ECSEL research project
DynTIM	thermal interface material thermal conductivity measurement equipment from Mentor Graphics [45]
DUT	device under test
FloTHERM	CFD simulator from Mentor Graphics
FVM	finite volume method
IVL	current-voltage-light output (e.g. radiant flux)
JEDEC	Joint Electron Device Engineering Council
JESD	JEDEC standards
LED	light emitting diode
LEP	light-emitting polymer
LightTools	a 3D optical engineering and design software by Synopsys Inc.
MCPCB	metal core printed circuit board
OLED	organic light emitting device
pc-WLED	phosphor-converted white LED
PDMS	polydimethylsiloxane
QE	quantum efficiency
SPD	spectral power distribution
SUNRED	Successive Network Reduction
T3Ster	thermal transient tester equipment from Mentor Graphics [17]
THERMINIC	International Workshop on Thermal Investigations of ICs and Systems
TracePro	optical engineering software by Lambda Research Corp.
YAG:Ce	yttrium aluminium garnet activated by cerium

Symbol	Definition	Unit
I_F	forward current (of an LED)	[A]
V_F	forward voltage (of an LED)	[V]
$V_{F\_ensemble}$	ensemble forward voltage of an LED array	[V]
V_{F_i}	forward voltage of the i-th individual LED chip of an LED array	[V]
$V_{F_{chip}}$	average forward voltage of the LED chips within an LED array	[V]
$T_{J\_ensemble}$	junction temperature associated with $V_{F\_ensemble}$ of an LED array	[°C]
T_J	junction temperature of an LED (see JEDEC JESD51-51 [36])	[°C]
T_{amb}	ambient temperature of the environment (e.g. the laboratory)	[°C]
T_C	temperature in the center of of a simulation grid cell of the FVM model	[°C]
Φ_e	(total emitted) radiant flux (of an LED), also known as emitted optical power (alternate notation: P_{opt})	[W]
Φ_V	(total emitted) luminous flux (of an LED)	[lm]
Φ_X	(total emitted) flux (of an LED); radiant (X=e) or luminous (X=V)	[W] or [lm]
$\Phi_{X\_ensemble}$	ensemble flux of an LED array; radiant (X=e) or luminous (X=V)	[W] or [lm]
Φ_{X_i}	(total emitted) flux of the i-th individual LED chip of an LED array; radiant (X=e) or luminous (X=V)	[W] or [lm]
$\Phi_{x_{chip}}$	average (total emitted) flux of the LED chips within an LED array; radiant (X=e) or luminous (X=V)	[W] or [lm]
λ	wavelength	[nm]
R_{th}	thermal resistance (of e.g. a grid cell of the FVM model)	[K/W]
$R_{th\_ensemble}$	ensemble thermal resistance of an LED array (see JEDEC JESD51-51 [36])	[K/W]
R_{th_i}	thermal resistance i-th individual LED chip of an LED array	[K/W]
C_{th}	thermal capacitance (of e.g. a grid cell of the FVM model)	[Ws/K]
P_H	heating power (of an LED chip / heat loss in the phosphor cell in the FVM model) calculated by the LED chip or phosphor multi-domain model	[W]
P_d	local heating power of a simulation grid cell of the FVM model	[W]
N	number serially connected LED chips of an LED array	[-]
N_e, N_{en}, N_B, N_Y	number of emitted, nodal emitted, blue and yellow photons	[-]

μ_b, μ_y	sum of the attenuation and conversion coefficients for the blue and yellow photons, respectively	[1/m]
μ_c	conversion coefficient	[1/m]
d.	thickness of the phosphor layer	[m]
x	distance (e.g. from the surface of a blue LED chip)	[m]
r	reflection coefficient (the ratio of reflected and absorbed photons)	[-]
E_{ba}, E_{ya}	energy of the blue and yellow photons, respectively	[J]
E_c	energy difference due to wavelength conversion of photons	[J]
Φ_{e_b}, Φ_{e_y}	radiant flux of the blue and yellow photons, respectively	[W]
Φ_{e_ny}	radiant flux of the *new* yellow photons	[W]
r	vector of location *r*	[m]
α_{blue}, α_{yellow}	blue and yellow attenuation coefficients	[1/m]
$\Phi_{e\_blue\_in}$, $\Phi_{e\_blue\_out}$	input and output radiant flux of blue photons in a simulation grid cell of the FVM model	[W]
$\Phi_{e\_yellow\_in}$, $\Phi_{e\_yellow\_out}$	input and output radiant flux of yellow photons in a simulation grid cell of the FVM model	[W]
$\Phi_{e\_yellow\_trans}$	radiant flux of the transmitted yellow photons in a simulation grid cell of the FVM model (part of yellow light not absorbed)	[W]
$\Phi_{e\_blue\_absorb}$, $\Phi_{e\_yellow\_absorb}$	absorbed blue and yellow radiant flux in a simulation grid cell of the FVM model	[W]
$\Phi_{e\_yellow\_conv}$	radiant flux of the yellow photons converted from blue photons in a simulation grid cell of the FVM model	[W]
$\Phi_{e\_yellow\_re}$	radiant flux of the yellow photons re-emitted from the absorbed yellow photons in a simulation grid cell of the FVM model	[W]
η_{conv}	blue-to-yellow conversion efficiency (proportion of the emitted yellow and absorbed blue radiant fluxes)	[-]
η_{yellow_re}	yellow-to-yellow conversion efficiency conversion efficiency (proportion of the emitted yellow and absorbed yellow radiant fluxes)	[-]
P_{loss_conv}	heating power in the phosphor due to the blue-to-yellow conversion loss	[W]
$P_{loss\_yellow\_trans}$	heating power in the phosphor due to the yellow transmission loss	[W]

References

1. Chen, L.; Lin, C.; Yeh, C.; Liu, R. Light converting inorganic phosphors for white light-emitting diodes. *Materials* **2010**, *3*, 2172–2195. [CrossRef]
2. Smet, P.F.; Parmentier, A.B.; Poelman, D. Selecting conversion phosphors for white light-emitting diodes. *J. Electrochem. Soc.* **2011**, *158*, R37. [CrossRef]
3. Meyer, J.; Tappe, F. Photoluminescent materials for solid-state lighting: State of the art and future challenges. *Adv. Opt. Mater.* **2015**, *3*, 424–430. [CrossRef]
4. McKittrick, J.; Shea-Rohwer, L.E. Review: Down conversion materials for solid-state lighting. *J. Am. Ceram. Soc.* **2014**, *97*, 1327–1352. [CrossRef]
5. Xia, Z.; Liu, Q. Progress in discovery and structural design of color conversion phosphors for LEDs. *Prog. Mater. Sci.* **2016**, *84*, 59–117. [CrossRef]
6. Song, Y.; Ji, E.; Jeong, B.; Jung, M.K.; Kim, E.Y.; Yoon, D.H. High power laser-driven ceramic phosphor plate for outstanding efficient white light conversion in application of automotive lighting. *Sci. Rep.* **2016**, *6*, 31206. [CrossRef] [PubMed]
7. Kang, D.; Wu, E.; Wang, D. Modeling white light-emitting diodes with phosphor layers. *Appl. Phys. Lett.* **2006**, *89*, 231102. [CrossRef]
8. Du, K.; Li, H.; Guo, K.; Wang, H.; Li, D.; Zhang, W.; Mei, T.; Chua, S.J. The rate equation based optical model for phosphor-converted white light-emitting diodes. *J. Phys. D* **2017**, *50*, 095101. [CrossRef]
9. Hu, R.; Cao, B.; Zou, Y.; Zhu, Y.; Liu, S.; Luo, X. Modeling the light extraction efficiency of bi-layer phosphors in white LEDs. *IEEE Photon. Technol. Lett.* **2013**, *25*, 1141–1144. [CrossRef]
10. Huang, M.; Yang, L. Heat generation by the phosphor layer of high-power white LED emitters. *IEEE Photon. Technol. Lett.* **2013**, *25*, 1317–1320. [CrossRef]
11. Li, J.; Tang, Y.; Li, Z.T.; Chen, J.X.; Ding, X.R.; Yu, B.H. Precise optical modeling of phosphor-converted LEDs with arbitrary concentration and thickness using bidirectional scattering distribution function. *IEEE Photonics J.* **2018**, *10*, 1–17. [CrossRef]

12. Huang, C.; Tien, C. Phosphor-converted LED modeling by bidirectional photometric data. *Opt. Express* **2010**, *18*, A261–A271. [CrossRef]

13. Tan, C.M.; Singh, P.; Zhao, W.; Kuo, H.C. Physical limitations of phosphor layer thickness and concentration for white LEDs. *Sci. Rep.* **2018**, *8*, 2452. [CrossRef] [PubMed]

14. Juntunen, E.; Tapaninen, O.; Sitomaniemi, A.; Heikkinen, V. Effect of phosphor encapsulant on the thermal resistance of a high-power COB LED module. *IEEE Trans. Compon. Packag. Manuf. Technol.* **2013**, *3*, 1148–1154. [CrossRef]

15. Zollers, M. Phosphor Modeling in LightTools Ensuring Accurate White LED Models, White Paper, Synopsys. 2011. Available online: https://www.synopsys.com/content/dam/synopsys/optical-solutions/documents/datasheets/modeling-phosphors-in-lighttools.pdf (accessed on 2 June 2020).

16. Alexeev, A.; Cassarly, W.; Hildenbrand, V.D.; Tapaninen, O.; Sitomaniemi, A.; Wondergem, A. Simulating Light Conversion in mid-power LEDs. In Proceedings of the 17th International Conference on Thermal, Mechanical and Multi-Physics Simulation and Experiments in Microelectronics and Microsystems (EuroSimE 2016), Montpellier, France, 18–20 April 2016. [CrossRef]

17. Mentor Graphics' T3Ster Equipment Product Homepage. Available online: https://www.mentor.com/products/mechanical/micred/t3ster/ (accessed on 2 June 2020).

18. Jeon, S.-W.; Noh, J.N.; Kim, K.H.; Kim, W.H.; Yun, C.; Song, S.B.; Kim, J.P. Improvement of phosphor modeling based on the absorption of Stokes shifted light by a phosphor. *Opt. Express* **2014**, *22*, A1237. [CrossRef]

19. Qian, X.; Zou, J.; Shi, M.; Yang, B.; Li, Y.; Wang, Z.; Liu, Y.; Liu, Z.; Zheng, F. Development of optical-thermal coupled model for phosphor-converted LEDs. *Front. Optoelectron.* **2019**, *12*, 249–267. [CrossRef]

20. Ye, H.; Koh, S.W.; Yuan, C.; van Zeijl, H.; Gielen, A.W.J.; Lee, S.-W.R.; Zhang, G. Electrical–thermal–luminous–chromatic model of phosphor-converted white light-emitting diodes. *Appl. Therm. Eng.* **2014**, *63*, 588–597. [CrossRef]

21. Chen, H.T.; Tan, S.-C.; Hui, S.Y.R. Analysis and modeling of high-power phosphor-coated white light-emitting diodes with a large surface area. *IEEE Trans. Power Electron.* **2015**, *30*, 3334–3344. [CrossRef]

22. RunHu, Y.M.; Yu, X.; Shu, W.; Luo, X. A modified bidirectional thermal resistance model for junction and phosphor temperature estimation in phosphor-converted light-emitting diodes. *Int. J. Heat Mass Transf.* **2017**, *106*, 1–6. [CrossRef]

23. Pohl, L.; Németh, M.; Hegedüs, J.; Hantos, G.; Kohári, Z.; Poppe, A. Multi-Domain Modelling and Simulation of White CoB LEDs. In Proceedings of the 25th International Workshop on THERMal INvestigation of ICs and Systems (THERMINIC'19), Lecco, Italy, 25–27 September 2019. [CrossRef]

24. Németh, M.; Kohári, Z.; Hegedüs, J.; Hantos, G.; Pohl, L.; Pálovics, P.; Poppe, A. Transient reduced order thermal model of LEDs with phosphor layer. In Proceedings of the Symposium on Design, Test, Integration and Packaging of MEMS/MOEMS (DTIP'19), Paris, France, 12–15 May 2019; pp. 64–69. [CrossRef]

25. Hegedüs, J.; Hantos, G.; Németh, M.; Pohl, L.; Kohári, Z.; Poppe, A. Multi-Domain Characterization of CoB LEDs. In Proceedings of the CIE 2019: 29th QUADRENNIAL SESSION, Washington, DC, USA, 14–22 June 2019; pp. 387–397. [CrossRef]

26. Delphi4LED Project Website. Available online: https://delphi4led.org (accessed on 14 April 2020).

27. Martin, G.; Marty, C.; Bornoff, R.; Poppe, A.; Onushkin, G.; Rencz, M.; Yu, J. Luminaire Digital Design Flow with Multi-Domain Digital Twins of LEDs. *Energies* **2019**, *12*, 2389. [CrossRef]

28. Poppe, A.; Farkas, G.; Gaál, L.; Hantos, G.; Hegedüs, J.; Rencz, M. Multi-domain modelling of LEDs for supporting virtual prototyping of luminaires. *Energies* **2019**, *12*, 1909. [CrossRef]

29. Bornoff, R. Extraction of Boundary Condition Independent Dynamic Compact Thermal Models of LEDs—A Delphi4LED Methodology. *Energies* **2019**, *12*, 1628. [CrossRef]

30. Poppe, A. Simulation of LED Based Luminaires by Using Multi-Domain Compact Models of LEDs and Compact Thermal Models of their Thermal Environment. *Microelectron. Reliab.* **2017**, *72*, 65–74. [CrossRef]

31. Marty, C.; Yu, J.; Martin, G.; Bornoff, R.; Poppe, A.; Fournier, D.; Morard, E. Design flow for the development of optimized LED luminaires using multi-domain compact model simulations. In Proceedings of the 24th International Workshop on Thermal Investigation of ICs and Systems (THERMINIC'18), Stockholm, Sweden, 26–28 September 2018. [CrossRef]

32. Kohári, Z.; Székely, V.; Rencz, M.; Páhi, A.; Dudek, V.; Höfflinger, B. Studies on the heat removal features of stacked SOI structures with a dedicated field solver program (SUNRED). *Microelectron. Rel.* **1998**, *38*, 1881–1891. [CrossRef]

33. Pohl, L.; Székely, V. A more flexible realization of the SUNRED algorithm. In Proceedings of the 12th International Workshop on THERMal INvestigation of ICs and Systems (THERMINIC'06), Nice, France, 27–29 September 2006; pp. 96–100. Available online: https://arxiv.org/ftp/arxiv/papers/0709/0709.1864.pdf (accessed on 31 July 2020).
34. Pohl, L.; Kollár, E.; Poppe, A.; Kohári, Z. Nonlinear electro-thermal modeling and field-simulation of OLEDs for lighting applications I: Algorithmic fundamentals. *Microelectron. J.* **2012**, *43*, 624–632. [CrossRef]
35. Kohári, Z.; Kollár, E.; Pohl, L.; Poppe, A. Nonlinear electro-thermal modeling and field-simulation of OLEDs for lighting applications II: Luminosity and failure analysis. *Microelectron. J.* **2013**, *44*, 1011–1018. [CrossRef]
36. JEDEC JESD51-51 Standard. *Implementation of the Electrical Test Method for the Measurement of the Real Thermal Resistance and Impedance of Light-Emitting Diodes with Exposed Cooling Surface;* JEDEC: Arlington, VA, USA, 2012.
37. JEDEC JESD51-52 Standard. *Guidelines for Combining CIE 127-2007 Total Flux Measurements with Thermal. Measurements of LEDs with Exposed Cooling Surface;* JEDEC: Arlington, VA, USA, 2012.
38. Zhang, Q.; Pi, Z.; Chen, M.; Luo, X.; Xu, L.; Liu, S. Effective thermal conductivity of silicone/phosphor composites. *J. Compos. Mater.* **2011**, *45*, 2465–2473. [CrossRef]
39. Yuan, C.; Luo, X. A unit cell approach to compute thermal conductivity of uncured silicone/phosphor composites. *Int. J. Heat Mass Transf.* **2013**, *56*, 206–221. [CrossRef]
40. Alexeev, A.; Martin, G.; Hildenbrand, V.; Bosschaart, K.J. Influence of Dome Phosphor Particle Concentration on Mid-Power LED Thermal Resistance. In Proceedings of the 32nd IEEE Thermal Measurement, Modeling & Management Symposium (SEMI-THERM), San Jose, CA, USA, 14–17 March 2016. [CrossRef]
41. Wenzl, F.P.; Fulmek, P.; Sommer, C.; Schweitzer, S.; Nemitz, W.; Hartmann, P.; Pachler, P.; Hoschopf, H.; Schrank, F.; Langer, G.; et al. Impact of extinction coefficient of phosphor on thermal load of color conversion elements of phosphor converted. *LEDs J. Rare Earths* **2014**, *32*, 201–206. [CrossRef]
42. Bachmann, V.M. Studies on Luminescence and Quenching Mechanisms in Phosphors for Light Emitting Diodes. Ph.D. Thesis, University of Utrecht, Utrecht, The Netherlands, 2007. Available online: https://dspace.library.uu.nl/bitstream/handle/1874/22761/full.pdf (accessed on 31 July 2020).
43. Poppe, A.; Farkas, G.; Szabó, F.; Joly, J.; Thomé, J.; Yu, J.; Bosschaartl, K.; Juntunen, E.; Vaumorin, E.; di Bucchianico, A.; et al. Inter Laboratory Comparison of LED Measurements Aimed as Input for Multi-Domain Compact Model Development within a European-wide R&D Project. In Proceedings of the Conference on "Smarter Lighting for Better Life" at the CIE Midterm Meeting 2017, Jeju, Korea, 23–25 October 2017; CIE x044:2017. pp. 569–579, ISBN 978-3-901906-95-4. [CrossRef]
44. Schubert, E.F. *Light-Emitting Diodes*, 2nd ed.; Cambridge University Press: Cambridge, UK, 2006. [CrossRef]
45. Mentor Graphics' DynTIM Equipment Product Homepage. Available online: https://www.mentor.com/products/mechanical/micred/dyntim/ (accessed on 2 June 2020).
46. Appel, A. Some techniques for shading machine renderings of solids. In Proceedings of the AFIPS Spring Joint Computing Conference, Atlantic City, NJ, USA, 30 April–2 May 1968; pp. 37–45. [CrossRef]
47. Lafortune, E. Mathematical Models and Monte Carlo Algorithms for Physically Based Rendering. Ph.D. Thesis, Faculty Of Engineering, KU Leuven, Belgium, 1996.
48. Alexeev, A.; Martin, G.; Onushkin, G. Multiple heat path dynamic thermal compact modeling for silicone encapsulated LEDs. *Microelectron. Reliab.* **2018**, *87*, 89–96. [CrossRef]
49. Alexeev, A.; Onushkin, G.; Linnartz, J.-P.; Martin, G. Multiple Heat Source Thermal Modeling and Transient Analysis of LEDs. *Energies* **2019**, *12*, 1860. [CrossRef]
50. Alexeev, A. Characterization of Light Emitting Diodes with Transient Measurements and Simulations. Ph.D. Thesis, TU Eindhoven, Eindhoven, The Netherlands, 2020. Available online: https://research.tue.nl/en/publications/characterization-of-light-emitting-diodes-with-transient-measurem (accessed on 21 July 2020).

Article

Lifetime Modelling Issues of Power Light Emitting Diodes

János Hegedüs, Gusztáv Hantos and András Poppe *

Department of Electron Devices, Budapest University of Technology and Economics, 1117 Budapest, Hungary; hegedus@eet.bme.hu (J.H.); hantos@eet.bme.hu (G.H.)
* Correspondence: poppe@eet.bme.hu; Tel.: +36-1-463-2721

Received: 27 May 2020; Accepted: 23 June 2020; Published: 1 July 2020

Abstract: The advantages of light emitting diodes (LEDs) over previous light sources and their continuous spread in lighting applications is now indisputable. Still, proper modelling of their lifespan offers additional design possibilities, enhanced reliability, and additional energy-saving opportunities. Accurate and rapid multi-physics system level simulations could be performed in Spice compatible environments, revealing the optical, electrical and even the thermal operating parameters, provided, that the compact thermal model of the prevailing luminaire and the appropriate elapsed lifetime dependent multi-domain models of the applied LEDs are available. The work described in this article takes steps in this direction in by extending an existing multi-domain LED model in order to simulate the major effect of the elapsed operating time of LEDs used. Our approach is based on the LM-80-08 testing method, supplemented by additional specific thermal measurements. A detailed description of the TM-21-11 type extrapolation method is provided in this paper along with an extensive overview of the possible aging models that could be used for practice-oriented LED lifetime estimations.

Keywords: power LED measurement and simulation; life testing; reliability testing; LM-80; TM-21; LED lifetime modelling; LED multi-domain modelling; Spice-like modelling of LEDs; lifetime extrapolation and modelling of LEDs

---

1. Introduction

The typical failure mode of LEDs, unlike fluorescent and incandescent light sources, is not catastrophic failure. The total luminous flux of solid state light sources (SSL) experiences a continuous decrease with the elapsed operating time. The most commonly applied end-of-life criterion of LEDs is related to this constantly declining nature. The most apparent manifestation of LED aging is the luminous flux decrease, or seen from another perspective, to what extent the initial value of the emitted total luminous flux of an LED package is maintained. (sloppy, every day terminology to denote this property of LEDs is called 'lumen maintenance' that we shall rigorously refrain from using; instead, we shall refer to this LED property as 'luminous flux maintenance'.) The IES LM-80 family of test methods have been developed and used in the SSL industry to measure LEDs' luminous flux maintenance. The experiment part of our work presented here is based partly on the provisions of the IES LM-80-08 test method [1]. Besides luminous flux measurements, measuring the continuously shifting forward voltage may be part of the life tests but it is still not required in LM-80 tests and therefore such measurements or the results reporting is often omitted. In addition, LM-80-08 (like any other common life testing method) is defined at predetermined ambient temperatures and does not consider any change in the dissipated power or the degradation of the heat flow path, i.e., the cooling capability of the LED, though such tests were already proposed as early as 2011 [2]. The maximum allowed depreciation of the total luminous flux depends on the exact field of the application; the end of the SSL product lifetime is considered to be at the time when its light output deteriorates to the critical value. If an LED-based light source operates with a fixed, constant drive current, its value shall be determined so that the level of illumination remains sufficient even at the critical light output level. This results in higher illumination than necessary during most of the life cycle, as well as a significant amount of extra energy consumption. A smart controlling scheme that keeps the light output at a constant level throughout the lifetime can therefore not only increase the visual comfort of the SSL product but also improves its luminous efficiency [3,4]. Some SSL vendors already provide LED drivers that can be pre-scheduled; by gradually increasing the forward current of the LEDs, effects of the continuous total luminous flux degradation can be compensated. The elapsed lifetime dependent controlling scheme is defined for each configuration that consists of the driver, the applied heatsink and LEDs etc. The available CLO solutions most probably rely on extensive life testing results, however, the exact technical details (above the theory of the approved lifetime testing and extrapolating methods) are not put to public.

The recent industrial trends are continuously pushing product development under digitalization to reduce time-to-market and development cost. This mostly means system level computer aided simulations with the so-called digital twins (computer simulation models) of the real life components, like light sources, optical parts, heatsinks etc. Power LED modelling is still an active research area; a recent European H2020 project on LED characterization and modelling (Delphi4LED) [5] undertook to fulfil the growing industrial needs and aimed to generate the measured-data based digital twins of power LEDs [6–9]. Besides many considerations like round-robin testing [10], product variability analysis [11–13], chip-on-board device modelling [14] etc., one could rise the question: how could the electrical, optical and thermal parameter degradation of the LEDs be modelled? There are several analytical models for different stress conditions and parameter changes, e.g., mechanical stresses of the wire bonds [15], termo-hygro-mechanical stresses inside the package [16], shift of parameters in the Shockley diode equation [17], the course and effects of electro migration [18,19], etc.

Our present work originates also from the Delphi4LED in two ways. On the one hand, the models developed (e.g., [7]) and the test methods (see e.g., [10]) used in that project have also been used in this work. On the other hand, we focused on mainstream LEDs of today's SSL industry (operating in the visible range) that were also the subject of investigation in Delphi4LED and in terms of classifying these devices as 'mid-power' or 'high-power', we use the same terminology that was also been used within Delphi4LED [9]. This also explains why in our study we did not consider novel LED structures or recent LEDs aimed for the display industry, despite the fact that top level publications on these devices also share significant amount of test data [20–23]; rather, we re-used some of our own archived data measured during earlier European collaborative R&D projects such as NANOTHERM [24] and we also used our own test data obtained recently.

Reliability testing and investigation of LEDs has long been a hot topic, just some examples are papers [25–40]. Still, there is a lack of a lifetime-lasting multi-physical digital pair of the already existing and widely used solid state lighting solutions, not to mention the novel devices in the research phase such as the LEDs described in papers [20–23]. This exceeded the goals of the aforementioned H2020 research project but the solution of this issue is of an increasing interest as it offers many new options in certain LED applications. The capability of modelling the parameter degradation under different environmental conditions allows to determine the controlling scheme that results in constant light output (CLO). Furthermore, accurate system level lifetime simulations could give appropriate feedback to the luminaire designers by the means of the operational pn-junction temperatures and the suitability of the cooling assembly. These altogether could provide improved reliability, lower power consumption and higher visual comfort during the whole lifetime of streetlighting luminaires.

Our initial concepts and the summary of some of our test data were provided in our prior conference publications [3,4]. In [3] the concept of stabilizing the total emitted luminous flux of LEDs for their foreseen total expected life-span was presented. In [4] actual test data were provided along with our first attempt to extend one of the multi-domain chip level LED models of the Delphi4LED project [7] with the elapsed LED lifetime. The work described in this article is a comprehensive extension of our theory and test data already presented at the THERMINIC workshops in the previous years [3,4].

2. Total Luminous Flux Maintenance Projections

Reliability and lifetime testing of electrical components is quite a diverse field of research. Due to the complex use of materials in the LED package, various types of the failure mechanisms are induced by the different ambient stress conditions, such as extremely low or high humidity and temperature or high speed change of these, off/on power switching etc. [28–34]. Depending on the field of application, the industry may require a wide range of various reliability tests from the light source manufacturers; to reduce the total testing time it is a common practice to accelerate the degradation mechanisms by increasing the test conditions. The so-called accelerated life tests are based on the Arrhenius model that is used to predict the aging progress under varying degrees of the environmental stress conditions. It is worth mentioning however, that besides the standard laboratory reliability tests widely used in the SSL industry, there are a couple of academic studies about possible new in-field, in-situ test methods aimed primarily for health monitoring, such as identification of LEDs' junction temperature form their emission spectra [35–37] or from changes of certain diode model parameters [38], or from certain dynamic operating characteristics such as the small-signal impedance, non-zero intercept frequency or the optical modulation bandwidth [39,40].

As our work is strongly related to our prior, SSL industry inspired projects, in terms of the considered test methods we aimed to stay as close to the already standardized methods as possible. These are recommendations, approved testing methods and standards, like the IES LM-80-08 and the JESD 22-A family of standards from JEDEC [41,42]. These documents contain requirements on the measurement devices, and specify the needed test conditions like the temperature and humidity [41], change of the stress conditions with time [42], as well as the needed accuracy level of the performed measurements and the set stress parameters [1] etc.

In LED-based lighting applications the main source of any lifetime approximation is provided by the IES LM-80-08 approved method and the IES TM-21-11 technical memorandum [43]. The work described in this paper is also based on these documents. Therefore, first we would like to provide a detailed insight to show the methodology and the roots of our concept.

2.1. IES LM-80

The IES LM-80-08 description does not provide detailed instructions on the proper sample size or the sample selection, it only states that the samples under test should adequately represent the overall population. It specifies the necessary case temperatures of 55 °C and 85 °C while the value of the third testing temperature is left to the choice of the manufacturer. The tolerance of the testing case temperatures is 2 °C during the burning time, and the temperature of the surrounding air in the chamber should remain within the ±5 °C range, which should be continuously monitored by a thermocouple measurement system. The relative humidity level is prescribed to be under 65%.

Duration of the life test should be documented at least with 0.5% accuracy and also the length of any possible power failure should be considered. The optical measurements have to be measured at least at every 1000 h, while the LEDs should be driven by the aging forward current and the ambient (or heatsink) temperature should be 25 °C ± 2 °C. The total length of the test should be at least 6000 h but it is preferred to reach the total time of 10,000 h. At each photometric measurement interval chromaticity shift should be measured, as well as any possible catastrophic failure of the samples should be monitored and recorded.

The LM-80 method also gives a recommendation on the format and content of the measurement report generated at the end of the life test. Nevertheless, it does not provide provisions to qualify the LED samples and does not state anything about their lifetimes, it provides a procedure only for the measurement of the total luminous flux maintenance.

In 2015 IES published the LM-80-15 approved method [44] which is the revision of LM-80-08. In the new version there are additional requirements towards the optical and colorimetric measurements, but the prescribed three case temperatures have been reduced to only two and even the minimum test duration of 6000 h has been abolished. Regarding our LED modelling concepts the extra colorimetric measurements are not that necessary at this stage while it is still advantageous to keep the 3 case temperatures, therefore, all of our aging tests were still based on the LM-80-08 document [1].

2.2. IES TM-21-11

The IES TM-21-11 technical memorandum provides a lifetime estimation method to the measurement results of the LM-80-08 testing. The November/December 2011 issue of LEDs Magazine provides a good overview [45] on the extrapolation method.

The extrapolation technique applied by the TM-21-11 method is based on exponential curve fittings to the measured optical data of the LM-80 test. Each case temperature is approximated individually in between which the Arrhenius-equation may be used for interpolations. The recommended sample size is established by 20 packaged LEDs (either on PCB or without it). 30 or more samples would not considerably improve the estimation capability, but there is an uncertainty of extrapolations based on test results of only 10 LEDs. Within this range the number of the samples also sets the limit to the time projection; below 10 tested LEDs the extrapolation method should not be applied, up to 19 tested pieces the document allows a 5.5 times while from 20 samples it admits a 6 times extrapolation of the total test duration. The amount of the measured data used for the exponential curve fitting depends on the total test duration: collected data of the last 5k hours is taken into account up to 10k hours of aging, above that data of the last half of the test is used.

The end of operating lifetime is then defined according to the LM-80 and the TM-21 results. If the pre-defined light output degradation cannot be reached within the extrapolation limit then the result is the maximum extrapolation time itself marked with a "less-than" sign (e.g., "L70 (10k) > 55,000 h" where the "10k" denotes that the LM-80 test lasted for 10,000 h and the 5.5 times rule is applied). If the life output degradation is reached using the TM-21 estimation then the result is reported with an "equals" sign. If the samples reach the minimum light output level during the LM-80 test then the result equals to the testing time in the general reporting formula (e.g., "L90 (5k) = 5100 h").

2.3. The Degradation Model Used by TM-21-11

The TM-21-11 technical memorandum applies an exponential curve fitting method to extrapolate the measured data in time and the Arrhenius-equation to interpolate between the three different case temperatures. The idea behind these techniques is well described in chemistry; in the following part we will use some of the basic concepts of reaction kinetics in order to give an analytical description of the LED degradation models. Our intention was to build-up and cover the fundamentals for the derivations in the later sections in the lack of such a textbook. The textbook-like manner not only introduces our efforts but also aims to provide a reliable baseline for researchers aiming to build on our work.

Reaction rate of a first-order reaction depends linearly on the reactant concentration [46]. The differential form of the rate law is:

$$\text{Rate} = -\frac{dc}{dt} = k \cdot c \tag{1}$$

where c is the changing reactant concentration, t is the elapsed time and k is the reaction rate coefficient. The separable differential equation can be solved by rearranging it and integrating both sides of the following equation:

$$\int_{c_0}^{c} \frac{1}{c}dc = -\int_{t_0}^{t} kdt \tag{2}$$

where c_0 is the initial concentration and t_0 is the initial time instant. The integration should be performed with the condition of $t_0 = 0$ s. After rearranging the achieved formula the rules of logarithm shall be applied. Finally, the integral form of the rate law is:

$$\frac{c}{c_0} = exp(-k \cdot t) \tag{3}$$

Considering the total luminous flux as the decreasing quantity of the homogenous aging process, where the initial value of the regression curve fit to the luminous flux is c_0 and c is the actual value at time t, then the TM-21-11 defined extrapolation of the total luminous flux maintenance curve is obtained:

$$\Phi(t) = \beta \cdot exp(-\alpha \cdot t) \tag{4}$$

where the normalized light output is Φ at the time t, β is a fitting parameter and α corresponds to the reaction rate coefficient k specified at Equation (1).

Homogeneous chemical processes proceeding in the solid phase involves various aging phenomena in plastic and glass, thermal changes induced transformations, recrystallization, transformation of alloys and metals throughout and following a thermal treatment etc. In the solid state these progresses need a lot more time than in gas or in liquid phase. The reaction rate coefficient (i.e., the speed of these reactions) is an exponential function of the absolute temperature. The exact formula is described by the Arrhenius-Equation:

$$k = A \cdot exp\left[\frac{-E_a}{k_B \cdot T}\right] \tag{5}$$

where A is a pre-exponential factor, E_a is the activation energy (the energy barrier below which the reaction in question does not proceed), k_B is Boltzmann's constant and T is the absolute temperature in kelvins. Concerning the TM-21 interpolations, values of A and E_a can be calculated if datasets of two or more temperatures are available.

The Arrhenius-equation is typically used to express the AF acceleration factor of the reaction rate coefficient at elevated temperatures:

$$AF = exp\left[\left(\frac{-E_a}{k_B \cdot T}\right) \cdot \left(\frac{1}{T_2} - \frac{1}{T_1}\right)\right] \tag{6}$$

where T_2 is the elevated temperature.

Upon taking into account the speeding-up effect of higher temperatures, the Arrhenius-equation has various expansions describing the effects of other stress conditions as well, like humidity or other non-thermal stresses. In case of LEDs, with respect to the LM-80 testing the most important non-thermal impact corresponds to the forward current [47]. The forward current dependent reaction rate coefficient can be described as:

$$k = A \cdot exp\left[\frac{-E_a}{k_B \cdot T}\right] \cdot I^n \tag{7}$$

where I is the forward current and n is the so called life-stressor slope [47]. With the help of Equation (7) one can perform the necessary interpolations between the measurement results belonging to the different case temperatures and forward current values captured during an LM-80 life testing of LEDs.

2.4. Further Possible Degradation Trends

The exponential decay of the total luminous flux output of an LED corresponds to the first-order reaction rate. The order of a reaction defines the relationship between the reaction rate (or the decay rate, in this case) and the changing concentration of the reactant(s) (or the decreasing total luminous flux, in this case). Practically, the order of a rate law is the sum of the exponents of the changing parameters. Accordingly, the reaction rates of the zero-, first- and second-order reactions are as follows:

$$\text{Rate} = -\frac{dc}{dt} = k \cdot c^0 = k \tag{8}$$

$$\text{Rate} = -\frac{dc}{dt} = k \cdot c^1 = k \cdot c \tag{9}$$

$$\text{Rate} = -\frac{dc}{dt} = k \cdot c^2 \tag{10}$$

For the sake of curiosity, a third-order reaction rate with two concentrations of species looks like:

$$\text{Rate} = -\frac{dc}{dt} = k \cdot c_1 + k \cdot c_2^2 \tag{11}$$

where c_1 and c_2 are the concentrations of the two different species (note that the upper indices indicate the order of the reaction while the lower indices refer to the different species). To get the integral form of the above rate laws, the same mathematical steps should be followed as described in case of Equations (1)–(3).

The reaction in which one chemical species is irreversibly transformed into more than one other species is the so called parallel reaction (see also other more complex reactions in [48]). In this case the rates of the parallel reactions add up. Supposing a zero- and a first-order parallel reaction with the coefficients of k_1 and k_2 respectively, the decay rate can be written as:

$$\text{Rate} = -\frac{dc}{dt} = k_1 + k_2 \cdot c \tag{12}$$

The separable differential equation can be solved by rearranging it and integrating both sides of the following equation:

$$\int_{c_1}^{c_2} \frac{1}{k_1 + k_2 \cdot c} dc = -\int_{t_1}^{t_2} dt \tag{13}$$

$$\frac{ln(|k_1 + k_2 \cdot c_2|)}{k_2} - \frac{ln(|k_1 + k_2 \cdot c_1|)}{k_2} = -(t_2 - t_1) \tag{14}$$

Rearranging Equation (14) and applying the logarithm rules, we get:

$$ln\left(\left|\frac{k_1 + k_2 \cdot c_2}{k_1 + k_2 \cdot c_1}\right|\right) = -k_2 \cdot (t_2 - t_1) \tag{15}$$

where we know that all terms are positive. Raising both sides of the equation to the power equal to the base of natural logarithm and further rearranging it we get a final version as follows:

$$c_2 = \left[c_1 + \frac{k_1}{k_2}\right] \cdot exp[-k_2 \cdot (t_2 - t_1)] - \frac{k_1}{k_2} \tag{16}$$

where c_1 and c_2 are the concentration values at the time instants t_1 and t_2, respectively. In case of LEDs, a description with parallel reactions should be appropriate when the root causes of the light output degradation can be separated. That is, for example, if the packaged blue LED, the light conversion phosphor material and the lens are aged and tested separately. These aging modes independently reduce the light output of the LED (by the means of decreasing radiant and light conversion efficiencies and light transmission), therefore, theoretically a parallel reaction model with the individual reaction rate coefficients and modes could be matched with the aging results of the complete white LED.

At the EPA ENERGY STAR Lamp Round Table held in San Diego (CA, USA) in 2011, a wider set of decay rate models describing the total luminous flux maintenance of LEDs was proposed by Miller, of the U.S. National Institute of Standards and Technology (NIST, Gaithersburg, MD, USA)—see Table 1 [49]. Among the different decay models one can find zero-, first- and second-order reaction rates, models that are inversely proportional to the elapsed time, and the parallel mixture of the previously listed ones. Miller also drew attention to the fact that the light output degradation trend of LEDs may change significantly during the operation time; Figure 1a,b indicate two relative total luminous flux maintenance curves where the estimated L_{70} lifetime from the 10,000-h results is halved or doubles compared to the estimation from the 6000-hour results.

Table 1. Aging models of different decay rates with the closed form solution, i.e., the integral form (after [49]).

#	Decay Rate	Integral Form	
1	$\frac{dI_V}{dt} = k_1$	$I_V = I_V^0 + k_1 \cdot \left(t - t^0\right)$	
2	$\frac{dI_V}{dt} = k_2 \cdot I_V$	$I_V = I_V^0 \cdot exp\left[k_2 \cdot \left(t - t^0\right)\right]$	
3	$\frac{dI_V}{dt} = k_1 + k_2 \cdot I_V$	$I_V = \left(I_V^0 + \frac{k_1}{k_2}\right) \cdot exp\left[k_2 \cdot \left(t - t^0\right)\right] - \frac{k_1}{k_2}$	Model 1 + Model 2
4	$\frac{dI_V}{dt} = \frac{k_3}{t}$	$I_V = I_V^0 + k_3 \cdot ln\left(\frac{t}{t^0}\right)$	
5	$\frac{dI_V}{dt} = k_1 + \frac{k_3}{t}$	$I_V = I_V^0 + k_1 \cdot \left(t - t^0\right) + k_3 \cdot ln\left(\frac{t}{t^0}\right)$	Model 1 + Model 4
6	$\frac{dI_V}{dt} = k_4 \cdot I_V^2$	$I_V = \frac{I_V^0}{1 + I_V^0 \cdot k_4 \cdot (t - t^0)}$	
7	$\frac{dI_V}{dt} = k_5 \cdot \frac{I_V}{t}$	$I_V = I_V^0 \cdot \left(t/t^0\right)^{k_5}$	
8	$\frac{dI_V}{dt} = k_2 \cdot I_V + k_5 \cdot \frac{I_V}{t}$	$I_V = I_V^0 \cdot exp\left[k_2 \cdot \left(t - t^0\right)\right] \cdot \left(t/t^0\right)^{k_5}$	Model 2 + Model 7
9		$I_V = I_V^0 \cdot exp\left[-\frac{(t - t^0)}{k_6}\right]^{k_7}$	

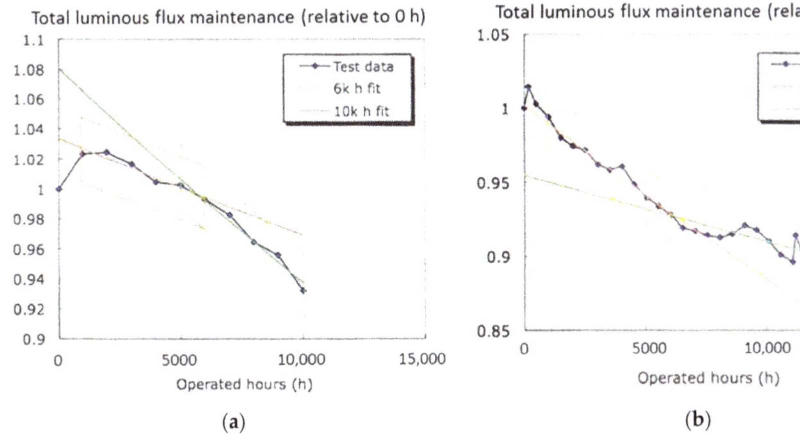

Figure 1. The change in aging trends: extrapolation of the 6k and 10k hours may differ dramatically (based on [49]) (a) $L_{70}(6k) = 60,000$ h vs. $L_{70}(10k) = 30,000$ h; (b) $L_{70}(6k) = 30,000$ h vs. $L_{70}(10k) > 60,000$ h.

3. The Pn-Junction Temperature During the LM-80 Test

The pn-junction temperature of LEDs may change significantly during an LM-80 testing procedure due to the increased electrical power consumption, to the decreased energy conversion efficiencies and even due to the possible degradation of the thermal interfaces. The temperature increase can range from only a few degrees Celsius to as high as 20–30 °C. Its exact value depends mostly on the testing forward current, the overall thermal resistance, the zero-hour radiant efficiency and on the luminous flux maintenance value reached during the test. Figure 2 shows a theoretical approximation for a high- (a) and a mid-power (b) LED, as the function of the main causes of the increase, assuming at this point, that the 3 V forward voltage and the thermal resistance remain constant. During the calculations we neglected any effects of the temperature sensitive radiant efficiency; the extremely high temperature dependence in case of red and amber LEDs is well-known and acts as a positive loopback to the junction temperature, further increasing the discussed effect.

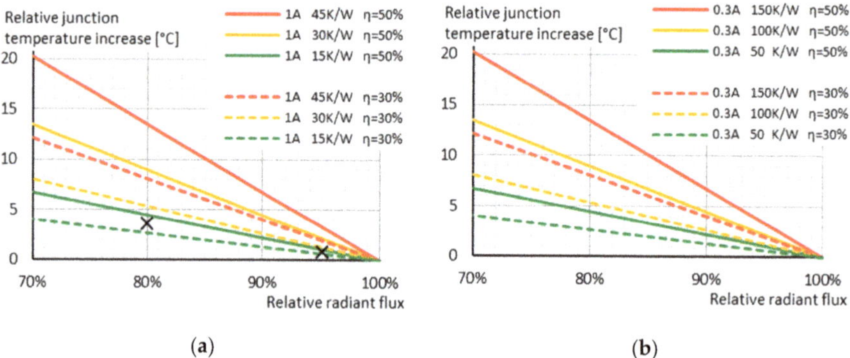

Figure 2. Theoretical pn-junction temperature increase during an LM-80 test in case of (**a**) a high-power and (**b**) a mid-power LED, as the function of the forward current, the thermal resistance, the zero-hour radiant efficiency and the reached luminous flux decay (assuming a constant thermal resistance and a Figure 3. V forward voltage). Note that the curves on (**a**) and (**b**) are identical implying that the junction temperature increase may affect high- and mid-power LEDs equally, depending on the thermal resistance.

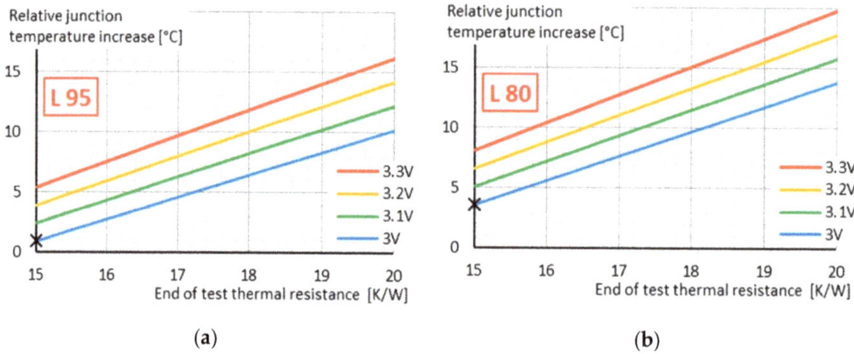

Figure 3. Effects of the increase in the forward voltage and in the thermal resistance on the pn-junction temperature increase during an LM-80 test. The initial parameters are: 1 A forward current, 40% radiant efficiency, 15 K/W thermal resistance, 3 V forward voltage. The figures show the junction temperature increase during the test as the function of the end of test thermal resistance and forward voltage. The end of test relative light outputs are (**a**) 95% and (**b**) 80%. The black crosses indicate values corresponding to that of in Figure 2a.

It is obvious that a few degrees Celsius change in the junction temperature is not an issue. In this article we are dealing with the >10 °C increases caused by high testing currents, high initial efficiency, poor thermal conductivity, significant increase of the forward current and/or the thermal resistance [50,51] etc. Figure 3a,b indicate the effects of the increase in the forward voltage and in the thermal resistance. During the calculations we considered a power LED with a test current of 1 A. The initial values of the radiant efficiency, the forward voltage and the thermal resistance were 40%, 3 V and 15 K/W in order and the relative junction temperature increase was calculated as the function of the end of test thermal resistance and forward voltage. The end of test relative light output was regarded to be 95% and 80%. Stated practically, Figure 3a,b are the extensions of Figure 2a; the indicated black crosses in the figures correspond to each other. As an example for the parameter increase, Figure 4 shows the aging related forward voltage shift of a Luxeon Z power LED; all the measurement points in the figure belong to the 80 °C junction temperature.

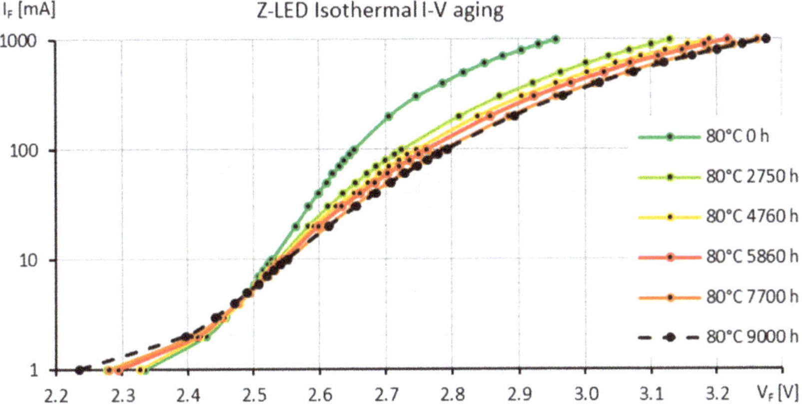

Figure 4. Shift of the forward voltage—forward current characteristics during the first 9000 h of a Luxeon Z LED sample (aged at 85 °C case temperature and 1 A forward current).

Most of the root causes of LED aging is closely connected to the pn-junction temperature. The temperature on which the die-, the interconnection-, the phosphor- and the lens-related aging processes undergo is practically much closer to the pn-junction temperature than the case- or the soldering point temperature. Therefore, we make an attempt to determine the parameter set of the Arrhenius-equation as the function of the changing junction-temperature. To do that first we need to determine the elapsed test-time dependent function of the junction temperature.

3.1. Analytical Calculation of the Pn-Junction Temperature

The T_J pn-junction temperature can be calculated from the T_A ambient temperature, the optically corrected real R_{th} thermal resistance and the P_{dis} dissipated power:

$$T_J = T_A + R_{th} \cdot P_{dis} \tag{17}$$

The dissipated power is the difference of the consumed electrical power and the radiant flux. For simplicity we consider R_{th} to be constant over the time, therefore:

$$T_J(t) = T_A + R_{th} \cdot \left(I_F \cdot V_F(t, T_J, I_F) - \Phi_e(t, T_J, I_F) \right) \tag{18}$$

where I_F and V_F are the forward current and the forward voltage, while Φ_e is the total radiant flux. The latter two parameters depend on the elapsed operation time t, and the actual junction-temperature and forward current. We assume an exponential decay model for the radiant flux function over time, so we apply Equations (4) and (7) and we also take into account the S_{Φ_e} temperature sensitivity of the radiant flux. For simplicity we consider S_{Φ_e} to be constant over the time. Therefore:

$$\Phi_e\big(t, T_J, I_F\big) = \Phi_{e0} \cdot exp\left[-t \cdot A_{\Phi_e} \cdot I_F^{n\Phi_e} \cdot exp\left(\frac{-E_a^{\Phi_e}}{k_B \cdot T_J}\right)\right] \cdot \left[1 + S_{\Phi_e} \cdot \big(T_J - T_{ref}\big)\right] \tag{19}$$

where Φ_{e0} and T_{ref} are the initial radiant flux and the reference operating junction temperature of the LED in the test environment at the zero-hour condition. Next, we assume a zero-order model for the increase of the forward voltage (i.e., a value linearly increasing with time) and we also consider the S_{VF} temperature sensitivity of the forward voltage. For simplicity we consider S_{VF} to be constant over the time:

$$V_F\big(t, T_J, I_F\big) = \left[V_{F0} + t \cdot A_{V_F} \cdot I_F^{nV_F} \cdot exp\left(\frac{-E_a^{V_F}}{k_B \cdot T_J}\right)\right] \cdot \left[1 + S_{V_F} \cdot \big(T_J - T_{ref}\big)\right] \tag{20}$$

After all this, the overall elapsed test-time dependent junction temperature can be written as:

$$\begin{aligned}
T_J(t) = \quad & T_A + R_{th} \cdot I_F \cdot \left[V_{F0} + t \cdot A_{V_F} \cdot I_F^{nV_F} \cdot exp\left(\frac{-E_a^{V_F}}{k_B \cdot T_J}\right)\right] \cdot \left[1 + S_{V_F} \cdot \big(T_J - T_{ref}\big)\right] \\
& - R_{th} \cdot \Phi_{e0} \cdot exp\left[-t \cdot A_{\Phi_e} \cdot I_F^{n\Phi_e} \cdot exp\left(\frac{-E_a^{\Phi_e}}{k_B \cdot T_J}\right)\right] \\
& \quad \cdot \left[1 + S_{\Phi_e} \cdot \big(T_J - T_{ref}\big)\right]
\end{aligned} \tag{21}$$

which is analytically not solvable by ordinary mathematical methods. At this point we gave up with the analytical attempt and tried a measurement based practical method.

3.2. Determination of the Pn-Junction Temperature by Measurement

The strong temperature dependencies of LEDs make it necessary to measure their optical, electrical and thermal parameters simultaneously. The JEDEC JESD 51-5x family of standards and the related CIE standards [52–61] provide a multi-domain characterization method, especially for LEDs, which includes the measurement of the pn-junction temperature by the help of thermal transient testing [62–64] and the calibrating process of the temperature sensitive parameter, i.e., the S_{VF}. Of course, there are other measurement techniques (e.g., as described in [65,66]) beside the mentioned JEDEC and CIE standards/recommendations, we still work with these; over the past 15 years our research team has made significant contributions to the development of the related instruments. The gained experiences and the opportunity to customize the instrumentation also provide a higher flexibility during our investigations. In addition to these, the JEDEC JESD 51-5x family of standards is also conform with the new CIE 225:2017 recommendation and it is also used by many leading SSL companies—the reason why the Delpi4LED consortium has also chosen these standards as the basis of the LED modelling methodology also cited and described in this article.

We have performed the LM-80-08 based life testing of a mid-power LED sample set at the Department of Electron Devices of the Budapest University of Technology and Economics (see details in Section 5). Beside the optical measurements at room temperature specified by the testing method we also performed thermal transient testing both at 25 °C and at the testing case temperature. What we had after the measurements were the forward voltages, the radiant fluxes and the junction temperatures at 25 °C and at 55 °C ambient temperatures. Figure 5 shows the measured pn-junction temperatures at 300 mA forward current. Also, a linear approximation was applied on the measured results after 340 h of aging.

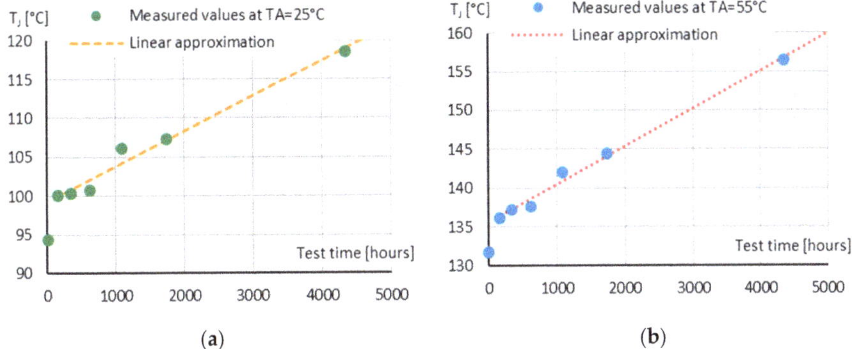

Figure 5. Pn-junction temperatures of the tested mid-power LED (#S11) over the testing time, measured at 300 mA forward current, (**a**) at 25 °C and (**b**) at 55 °C ambient temperatures.

The S_{Φ_e} temperature sensitivity of the radiant flux and the S_{VF} temperature sensitivity of the forward voltage can be calculated from the measurement results belonging to different junction temperatures. By having these sensitivity values it is possible to make approximate calculations for the operating parameters at any arbitrary junction temperature value.

3.3. The Transient Testing Based Calculation of the Arrhenius-Equation

The method that we propose consists of four main steps:

(1). Determine the pn-junction temperature during the LM-80 process, at the test current and case temperature.

(2). Determine the pre-exponential factor A and the activation energy E_a of the Arrhenius-equation from the continuously increasing junction temperature values of each measurement time and the corresponding measured radiant flux values.

(3). Determine the luminous flux maintenance curve belonging to a fixed junction-temperature value, applying an arbitrary time profile of the junction temperature.

(4). Determine the S_{Φ_e} temperature sensitivity of the radiant flux and calculate the light output parameters at any arbitrary junction temperature and at any time of the aging process.

3.3.1. Determine the Pn-Junction Temperature

The first step was introduced in the previous subsection (see the listed references).

3.3.2. Determine the Pre-Exponential Factor and the Activation Energy

The second step is based on the LM-80 results and on the recorded junction temperature values during the test. Let us recall the differential form of the rate law (Equation (1)) and insert the Arrhenius-equation in place of the reaction rate coefficient k (Equation (5)):

$$\text{Rate} = -\frac{dc}{dt} = c_t \cdot A \cdot exp\left[\frac{-E_a}{k_B \cdot T_J(t)}\right] \tag{22}$$

where c_t and $T_J(t)$ are the changing concentration and the junction temperature at the time instance of t; c_t corresponds to Φ_e (or Φ_V). Equation (22) clearly indicates the relationship between the parameters of the Arrhenius-equation and the reaction rate. Let us denote the reaction rate by D (representing the fact that it is the derivative of the c over t function):

$$D = \frac{dc}{dt} \tag{23}$$

At any two measurement times we can write the above equations with the actual parameters:

$$-D_{t_1} = c_{t_1} \cdot A \cdot exp\left[\frac{-E_a}{k_B \cdot T_J(t_1)}\right] \tag{24}$$

$$-D_{t_2} = c_{t_2} \cdot A \cdot exp\left[\frac{-E_a}{k_B \cdot T_J(t_2)}\right] \tag{25}$$

which is a set of equations with two variables (since D_{t_i}, c_{t_i} and $T_J(t_i)$ are measured values). After rearranging it we get that:

$$A = -\frac{D_{t_1}}{c_{t_1}} \cdot exp\left[\frac{E_a}{k_B \cdot T_J(t_1)}\right] \tag{26}$$

$$E_a = \frac{k_B \cdot T_J(t_1) \cdot T_J(t_2)}{T_J(t_2) - T_J(t_1)} \cdot ln\left(\frac{D_{t_2} \cdot c_{t_1}}{c_{t_2} \cdot D_{t_1}}\right) \tag{27}$$

The value of D_{t_i} can be calculated as the derivative of the experimentally acquired luminous flux maintenance curve (see Equation (4)) at the time instance of t_i:

$$D_{t_i} = \frac{d[\beta \cdot exp(-\alpha \cdot t_i)]}{dt} \tag{28}$$

$$D_{t_i} = -\alpha \cdot \beta \cdot exp(-\alpha \cdot t_i) \tag{29}$$

Although, theoretically Equations (26), (27) and (29) should unambiguously assign the values of A and E_a, still several orders of magnitudes differences may occur among the obtained results. To dissolve this issue it could be a good practice to calculate E_a for each measurement time with an arbitrarily fixed A value, then sweep A until the smallest difference amongst the calculated E_a values is reached.

Even if the junction temperature is continuously increasing during aging, its effects on the slope (i.e., the derivative) of the luminous flux maintenance curve (from which A and E_a were calculated) are negligible in most cases compared to the aging related changes of the light output parameters. Although, the following example will make it clear, that the temperature sensitivity of the optical parameters can be extremely high for red and amber LEDs. The S_{Φ_e} temperature sensitivity of the radiant flux of a red power LED from a well-recognized vendor was measured to be −4.3 mW/°C at 1 A forward current whereas the optical power was found to be 1.1 W with a radiant efficiency of 41%. Mounted on a cooling assembly with a 25 K/W junction-to-ambient thermal resistance and supposing a 0.1 V and 1 K/W increase in the forward voltage and thermal resistance respectively, roughly a 7 °C increase occurs in the pn-junction temperature until the time of the 10% light output degradation, causing another thermally induced 2.8% drop in the radiant flux—which is not negligible any more. In such cases Equation (23) should be corrected in the following format:

$$D_{corr} = \frac{dc}{dt} - S_{\Phi_e} \cdot \frac{dT_J}{dt} \tag{30}$$

Another possible compensation method of this effect is to determine the optical flux values that would be emitted at the reference junction temperature and use the gained data set as the new maintenance curve:

$$\Phi_e(t)\big|_{T_{ref}} = \Phi_e(t) \cdot \left[1 + S_{\Phi_e} \cdot \left(T_J - T_{ref}\right)\right] \tag{31}$$

We must note, that it could seem to be a good idea to compare the values achieved by the proposed method with the results of an LM-80 test sequence performed on multiple case temperatures. In fact, the technique discussed here only makes sense if the junction temperature increase during the test is significant, but this also means that the parameters calculated from different aging case temperatures would not be well correlated with the junction temperature, therefore the latter method gives a completely different result in principle. From this reasoning we can tell that only testing at multiple case temperatures provides the needed data if the junction temperature rise is negligible, but otherwise consistent pn-junction temperature based test results can be reached only if the experiment is supported by accurate junction temperature measurements (e.g., thermal transient testing).

3.3.3. Determine the Luminous Flux Maintenance Curve at a Fixed T_J

Our base concept is that LEDs have a kind of "lifetime budget", which (under nominal operating conditions) is consumed at a rate most dependent on the pn-junction temperature. We model the lifetime budget as a junction temperature, forward current and elapsed operating time dependent efficiency η_t (eta t) which is to be multiplied by the zero-hour value of the radiant efficiency η_e or the luminous efficacy η_v to get the prevailing light output parameters. It contains the effects of any aging phenomenon, at this point even including the change of the electrical consumption through the change of the forward voltage.

According to our theory the current value of the budget is not dependent of current value of temperature (in such a way maintaining causality). Also, it does not carry any information on the temperature sensitivity of the parameters, therefore the temperature sensitivity of the optical parameters should be applied when the junction temperature is out of the reference value. If the junction temperature remains constant during the test then the lifetime budget is identical to the luminous flux maintenance curve (described by Equation (4)) normalized to 100%.

To calculate the change of the lifetime budget (or the normalized luminous flux maintenance at the reference T_J) we need to recall again the differential form of the rate law. Assuming that the exact time function of the temperature change is known, we need to substitute it into:

$$\Delta \eta_t = \int_{c_1}^{c_2} \frac{1}{c} dc = - \int_{t_1}^{t_2} A \cdot exp\left[\frac{-E_a}{k_B \cdot T_J(t)} \right] dt \tag{32}$$

The analytical solution of which is not trivial, even in case of a linearly changing temperature value. If the necessary mathematical tools are not available, then a practical solution could be to discretize the problem that way converting the integration to a sum calculation (a series of additions on very short time intervals). This means that the actual value of k can be calculated at every time instance knowing the mean value of the temperature, then the differential form of the rate law can be used with the difference that Δc change is calculated during the short time interval Δt. Adding up c and Δc will result in the total value of the next time interval.

It should be emphasized that during this step we calculated the theoretical light output parameters of the LED at the reference junction temperature but accounting for the real temperature data, which is not always equal. For example, the operating junction temperature of an LED sample is 85 °C at the 55 °C case temperature at the beginning of the test. Let us assume that after 10,000 h the operating temperature is 105 °C at the same 55 °C case temperature. In this case the value substituted into Equation (32) is the real and continuously increasing value (105 °C which is 378.15 K after 10k h) but the radiant or luminous flux provided by Equation (32) is the value the LED would emit at the reference 85 °C junction temperature. First it could be confusing but we must not forget that we calculated the pre-exponential factor A and the activation energy E_a as the function of the prevailing junction temperature (therefore the degradation itself is calculated after the aging T_J profile). Still, these values describe only the aging effects and they do not carry any information about the temperature sensitivity of the optical parameters. If they would do so, then the method described for Equation (30) or (31) should be applied.

3.3.4. Calculating the Light Output Parameters at any T_J

In the previous subsection we have determined the lifetime budget as the function of the elapsed lifetime and junction temperature profile in the meantime. The calculations result in the actual radiant flux value that would be emitted at the reference junction temperature, after the elapsed operating lifetime t. The next step is to apply the S_{Φ_e} temperature sensitivity of the radiant flux if the T_J value increases/changes significantly during aging:

$$\Phi_e(t, T_J) = \Phi_{e0} \cdot \eta_t(t) \cdot \left[1 + S_{\Phi_e} \cdot \left(T_J - T_{ref}\right)\right] \tag{33}$$

where Φ_{e0} is the zero-hour radiant flux, η_t (eta t) is the lifetime budget number at time t. The value of S_{Φ_e} can be determined at zero-hour and used during the whole lifetime as a constant or it can be re-determined at each control measurement. In practice according to the proposed method thermal transient testing should be performed both at case temperature and at the LM-80 prescribed 25 °C ambient temperature. From these measurements a linear approximation of all temperature sensitivity parameters can be determined.

3.4. Case Study

The procedure suggested in Section 3.3 was performed on the measurement results of the LED sample presented in Section 3.2. The steps taken are as follows:

(1). T_J and Φ_e were measured at $T_A = 25\,°C$ and at $T_A = 55\,°C$ (Figure 5 and the blue and green dots in Figure 6). Measurement results before 340 h were omitted.
(2). S_{Φ_e} was calculated.
(3). Φ_e ($T_J = 85\,°C$) was calculated for all the control measurements (Equation (31); the red dots in Figure 6).
(4). A logarithmic trend line was fit to the calculated Φ_e ($T_J = 85\,°C$) values (Model #4 from Table 1; the three continuous lines in Figure 6).
(5). A and E_a were determined; the measured T_J values at $T_A = 55\,°C$ and the maintenance curve determined in step 4 were used. Applying the same method described in Section 3.3.2 the formulas for the logarithmic model are:

$$A = -D_{t_1} \cdot t_1 \cdot exp\left[\frac{E_a}{k_B \cdot T_J(t_1)}\right] \tag{34}$$

$$E_a = \frac{k_B \cdot T_J(t_1) \cdot T_J(t_2)}{T_J(t_2) - T_J(t_1)} \cdot ln\left(\frac{D_{t_2} \cdot t_2}{D_{t_1} \cdot t_1}\right) \tag{35}$$

(1). To check the accuracy of the achieved model we calculated the maintenance curve belonging to $T_J = 85\,°C$ (dashed dark red line in Figure 6). The simulation run with a time increments of 1 h.
(2). We also calculated the maintenance curves belonging to $T_A = 25\,°C$ and to $T_A = 55\,°C$ (dashed dark blue and green lines in Figure 6).

To represent the appropriateness of the proposed technique the R-square values were determined with respect to the measured data. The R^2 values of the simulation are 0.992 and 0.988 for the 25 °C and 55 °C measurements while that of the logarithmic approximation are 0.993 and 0.983. These values show that the accuracy of the new aging model and the classical curve fitting method is practically the same.

Figure 6 also indicates that the simulated results and the fitted curves have different curvatures and their separation becomes quite significant after around 5000 h. Obviously, neither approximation of the measured values is more accurate than the other. The main cause of this misfit is probably the low statistical power of the measurement results of one single sample (and perhaps the changing aging rate). The purpose of this short case study was only to demonstrate the potential and feasibility of the theory.

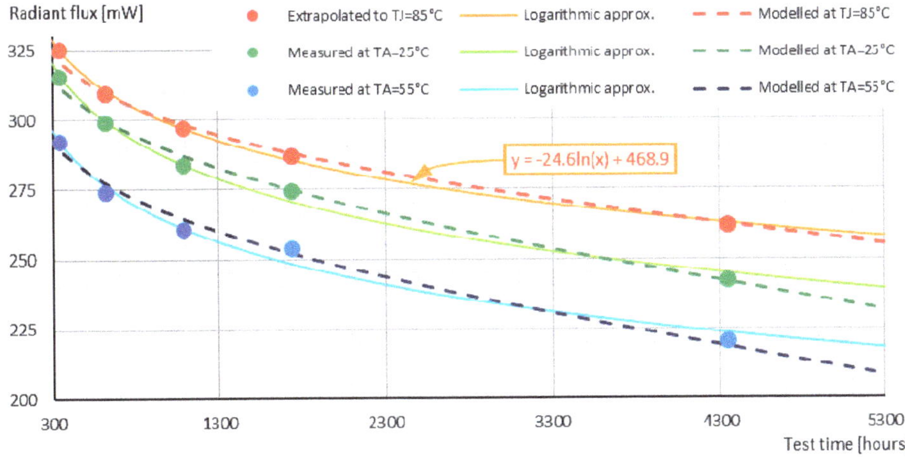

Figure 6. Measurement and simulation results of the aged mid-power blue LED; the logarithmic regression fitting parameters to the 85 °C values are indicated just as an example.

4. LM-80 Based Lifetime Modelling of Power LEDs

Once the fitting parameters of the total luminous flux maintenance curves (and the pre-exponential factor(s) along with the activation energy (or energies)) are available, it becomes possible to estimate the in-situ light output parameters of an LED at any time (by Equations (7) and (32)), provided that the prevailing junction temperature and forward current values are always known.

In case of a streetlighting luminaire the forward current is either kept constant or it is controlled by a smart device. In fact, aging of the LED driver may cause aging related deviations in the set current but discussion of such issues is out of the scope of this article.

In-situ measurement of the pn-junction temperatures is not impossible, but it requires specialized laboratory equipment. Another solution is to reveal the operating temperature map of the luminaire can be achieved by system level simulations [67–70] that way shifting the junction temperature measurements to a predetermined point of the luminaire. System level simulations with multi-domain LED models make it possible to tell the operating parameters just by measuring the luminaire case temperature which is then on the field, mostly depends on the actual weather conditions.

4.1. Continuous In-Situ Lifetime Modelling of LEDs

Generating a time-dependent absolute temperature function from years of weather conditions is not realistic, but in the short period of time (e.g., within an hour) it can be assumed that the temperature changes linearly over the time. This approximation makes it possible to solve Equation (32) analytically, but despite of it, the computing capacity of the intelligent control unit of the luminaire may still be insufficient for the required calculations (or it is not acceptable for the unit to be kept busy by these calculations—also not counting with the power consumption of the CPU). If the value of reaction (or decay) rate coefficient is recalculated periodically at short intervals (e.g., every 5–10 min) and approximating the temperature to be constant in the meantime, then Equation (32) is greatly simplified:

$$c_2 = c_1 \cdot exp\{k(T_J) \cdot [t_2 - t_1]\} \tag{36}$$

where c is the actual light output value of the LED and the lower indices 1 and 2 denote the initial and the terminating time in-between the LED aging is calculated.

A case study was carried out to present the capabilities of the theory. For this purpose, the LM-80-08 based measurement results of Luxeon Z LEDs aged in our department were used (see the results in Figure 7a). The natural white high-power LEDs had a junction-to-ambient thermal resistance of 35 K/W. In Figure 7b the continuous red curve indicates the exponential fit to the measurement results (the same as the dashed orange curve in Figure 7a). Further eight theoretical aging curves were added in order to make the necessary calculations feasible, corresponding to T_J = 85 °C and 50 °C and to I_F = 850 mA and 700 mA.

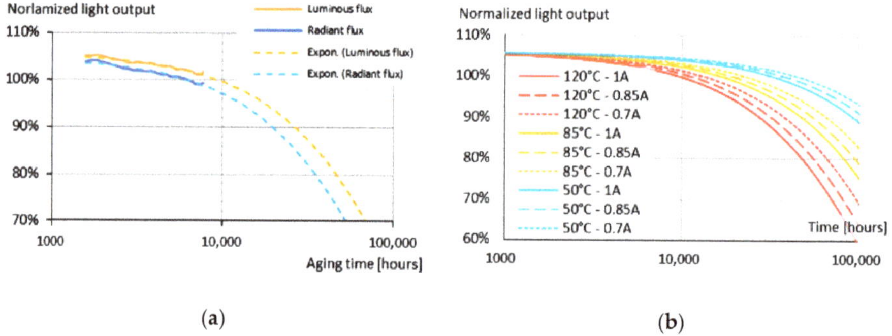

Figure 7. (**a**) Averaged LM-80 measurement results of 9 pieces of Luxeon Z samples and extrapolation until 50k hours (T_J = 120 °C, I_F = 1 A); (**b**) Exponential curve fit to the LM-80 measurement set along with the eight assumed aging trends.

Figure 8 shows an illustrative example of the theory, applying Equation (36): the stress conditions are abruptly changed at 30k hours of aging from T_J = 120 °C and I_F = 1 A to T_J = 50 °C, I_F = 0.85 A.

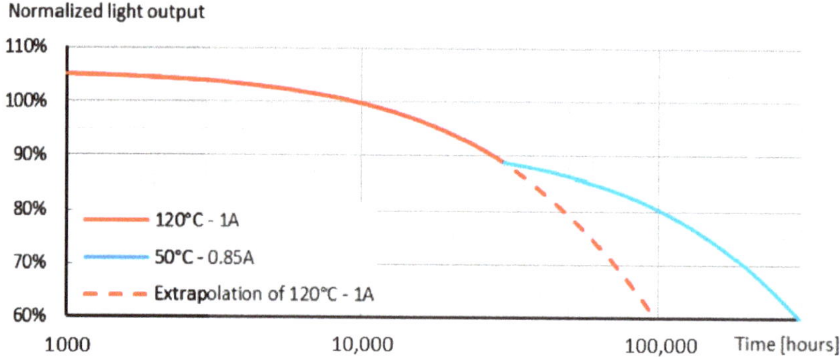

Figure 8. Illustration of the aging model; the simulation conditions are changed after 30,000 h.

4.2. Lifetime Modelling of Iso-Flux Operation; a Case Study

By keeping the total luminous flux of solid-state light sources constant, not only the visual comfort can be improved, but the reliability and energy efficiency of the luminaires could also be greatly increased. The advantage of a design method that compensates for the effects of temperature changes has been presented in previous articles [68–70]. The previously proposed methodology has not yet taken into account the aging of LEDs, which in the long time range may be even more significant than the effects of temperature changes.

The total lifespan of LEDs is typically referred to be in the range of 50–100k hours. This period can be easily converted to lifetime in years for light sources that operate without interruption (e.g., tunnel lighting or lamps at industrial facilities), but in case of streetlighting luminaires the conversion is not that straightforward as the on and off time depends on the length of days that alternate continuously throughout the year. In addition, switching on and off the lights does not necessarily depend on the exact occurrence of the twilights; the operating time measured in years should be considered based on astronomical data, taking into account the changing length of the nights. Table 2 illustrates the typical practical lifetime of a luminaire (assuming a rated LED lifetime of 50k hours) in Hungary.

Table 2. Elapsed time till 50k hours of operation of a streetlighting luminaire in years, in Hungary.

Dark-Hours in-Between		50k Dark-Hours (Years)	Dark-Hours during a Year (Hours)
Sunset to Sunrise		12 years	4291 h
Civil		13 years	3875 h
Nautical	Twilights	15 years	3373 h
Astronomical		18 years	2802 h

A case study was carried out on the real and theoretical LED aging results in order to demonstrate the range of differences between a constant current and a constant luminous flux operation. In order to count with realistic temperature variations the daily temperature values of the past decade of the Hungarian city of Szombathely (47.23512° N 16.62191° E) available at the Hungarian Meteorological Service, have been used, for which the constant current mode and the constant luminous flux operation was compared. The difference between the length of cold winter nights and warm summer nights was also considered by taking into account the annual change in the time differences between the civil twilight. At this stage we dealt with only one single LED, with the junction-to-ambient thermal resistance of 35 K/W.

The results of the case study are shown in Figures 9–11: the consumed electricity can be considerably decreased, particularly in the first year of the operation which means that a significant portion of the cost of a new luminaire installation is recovered within a relatively short time. Figure 10 clearly indicates, that using the smart controlling scheme the light output can be kept constant against the effects of temperature changes in the short run as well as against the LED aging in the time scale of decades (considering only the LED light sources–any other aging effects form a different issue). Due to the more favorable operating conditions (lower forward current and junction temperature values), the expected product lifetime could also be increase significantly.

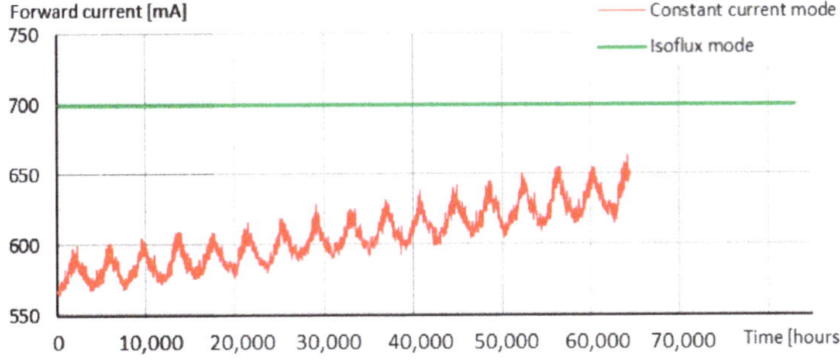

Figure 9. The forward currents applied during the simulation (the absolute maximum DC forward current of the Luxeon Z LEDs is 1 A).

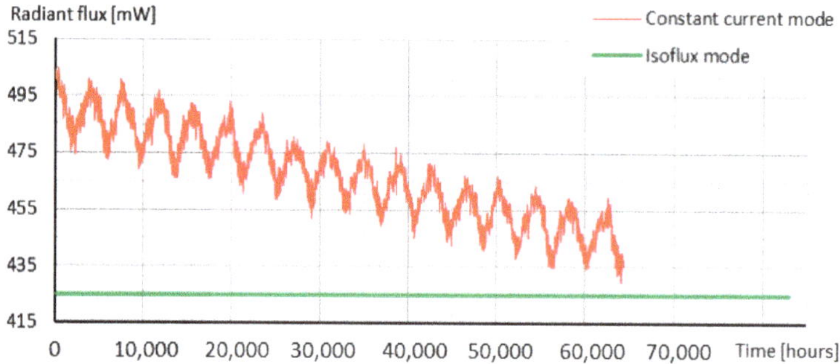

Figure 10. Simulated light output of a Luxeon Z LED.

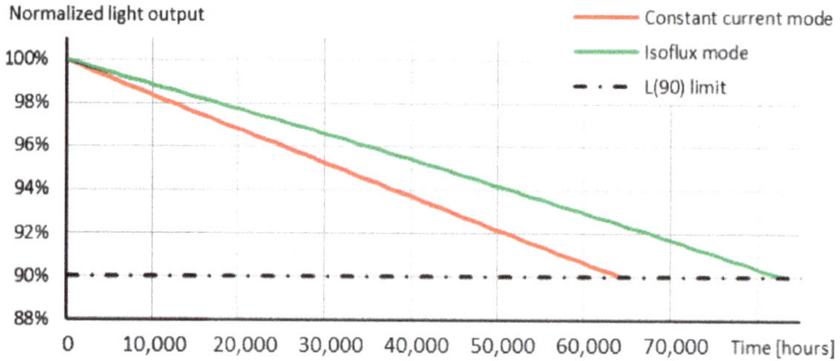

Figure 11. Simulated lifetime budget of a Luxeon Z LED.

In case of the smart control that realizes constant luminous flux output throughout the whole product lifetime, it is obvious that the normalized total luminous flux maintenance percentage can hardly be used to determine the end of the product lifetime. For that purpose we propose the lifetime budget efficiency η_t (eta t) as the new end-of-life metric for adaptively controlled LED luminaires. The final results are summarized in Table 3.

Table 3. Comparison of operation with constant forward current (700 mA) and with constant light output (425 mW), simulated with the help of our LED aging theory.

Examined Parameter	$I_F = const.$	$\Phi_e = const.$	Advantages of the Proposed CLO
Time to L(90) (hours)	64.4k h	83k h	+29%
Working years till L(90)	16.7 years	21.4 years	+4.7 years
Electricity consumed till 64.4k hours	130.8 kWh	112.8 kWh	−13.7%
Used energy in the 1st operational year	7.9 kWh	6.5 kWh	−17.7%

5. Lifetime Modelling Based on Multi-Domain Modelling of the LEDs

The purpose of our multi-domain LED modelling concept is the simultaneous and combined simulation of the optical, electrical and thermal operation parameters. The proposed modelling technique is continuously revised and enhanced in order to achieve higher accuracy, better industrial applicability and also to add further capabilities to the model. Significant improvements have been achieved during the Delphi4LED H2020 European R&D project and the next step shall be modelling the whole lifetime of LED based light sources by the help of a Spice compatible multi-physics model.

Multi-physics LED models are generated by the isothermal LED characteristics which consist of the electrical and optical data as a function of the forward current, measured at a fixed junction temperature. Such measurements can be performed by the help of the T3Ster and TeraLED [71,72] measuring instrument setup; to speed up the rather time-consuming characterization process the vendor also provides an automated control software to the instrumentation. Then the multi-domain LED model can be generated by a global parameter fitting process; a detailed description on the model and its variants, benefits and various properties is provided in the recently published paper in this journal [7].

Combining the existing multi-domain LED model with the measurement results of an LM-80 based life test is a promising attempt. It means that a complete isothermal characterization of the LEDs should be performed as the testing time passes, which is the main drawback of this technique, compared to the common measurement methods during life testing. Its benefit is, however, that the elapsed operating time dependent model parameters can be revealed.

5.1. Evaluation of Precious Life Testing Results

A sample set of 30 HP LEDs has already been aged under an LM-80-08 based life test, during which the measurements were also extended by the abovementioned isothermal characterization technique. In the earlier study the samples were exposed to the case temperature of 85 °C while they were driven by the forward current of 1 A. The whole test lasted for 8k hours (Figure 12a shows the normalized total radiant and luminous maintenance curves). Detailed isothermal characterization was performed in the first 6k hours on 5 LED samples from which the model parameters were determined as the function of the elapsed time (Figure 12b shows an example) [73].

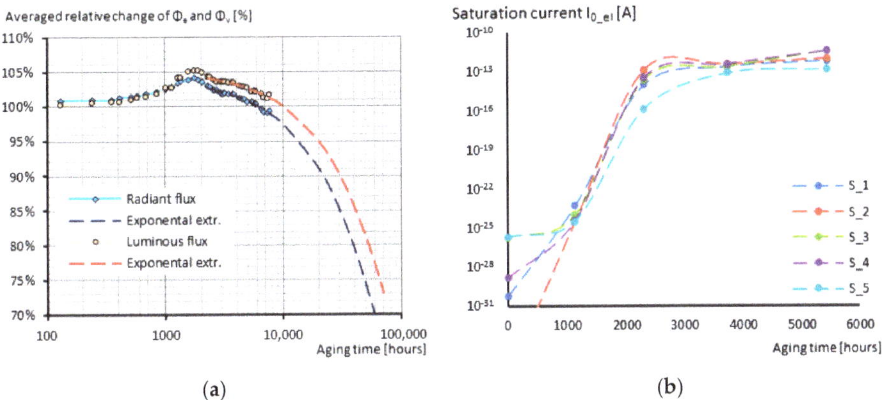

Figure 12. (a) Total luminous and radiant flux maintenance curves with their extrapolation [73]; (b) Time dependence of the model parameters: change of the saturation currents [73].

In case of all the characterized samples the obtained model parameters showed high similarity, and also the time dependent trends were consistent with each other. Still, a proper elapsed lifetime dependent model could not be generated from these. The difference in the ideality factors were found to be relatively small on a linear scale, but due to the exponential form of the Shockley diode equation even very tiny misfits can cause large errors in the output electrical and optical values. Besides this, there is an inflection point at around 1k–1.5k h of aging which prevents any modelling attempts with a simple time-function of the parameters; the use of complicated complex functions would not necessary describe the real physical aging processes but it would significantly increase the difficulty and the needed time of the global parameter fitting process. It is also obvious that various aging phenomena took place in the early "burn-in" part of the LEDs' lifetime and also their significance were changing with time; the collected data before 2.3k h of aging should be omitted during the model generation process. The appropriate time dependent model, however, cannot be created from only 3 characterization points.

5.2. Launch of a Targeted LM-80 Based Test Sequence

Instead of fixing the total testing time in advance, it could be more advantageous to pre-define a targeted total luminous flux depreciation level in order to have a better overview on the time evolution of the time evolution of the modelling parameters. Based on this idea a new LM-80 like test was conducted on 18 mid-power LED samples from a well-recognized vendor. The LED type selection was made according to our previously performed tests and detailed measurements. Testing a set of high-power LEDs would also been interesting; preparation of such tests is already underway.

The samples were exposed to the case temperatures of the specified 55 °C and 85 °C and at the arbitrarily chosen 70 °C. At each case temperature, three different forward currents were applied: 220, 260 and 300 mA, the latter one as the absolute maximum allowed forward current of this LED type (the nominal current value is 150 mA). This means only two samples per each aging condition (case temperature/forward current), which is insufficient for TM-21-11 extrapolations but the purpose of the test was much rather to support our theoretical assumptions than to collect statistical data for industrial applications and needs. During the test we did not only measure the necessary parameters prescribed by the LM-80-08 standard but we also performed a complete isothermal forward current–forward voltage–radiant flux characterization of the samples in a 500 mm integrating sphere. These captured iso I–V–L curves were expected to provide enough input data to reveal the effects of the different aging tendencies.

The tested LEDs arrived on 0.8 mm FR4 stripes that had extended thermal pads both on the top and on the bottom sides (see in Figure 13). The samples were electrically connected in series in the chains of 6 LEDs. In order to make the JEDEC JESD 51 compliant individual measurements of the samples the stripes were chopped between the LEDs. Prior to the LM-80 aging test the samples had been pre-characterized during which the radiant flux was measured to be around 50% at the nominal forward current, however, the real thermal resistance was measured to be around 100–150 K/W; from the integral structure functions it was obvious that the most significant part of it belonged to the LED package itself. The exact causes of the unexpectedly high thermal resistance values were not examined in more depth, but according to our assumptions the pre-bake technological step may have been omitted before soldering, therefore the humidity accumulated in the LED package may have caused delamination of the internal mechanical layers during the reflow process.

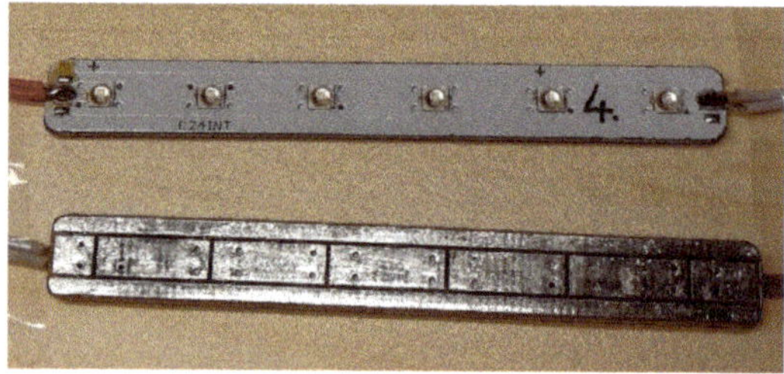

Figure 13. The sample mid-power LEDs mounted on FR4 strips; the printed circuit boards were provided with increased thermal interfaces for better cooling capabilities.

5.3. Results of the LM-80 Based Test

The LM-80-based investigation of the mid-power LEDs was ended after the elapsed time of 1735 h because of the high failure rate of the LEDs; till the termination of the test 14 LEDs suffered catastrophic failure and two further pieces had contact failure. The rapid and early failure of the samples was unfortunate but not unexpected; in only a few cases the testing junction temperature of the LEDs was close to the allowed maximum value, but in most cases it was far above that. Despite the fact that the testing case temperatures were specified according to the LM-80-08 description, the lifetime testing still went in an accelerated manner (reaching the L70 level was one of our goals anyway).

Figure 14 compares a faulty and an unaged sample. From the figure it can be clearly seen that one of the root causes of the failure may be traced back to the extremely high temperature around the chip that inflicted carbonization of the encapsulant material, causing discoloration and bubbling. Due to the reduced light transmission of the lens higher amount of the blue light was absorbed that further increased the self-heating effect inside the LED package. The catastrophic failure most probably occurred as a result of a thermal runaway.

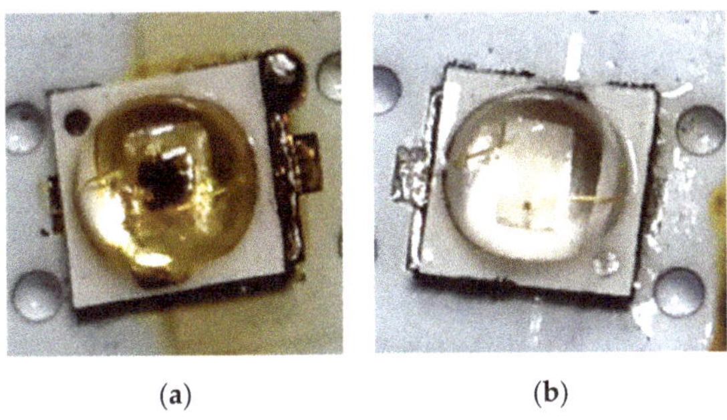

(**a**)　　　　　　　(**b**)

Figure 14. (**a**) A failed and (**b**) an unburnt sample of the investigated mid-power LEDs.

Figure 15 shows the attained total radiant flux maintenance curves; the case temperatures are indicated by different colors while the different forward current values are marked with different line types. Interestingly, the LEDs exposed to the lowest case temperature not only aged faster, but the trend over the time is also different from the others. Deeper investigations were not conducted to reveal the proper reasons and phenomena, but the main reason for the differences may be the higher RH formed at the lower temperature; this assumption is also supported by a previous study of a moisture resistance test on the same LED type, presented in paper [4].

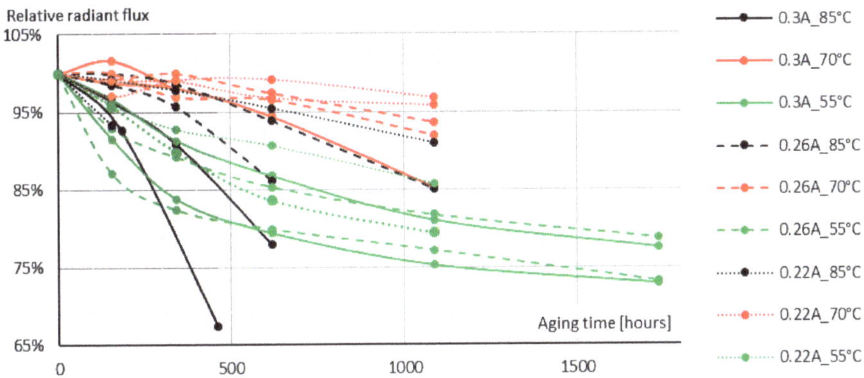

Figure 15. The attained total radiant flux maintenance results of the mid-power blue LEDs, sorted by case temperature and forward current.

All in all unfortunately, the captured LM-80 compliant measurement data altogether did not make it possible to further analyze the temperature and forward current dependence of the aging rate. In Section 3.4 however, an attempt was made to achieve the Arrhenius equation parameters, but in this exact case there is not enough measurement result of the other two case temperatures to support our theory; the resources required for such investigations far exceed the academic capabilities.

5.4. The Elapsed Lifetime Dependent Multi-Domain LED Model

Using the obtained measurement results an attempt was made in order to investigate the lifetime modelling possibilities and the extrapolation capabilities; for this purpose the multi-domain models of the still functional #S07 and #S11 samples were created at first. Various functions were tried out during a global parameter fitting process in order to set up the lifetime LED models with the best match to the measured characteristics. For this purpose, rudimentary parameter matching software was also developed, which ran on a mid-range 4-core processor for about 1 week (which was about one-tenth of the total software development time). The obtained Shockley model parameters and the value of the series resistance have significant elapsed time dependence. An improved version of the fitting application was re-run iteratively three times in order to achieve the first attempt of our elapsed lifetime dependent multi-physics LED model; Figure 16 indicates examples of the attained model parameters.

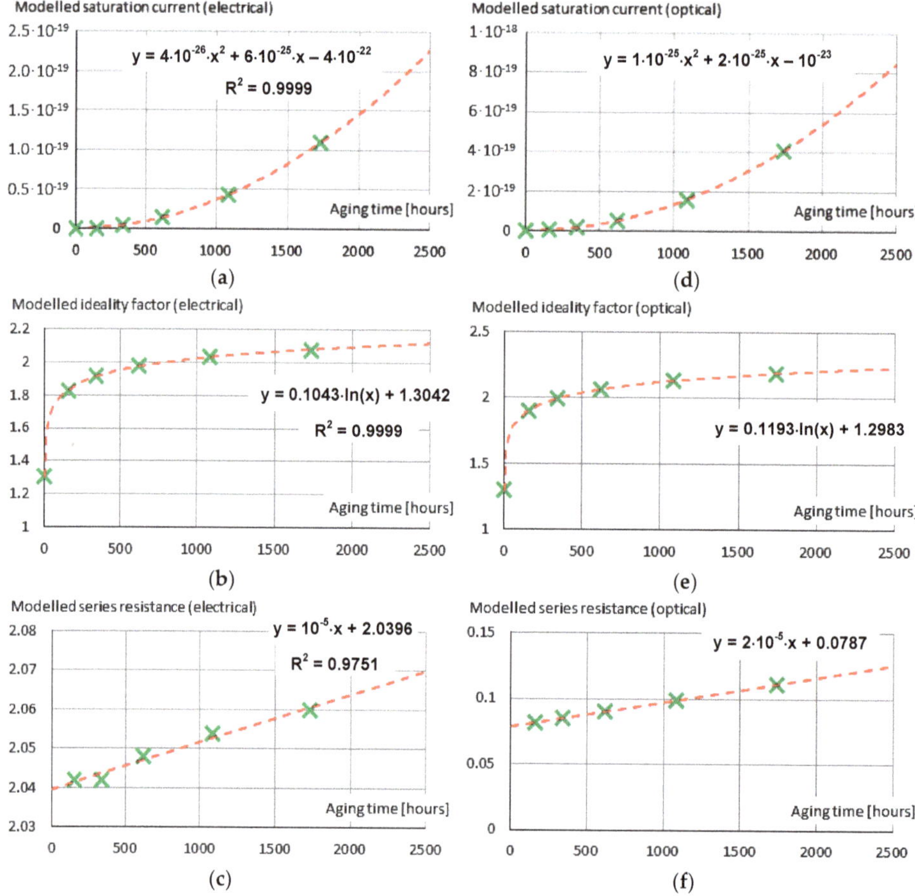

Figure 16. Electrical (**a–c**) and optical (**d–f**) model parameters of #S07.

After defining the new lifetime LED model a simulation test bench was created by which the LM-80 based test data could be compared with the modelled values (see the results in Figure 17a). The average absolute inaccuracy of the simulations is 0.5% while the maximum deviation is 1.2%. Figure 17a also shows that due to the quadratic and logarithmic time functions of the fitting parameters of the Shockley diode equation (see in Figure 16a,b) the model fits very well even in the early burn-in period—at least at the time of the measurements. Otherwise, further simulations with much higher time resolution has shown that the model becomes totally inconsistent between 1 and 100 h of aging (see in Figure 17b). These results strongly highlight the risk of using high-degree parameter matching algorithms. The extrapolation applicability of the new model still had to be tested therefore samples #S07 and #S11 were reinstated to the test chamber. In the meantime, the time functions of the fitting parameters were revised in order to eliminate the anomaly of the early aging time.

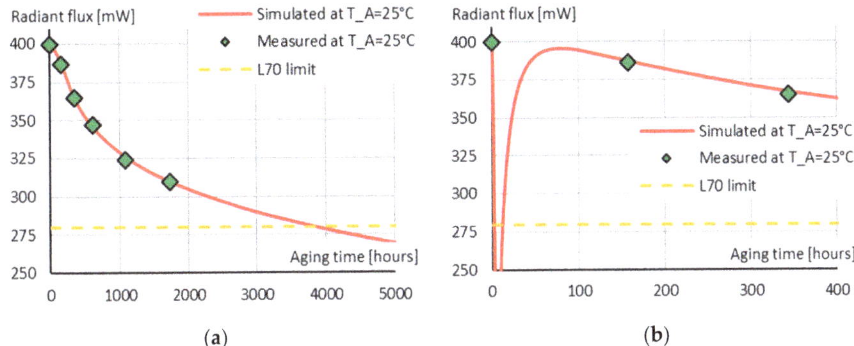

Figure 17. (**a**) Comparison of the simulated and measured total radiant flux maintenance curves of #S07; (**b**) The simulation results of #S07 with higher time resolution—the discontinuity can be clearly seen.

5.5. The Enhanced Time Functions

While the test continued on the two samples, we reconsidered the time functions that were previously found to provide the best match. After an extensive "trial and error" type investigation of the possibilities it was decided to apply only such functions that push the results only in the same direction while the rest of the parameters were kept constant. The biggest error in each case occurred in the initial burn-in stage, so we decided not to deal with it in the first round. That way the multi-domain model simplified remarkably: the shift of the forward voltage can be modelled by a linearly increasing series resistance, while the saturation current and the ideality factor of the electrical model are constant values. That way the electrical and the optical degradation of the LED can be modelled completely separately: the time function of the light output decay can be applied for the saturation current of the optical branch while its ideality factor and series resistance remain constant.

Considering the initial 100 h of aging, an additive exponential decay with a very short time constant describes well the forward voltage of the LED. The electrical series resistance therefore is formed this way:

$$R_{ser\_el}(t) = R_0 + a \cdot t + b \cdot [1 - exp(-t \cdot \tau_{Rser})] \tag{37}$$

where t is the elapsed operation time R_0 is the zero-hour electrical series resistance, a and b are fitting parameters and τ_{Rser} is the time constant of the initial exponential deviation.

Regarding the radiant flux, so far no correction function was found to describe the burn-in time. The saturation current of the optical branch is therefore:

$$I_{0\_rad}(t) = I_0 + d \cdot ln(t) \tag{38}$$

where I_0 is the zero hour saturation current of the optical branch and d is a fitting parameter. According to this function the model is not applicable if $t < 1$ h and also gives inaccurate results if $t < 100$ h.

5.6. Extrapolation Capabilities of the Model

At 4340 h of the total aging time the two samples #S07 and #S11 were again unloaded from the aging chamber and were re-measured in our integrating sphere. The obtained measurement results (the radiant flux and the forward voltage values) are shown in Figure 18a–d along with the lifetime extrapolation simulation curves. The simulations were performed by the LED models based on the measurement results obtained up to 1000 h.

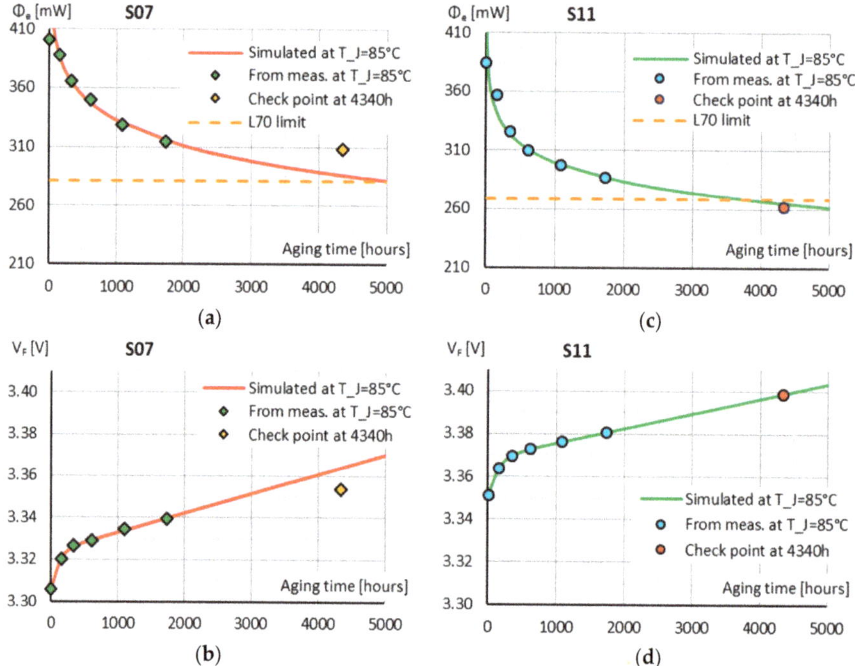

Figure 18. Simulated and measured total radiant flux maintenance of #S07 (**a**) and sample #S11 (**c**); Measured and simulated time function of the forward voltage of sample #S07 (**b**) and sample #S11 (**d**).

For sample #S11, the estimations acquired from the simulations are quite accurate, while the model of sample #S07 overestimates the extent of changes (see the mismatch values in Table 4). A possible explanation for the error could be the fact that the samples had to be removed from the aging environment for each measurement. Re-fixing the samples with different strengths may affect the value of thermal resistance. This possible reason is consistent with the previous assumptions that the high thermal resistance may have been caused by hygro-mechanical stresses that induced delamination of the mechanical layers in the LED packages.

It has to be noted that at this point the model is only valid for the 300 mA aging circumstances. Regarding the temperature issues, according to our theory the Arrhenius-equation parameters determined in Section 3.4 can be directly inserted into the model of #S11 sample:

$$I_{0\_rad}(t) = I_{0\_rad} - \int_{t_0}^{t} \frac{1}{t} \cdot A \cdot exp\left[\frac{-E_a}{k_B \cdot T_J(t)}\right] dt \tag{39}$$

by which the saturation current of the optical branch can be calculated at any time t. Dealing with discretized time steps of very short intervals and considering the temperature to be constant in the meanwhile we can calculate the I_{0_rad} value of the next period:

$$I_{0\_rad}(t_2) = I_{0\_rad}(t_1) - A \cdot exp\left[\frac{-E_a}{k_B \cdot T_J}\right] \cdot ln\left(\frac{t_2}{t_1}\right) \tag{40}$$

With the same considerations it is possible to sensitize the time-dependent electrical series resistance value to the junction temperature. In case of this LED type it should be based on the zero-order model of the differential rate law.

Table 4. The error made by the model compared to the measurement results at 4340 h.

Sample	V_F Mismatch (mV)	V_F Mismatch Compared to the Zero-Hour Results (%)	Φ_e Mismatch (mW)	Φ_e Mismatch Compared to the Zero-Hour Results (%)
#S07	10.2 mV	0.3%	−21.7 mW	5.4%
#S11	0.1 mV	0.003%	3.2 mW	0.8%

5.7. The Required Measurement and Testing Time

In case of the LM-80-08 based life testing of the blue mid-power LED set the isothermal characterization process was optimized to the needed time: the measured set of operating points was reduced to the minimally sufficient range while the small size of the LED packages and the relatively short thermal time constants (around 30–60 s) were also advantageous to significantly speed up the measurements. In spite of these facts, the required measurement time per each sample was around 180 min at each control event. Such characterization was performed 87 times during the test which altogether amounts a total measurement time of 260 h; an additional 15% of the original 1735 h of LED aging.

The minimal LM-80-08 and TM-21-11 compliant sample set consists of 90 LEDs as the testing process prescribes three case temperatures and at least three testing currents are necessary for proper interpolations in between while the extrapolation technique may be applied in the presence of at least 10 samples per aging conditions. Supposing a typical high-power LED the full characterization time may take even up to 6–8 h which means a total measurement time of 500–700 h at every control event that (according to LM-80-08) must be performed at every 1000 h. Although the measurements could be fully parallelized, it is still not realistic for academy in terms of the price of the currently available instruments required for the isothermal LED characterization. Therefore, supporting the theory described in this paper by appropriate statistical background remains an opportunity much rather for the major industry partners. Also, reducing the necessary measurement time of the current LED characterization system could also help to put this method into common practice.

6. Conclusions

In this work the LM-80-08 and TM-21-11 documents were briefly introduced after which an extensive description of the applied decay models were provided. In addition to the models used in the accepted methods, we also presented other aging models and the mathematical basis of their application.

Applying the theoretical basics of the light output degradation of LEDs we have introduced a novel method to determine the pre-exponential factor and the activation energy of the Arrhenius-equation only by measuring a sample set of only one aging case temperature instead of the prescribed three. We have also pointed out, that consistent pn-junction temperature based test results can be reached only if the experiment is supported by accurate junction temperature measurements in cases where the junction temperature rise during the life testing is considerable. Our theory was also supported by the evaluation of real measurement results. The case study showed that the LM-80 based measurement results of a power LED sample set and our LED aging theory of the "lifetime budget" were applied in a case study based on archive meteorological data. The case study showed that the iso-flux (or constant light output) operation mode has very significant benefits in terms of both electrical consumption and lifetime expectancy.

In order to create the elapsed lifetime dependent multi-domain LED model an LM-80-08-based test was performed on 18 blue mid-power LEDs of a well-recognized vendor. During the test even the isothermal characteristics of the samples were captured. Extremely high operating junction temperature of the samples even at the prescribed case temperatures caused fast and early failure of the LEDs. The speed and the trends of the LED aging also showed significant anomalies: the highest aging rate belonged to the samples of the lowest test temperature and even their total luminous flux maintenance curves follow a logarithmic trend instead of the expected exponential one. In the absence of the sufficient aging data ant due to the experienced anomalies it was not possible to determine the forward current dependent LED aging model but a theory was set up to specify the Arrhenius equation's parameters from only one testing temperature, by the help of thermal transient testing. In this specific case the needed extra measurements of the proposed method added a 15% surplus to the life testing duration which is estimated to be one half or one third of the extra time ordinarily required.

As the first approach of the Spice-compatible LED lifetime multi-physics modelling a mid-power LED was modelled from its captured aging results up to the L(78) level. The created LED model matches the measured values with a misfit less than 1.2%.

Simulations with higher time resolution had shown that the achieved model became inconsistent in the very early burn in period, therefore a new aging model was set up. In the meantime two LED samples were reinstated to the test in order to show extrapolation abilities of the lifetime multi-domain LED model. The model in case of sample #S11 performed over expectations, although, in case of #S07 the extrapolations proved to be fairly inaccurate. The cause of the modelling mismatch may be the fact that at this development and research state the LEDs have to be displaced from the aging chamber to perform the necessary measurement in an integrating sphere—it is still an issue that should be solved by a new, appropriate combination of the characterization and life testing methods.

At the academic level the currently running national R&D project allows resources only to such scale of studies. The theory described in this paper should be supported by testing and measurement results of much higher number of LED samples in order to represent appropriately the whole LED population and also the general aging physics of LEDs. Increasing the throughput of the presently applied LED characterization methods would also be needed. We are currently making efforts to develop new procedures to reduce measurement time and also to set up an international joint project consortium to enhance the statistical background of our theory.

Author Contributions: Conceptualization, A.P., G.H. and J.H.; Data curation, G.H. and J.H.; Formal analysis, J.H., Funding acquisition, A.P.; Investigation, G.H. and J.H.; Methodology, G.H. and J.H.; Project administration, A.P.; Software, G.H. and J.H.; Supervision, A.P.; Writing—original draft, J.H.; Writing—review and editing, A.P., G.H. All authors have read and agreed to the published version of the manuscript.

Funding: This research was funded by the K 128315 grant of the National Research, Development and Innovation Fund and was also supported by the Higher Education Excellence Program of the Ministry of Human Capacities in the frame of Artificial Intelligence research area (BME FIKP-MI/SC) and the Nanotechnology research area (BME FIKP-NANO) of the Budapest University of Technology and Economics. The support of the "TKP2020, National Challenges Program" of the National Research Development and Innovation Office (BME NC TKP2020) is also acknowledged.

Acknowledgments: Support from HungaroLux Light Ltd. (especially from A. Szalai and T. Szabó) is gratefully acknowledged.

Conflicts of Interest: The authors declare no conflict of interest. The funders had no role in the design of the study; in the collection, analyses, or interpretation of data; in the writing of the manuscript, or in the decision to publish the results.

References

1. IESNA. *IES Approved Method: Measuring Lumen Maintenance of LED Light Sources*; IES LM-80-08; Illuminating Engineering Society of North America: New York, NY, USA, 2008; p. 7.
2. Poppe, A.; Gábor, M.; Csuti, P.; Szabó, F.; Schanda, J. Ageing of LEDs: A Comprehensive Study Based on the LM80 Standard and Thermal Transient Measurements. In Proceedings of the 27th Session of the CIE, Sun City, South Africa, 9–16 July 2011; pp. 467–477.
3. Hegedüs, J.; Hantos, G.; Poppe, A. Lifetime Iso-flux Control of LED based Light Sources. In Proceedings of the 23rd International Workshop on Thermal Investigation of ICs and Systems (THERMINIC'17), Amsterdam, The Netherlands, 27–29 September 2017. [CrossRef]
4. Hegedüs, J.; Hantos, G.; Poppe, A. Reliability Issues of Mid-Power LEDs. In Proceedings of the 25th International Workshop on Thermal Investigation of ICs and Systems (THERMINIC'19), Lecco, Italy, 25–27 September 2019. [CrossRef]
5. Delphi4LED Project Website. Available online: https://delphi4led.org (accessed on 14 April 2020).
6. Martin, G.; Marty, C.; Bornoff, R.; Poppe, A.; Onushkin, G.; Rencz, M.; Yu, J. Luminaire Digital Design Flow with Multi-Domain Digital Twins of LEDs. *Energies* **2019**, *12*, 2389. [CrossRef]
7. Poppe, A.; Farkas, G.; Gaál, L.; Hantos, G.; Hegedüs, J.; Rencz, M. Multi-domain modelling of LEDs for supporting virtual prototyping of luminaires. *Energies* **2019**, *12*, 1909. [CrossRef]
8. Hantos, G.; Hegedüs, J.; Bein, M.C.; Gaál, L.; Farkas, G.; Sárkány, Z.; Ress, S.; Poppe, A.; Rencz, M. Measurement issues in LED characterization for Delphi4LED style combined electrical-optical-thermal LED modeling. In Proceedings of the 19th IEEE Electronics Packaging Technology Conference (EPTC'17), Singapore, 6–9 December 2017. [CrossRef]
9. Alexeev, A.; Bornoff, R.; Lungten, S.; Martin, G.; Onushkin, G.; Poppe, A.; Rencz, M.; Yu, J. Requirements specification for multi-domain LED compact model development in Delphi4LED. In Proceedings of the 2017 18th International Conference on Thermal, Mechanical and Multi-Physics Simulation and Experiments in Microelectronics and Microsystems (EuroSimE), Dresden, Germany, 3–5 April 2017; pp. 1–8. [CrossRef]
10. Poppe, A.; Di Bucchianico, A.; Vaumorin, E.; Juntunen, E.; Bosschaartl, K.; Yu, J.; Thomé, J.; Joly, J.; Szabo, F.; Merelle, T.; et al. Inter Laboratory Comparison of LED Measurements Aimed as Input for Multi-Domain Compact Model Development within a European-wide R&D Project. In Proceedings of the Conference on "Smarter Lighting for Better Life" at the CIE Midterm Meeting, Jeju, Korea, 23–25 October 2017; pp. 569–579. [CrossRef]
11. Bornoff, R.; Mérelle, T.; Sari, J.; Di Bucchianico, A.; Farkas, G. Quantified Insights into LED Variability. In Proceedings of the 24th International Workshop on Thermal Investigation of ICs and Systems (THERMINIC'18), Stockholm, Sweden, 26–28 September 2018. [CrossRef]
12. Mérelle, T.; Bornoff, R.; Onushkin, G.; Gaál, L.; Farkas, G.; Poppe, A.; Hantos, G.; Sari, J.; Di Bucchianico, A. Modeling and quantifying LED variability. In Proceedings of the 2018 LED Professional Symposium (LpS 2018), Bregenz, Austria, 25–27 September 2018; Luger Research e.U.—Institute for Innovation & Technology: Dornbirn, Austria, 2018; pp. 194–207, ISBN 978-3-9503209-9-2.
13. Merelle, T.; Sari, J.; Di Bucchianico, A.; Onushkin, G.; Bornoff, R.; Farkas, G.; Gaal, L.; Hantos, G.; Hegedus, J.; Poppe, A. Does a single LED bin really represent a single LED type? In Proceedings of the 29th Session of the CIE, Washington, WA, USA, 14–22 June 2019; pp. 1204–1214. [CrossRef]
14. Hegedüs, J.; Hantos, G.; Nemeth, M.; Pohl, L.; Kohári, Z.; Poppe, A. Multi-domain characterization of CoB LEDs. In Proceedings of the 29th Session of the CIE, Washington, WA, USA, 14–22 June 2019; pp. 387–397. [CrossRef]
15. Zhang, S.U.; Lee, B.W. Fatigue life evaluation of wire bonds in LED packages using numerical analysis. *Microelectron. Reliab.* **2014**, *54*, 2853–2859. [CrossRef]
16. Hu, J.; Yang, L.; Shin, M.W. Mechanism and thermal effect of delamination in light-emitting diode packages. *Microelectron. J.* **2007**, *38*, 157–163. [CrossRef]
17. Schubert, E.F. *Light Emitting Diodes*, 2nd ed.; Cambridge University Press: Cambridge, UK, 2006.
18. Kim, H.; Yang, H.; Huh, C.; Kim, S.W.; Park, S.J.; Hwang, H. Electromigration-induced failure of GaN multi-quantum well light emitting diode. *Electron. Lett.* **2000**, *36*, 908–910. [CrossRef]
19. de Orio, R.L.; Ceric, H.; Selberherr, S. Physically based models of electromigration: From Black's equation to modern TCAD models. *Microelectron. Reliab.* **2010**, *50*, 775–789. [CrossRef]

20. Pradhan, S.; Di Stasio, F.; Bi, Y.; Gupta, S.; Christodoulou, S.; Stavrinadis, A.; Konstantatos, G. High-efficiency colloidal quantum dot infrared light-emitting diodes via engineering at the supra-nanocrystalline level. *Nat. Nanotechnol.* **2019**, *14*, 72–79. [CrossRef]

21. Vasilopoulou, M.; Kim, H.P.; Kim, B.S.; Papadakis, M.; Gavim, A.E.X.; Macedo, A.G.; Da Silva, W.J.; Schneider, F.K.; Teridi, M.A.M.; Coutsolelos, A.G.; et al. Efficient colloidal quantum dot light-emitting diodes operating in the second near-infrared biological window. *Nat. Photonics* **2020**, *14*, 50–56. [CrossRef]

22. Won, Y.-H.; Cho, O.; Kim, T.; Chung, D.-Y.; Kim, T.; Chung, H.; Jang, H.; Lee, J.; Kim, D.; Jang, E. Highly efficient and stable InP/ZnSe/ZnS quantum dot light-emitting diodes. *Nature* **2019**, *575*, 634–638. [CrossRef]

23. Gao, L.; Na Quan, L.; De Arquer, F.P.G.; Zhao, Y.; Munir, R.; Proppe, A.H.; Quintero-Bermudez, R.; Zou, C.; Yang, Z.; Saidaminov, M.I.; et al. Efficient near-infrared light-emitting diodes based on quantum dots in layered perovskite. *Nat. Photonics* **2020**, *14*, 227–233. [CrossRef]

24. NANOTHERM Poject Website. Available online: http://project-nanotherm.com/ (accessed on 21 June 2020).

25. Lago, M.D.; Meneghini, M.; Trivellin, N.; Mura, G.; Vanzi, M.; Meneghesso, G.; Zanoni, E. "Hot-plugging" of LED modules: Electrical characterization and device degradation. *Microelectron. Reliab.* **2013**, *53*, 1524–1528. [CrossRef]

26. Meneghini, M.; Podda, S.; Morelli, A.; Pintus, R.; Trevisanello, L.; Meneghesso, G.; Vanzi, M.; Zanoni, E. High brightness GaN LEDs degradation during dc and pulsed stress. *Microelectron. Reliab.* **2006**, *46*, 1720–1724. [CrossRef]

27. van Driel, W.D.; Fan, X.J. (Eds.) *Solid State Lighting Reliability*; Springer: New York, NY, USA, 2013; ISBN 978-1-4614-3066-7. [CrossRef]

28. van Driel, W.D.; Fan, X.; Zhang, G.Q. *Solid State Lighting Reliability Part 2 (Solid State Lighting Technology and Application Series)*; Springer International Publishing: Cham, Switzerland, 2018; ISBN 978-3-319-58174-3. [CrossRef]

29. Chang, M.H.; Das, D.; Varde, P.V.; and Pecht, M. Light emitting diodes reliability review. *Microelectron. Reliab.* **2012**, *52*, 762–782. [CrossRef]

30. Koh, S.; van Driel, W.D.; Zhang, G.Q. Thermal and moisture degradation in SSL system. In Proceedings of the 2012 13th International Thermal, Mechanical and Multi-Physics Simulation and Experiments in Microelectronics and Microsystems, EuroSimE 2012, Cascais, Portugal, 16–18 April 2012. [CrossRef]

31. Singh, P.; Tan, C.M. Degradation Physics of High Power LEDs in Outdoor Environment and the Role of Phosphor in the degradation process. *Sci. Rep.* **2016**, *6*, 24052. [CrossRef]

32. Trevisanello, L.; Meneghini, M.; Mura, G.; Vanzi, M.; Pavesi, M.; Meneghesso, G.; Zanoni, E. Accelerated life test of high brightness light emitting diodes. *IEEE Trans. Device Mater. Reliab.* **2008**, *8*, 304–311. [CrossRef]

33. Csuti, P.; Kránicz, B.; Krüger, U.; Schanda, J.; Schmidt, F. Photometric and Colorimetric Stability of LEDs. In Proceedings of the CIE Expert Symposium on Advances in Photometry and Colorimetry, Turin, Italy, 7–8 July 2008.

34. Paisnik, K.; Rang, G.; Rang, T. Life-time characterization of LEDs. *Est. J. Eng.* **2011**, *17*, 241–251. [CrossRef]

35. Ikonen, E.; Vaskuri, A.; Baumgartner, H.; Pulli, T.; Poikonen, T.; Kantamaa, O.; Kärhä, P. Online measurement of LED junction temperature for lifetime prediction. In Proceedings of the Conference at the CIE Midterm Meeting, Jeju, Korea, 23–25 October 2017; pp. 36–37.

36. Vaskuri, A.; Kärhä, P.; Baumgartner, H.; Kantamaa, O.; Pulli, T.; Poikonen, T.; Ikonen, E. Relationships between junction temperature, electroluminescence spectrum and ageing of lightemitting diodes. *Metrologia* **2018**, *55*, S86–S95. [CrossRef]

37. Vaskuri, A. Spectral Modelling of Light-Emitting Diodes and Atmospheric Ozone Absorption. Ph.D. Thesis, Aalto University, Espoo, Finland, 2018; ISBN 978-952-60-8442-4. Available online: https://aaltodoc.aalto.fi/handle/123456789/34252 (accessed on 19 June 2020).

38. Vaitonis, Z.; Miasojedovas, A.; Novičkovas, A.; Sakalauskas, S.; Zukauskas, A. Effect of long-term aging on series resistance and junction conductivity of high-power resistance and junction conductivity of high-power. *Lith. J. Phys.* **2009**, *49*, 69. [CrossRef]

39. Alexeev, A.; Linnartz, J.-P.; Onushkin, G.; Arulandu, K.; Martin, G. Dynamic response-based LEDs health and temperature monitoring. *Measurement* **2020**, *156*, 107599. [CrossRef]

40. Alexeev, A. Characterization of Light Emitting Diodes with Transient Measurements and Simulations. Ph.D. Thesis, Eindhoven University of Technology, Eindhoven, The Netherlands, 2020; ISBN 978-90-386-5035-7. Available online: https://research.tue.nl/en/publications/characterization-of-light-emitting-diodes-with-transient-measurem (accessed on 19 June 2020).
41. JEDEC. *JESD22-A101C Standard, Steady State Temperature Humidity Bias Life Test*; JEDEC: Arlington, VA, USA, 2009.
42. JEDEC. *JESD22-A104D Standard, Temperature Cycling*; JEDEC: Arlington, VA, USA, 2009.
43. IESNA. *IES TM-21-11: Projecting Long Term Lumen Maintenance of LED Light Sources*; Illuminating Engineering Society of North America: New York, NY, USA, 2013.
44. IESNA. *IES Approved Method: Measuring Luminous Flux and Color Maintenance of LED Packages, Arrays and Modules*; IES LM-80-15; Illuminating Engineering Society of North America: New York, NY, USA, 2015; p. 8.
45. Richman, E. *The Elusive Life of LEDs: How TM21 Contributes to the Solution*; Pacific Northwest National Lab: Richland, WA, USA, 2011.
46. Chemistry LibreTexts: First-Order Reactions. Available online: https://chem.libretexts.org/Core/Physical_and_Theoretical_Chemistry/Kinetics/Reaction_Rates/First-Order_Reactions (accessed on 14 April 2020).
47. van Driel, W.D.; Schuld, M.; Jacobs, B.; Commissaris, F.; Van Der Eyden, J.; Hamon, B. Lumen maintenance predictions for LED packages. *Microelectron. Reliab.* **2016**, *62*, 39–44. [CrossRef]
48. Chemistry LibreTexts: More Complex Reactions. Available online: https://chem.libretexts.org/Bookshelves/Physical_and_Theoretical_Chemistry_Textbook_Maps/Map%3A_Physical_Chemistry_for_the_Biosciences_(Chang)/09%3A_Chemical_Kinetics/9.04%3A_More_Complex_Reactions (accessed on 14 April 2020).
49. Miller, C. IES TM-21-11 Overview, History and Q&A Session. EPA ENERGY STAR Lamp Round Table. National Institute of Standards & Technology, Sensor Science Division, San Diego, CA, USA, 24 October 2011. Available online: https://www.energystar.gov/sites/default/files/specs/TM-21%20Discussion_0.pdf (accessed on 14 April 2020).
50. Hantos, G.; Hegedüs, J.; Rencz, M.; Poppe, A. Aging Tendencies of Power MOSFETs—A Reliability Testing Method Combined with Thermal Performance Monitoring. In Proceedings of the 22nd International Workshop on Thermal Investigation of ICs and Systems (THERMINIC'16), Budapest, Hungary, 21–23 September 2016; pp. 220–223. [CrossRef]
51. Hantos, G.; Hegedüs, J.; Rencz, M. An efficient reliability testing method combined with thermal performance monitoring. *Microelectron. Reliab.* **2017**, *78*, 126–130. [CrossRef]
52. JEDEC. *JESD51-50 Standard. Overview of Methodologies for the Thermal Measurement of Single- and Multi-Chip, Single- and Multi-PNJunction Light-Emitting Diodes (LEDs)*; JEDEC: Arlington, VA, USA, 2012.
53. JEDEC. *JESD51-51 Standard. Implementation of the Electrical Test Method for the Measurement of Real Thermal Resistance and Impedance of Light-Emitting Diodes with Exposed Cooling*; JEDEC: Arlington, VA, USA, 2012.
54. JEDEC. *JESD51-52 Standard. Guidelines for Combining CIE 127-2007 Total Flux Measurements with Thermal Measurements of LEDs with Exposed Cooling Surface*; JEDEC: Arlington, VA, USA, 2012.
55. JEDEC. *JESD51-53 Standard. Terms, Definitions and Units Glossary for LED Thermal Testing*; JEDEC: Arlington, VA, USA, 2012.
56. *CIE 127:2007 Technical Report, "Measurement of LEDs"*; CIE: Vienna, Austria, 2007; ISBN 978-3-901-906-58-9.
57. *CIE 225: 2017 Technical Report, "Optical Measurement of High-Power LEDs"*; CIE: Vienna, Austria, 2017; ISBN 978-3-902842-12-1. [CrossRef]
58. Poppe, A.; Farkas, G.; Székely, V.; Horváth, G.; Rencz, M. Multi-domain simulation and measurement of power LED-s and power LED assemblies. In Proceedings of the 22nd IEEE Semiconductor Thermal Measurement and Management Symposium (SEMI-THERM'06), Dallas, TX, USA, 14–16 March 2006; pp. 191–198. [CrossRef]
59. Poppe, A.; Gábor, M.; Temesvölgyi, T. Temperature dependent thermal resistance in power LED assemblies and a way to cope with it. In Proceedings of the 26th IEEE Semiconductor Thermal Measurement and Management Symposium (SEMI-THERM'10), Santa Clara, CA, USA, 21–25 February 2010; pp. 283–288. [CrossRef]
60. Hantos, G.; Hegedüs, J. K-factor calibration issues of high power LEDs. In Proceedings of the 23rd International Workshop on Thermal Investigations of ICs and Systems (THERMINIC'17), Amsterdam, The Netherlands, 27–29 September 2017. [CrossRef]

61. Hantos, G.; Hegedüs, J.; Poppe, A. Different questions of today's LED thermal testing procedures. In Proceedings of the 34th IEEE Thermal Measurement, Modeling & Management Symposium (SEMI-THERM'18), San Jose, CA, USA, 19–23 March 2018; pp. 63–70. [CrossRef]

62. JEDEC. *JESD51-1 Standard. Integrated Circuits Thermal Measurement Method—Electrical Test Method (Single Semiconductor Device)*; JEDEC: Arlington, VA, USA, 1995.

63. JEDEC. *JESD51-14 Standard. Tranisent Dual Interface Test Method for the Measurement of the Thermal Resistance Junction-To-Case of Semiconductor Devices with Heat Flow through a Single Path*; JEDEC: Arlington, VA, USA, 2010.

64. Székely, V.; Bien, T.V. Fine structure of heat flow path in semiconductor devices: A measurement and identification method. *Solid State Electron.* **1988**, *31*, 1363–1368. [CrossRef]

65. Vaitonis, Z.; Pranciškus, V.; Zukauskas, A. Measurement of the junction temperature in high-power light-emitting diodes from the high-energy wing of the electroluminescence band. *J. Appl. Phys.* **2008**, *103*, 093110. [CrossRef]

66. Smirnov, V.I.; Sergeev, V.A.; Gavrikov, A.A.; Shorin, A.M. Modulation method for measuring thermal impedance components of semiconductor devices. *Microelectron. Reliab.* **2018**, *80*, 205–212. [CrossRef]

67. Poppe, A. Simulation of LED Based Luminaires by Using Multi-Domain Compact Models of LEDs and Compact Thermal Models of their Thermal Environment. *Microelectron. Reliab.* **2017**, *72*, 65–74. [CrossRef]

68. Hegedüs, J.; Hantos, G.; Poppe, A. Light output stabilisation of LED based streetlighting luminaires by adaptive current control. *Microelectron. Reliab.* **2017**, *79*, 448–456. [CrossRef]

69. Hegedüs, J.; Horváth, P.; Hantos, G.; Szabó, T.; Szalai, A.; Poppe, A. A New Dimming Control Scheme of LED Based Streetlighting Luminaires Using an Embedded LED Model Implemented on an IoT Platform to Achieve Constant Luminous Flux at Different Ambient Temperatures. In Proceedings of the Lux Europa 2017, Lubljana, Slovenia, 18–20 September 2017; pp. 87–92. Available online: https://pdfs.semanticscholar.org/f4a2/2878f59aa0a87a886402f638d3fb516dc710.pdf (accessed on 14 April 2020).

70. Hegedüs, J.; Horváth, P.; Szabó, T.; Szalai, A.; Poppe, A. A New Dimming Control Scheme of LED Streetlighting Luminaires Based on Multi-Domain Simulation models of LEDs in order to Achieve Constant Luminous Flux at Different Ambient Temperatures. In Proceedings of the Conference on "Smarter Lighting for Better Life" at the CIE Midterm Meeting, Jeju, Korea, 23–25 October 2017; pp. 267–276. [CrossRef]

71. Mentor Graphics T3Ster Product Website. Available online: https://www.mentor.com/products/mechanical/micred/t3ster/ (accessed on 14 April 2020).

72. Mentor Graphics TeraLED Product Website. Available online: https://www.mentor.com/products/mechanical/micred/teraled/ (accessed on 14 April 2020).

73. Hegedüs, J.; Hantos, G.; Poppe, A. A step forward in lifetime multi-domain modelling of power LEDs. In Proceedings of the 29th Session of the CIE, Washington, WA, USA, 14–22 June 2019; pp. 1154–1161. [CrossRef]

Article

Influence of a Thermal Pad on Selected Parameters of Power LEDs [†]

Krzysztof Górecki [1,*], Przemysław Ptak [1], Tomasz Torzewicz [2] and Marcin Janicki [2]

[1] Department of Marine Electronics, Gdynia Maritime University, Morska 81-87, 81-225 Gdynia, Poland; p.ptak@we.umg.edu.pl

[2] Department of Microelectronics and Computer Science, Technical University of Łódź, Wólczańska 221, 90-924 Łódź, Poland; torzewicz@dmcs.pl (T.T.); janicki@dmcs.pl (M.J.)

[*] Correspondence: k.gorecki@we.umg.edu.pl

[†] This paper is an extended version of our paper published in Proceedings of 25th International Workshop on Thermal Investigations of ICs and Systems Therminic 2019, Lecco, Italy, September 27–29 2019, doi:10.1109/THERMINIC.2019.8923795.

Received: 24 June 2020; Accepted: 16 July 2020; Published: 20 July 2020

Abstract: This paper is devoted to the analysis of the influence of thermal pads on electric, optical, and thermal parameters of power LEDs. Measurements of parameters, such as thermal resistance, optical efficiency, and optical power, were performed for selected types of power LEDs operating with a thermal pad and without it at different values of the diode forward current and temperature of the cold plate. First, the measurement set-up used in the paper is described in detail. Then, the measurement results obtained for both considered manners of power LED assembly are compared. Some characteristics that illustrate the influence of forward current and temperature of the cold plate on electric, thermal, and optical properties of the tested devices are presented and discussed. It is shown that the use of the thermal pad makes it possible to achieve more advantageous values of operating parameters of the considered semiconductor devices at lower values of their junction temperature, which guarantees an increase in their lifetime.

Keywords: power LEDs; thermal pads; thermal resistance; measurements; optical efficiency; self-heating; electronics cooling

---

1. Introduction

Power LEDs are today the most important components of solid-state lighting sources [1–3]. Temperature strongly influences properties of all semiconductor devices, including power LEDs, [1,2,4–6]. For a single semiconductor device, the value of its junction temperature depends on both the ambient temperature T_a and the excess ΔT of the device internal junction temperature, which is caused by the self-heating phenomenon [7–13]. Thus, the device temperature rise depends on power dissipated in a considered semiconductor device and on the efficiency of heat removal characterized by thermal parameters. In typically used compact thermal models of semiconductor devices, at the steady state, thermal resistance can be used for this purpose [14,15].

The thermal resistance between the p–n junction of a power LED and the ambient is used to describe the total influence of all components included in the heat flow path, e.g., the package of the device, the printed circuit board (PCB), and the heatsink, on the heat transfer efficiency [4,16]. Thus, the manner of assembly of a power LED can influence its thermal properties and, consequently, electric and optical properties of this device [17,18].

In our previous papers, [18–20], we investigated the dependences of the electric, optical, and thermal parameters of power LEDs on cooling conditions. In particular, these papers show that the influence of cooling conditions on mentioned parameters could be noticeable, e.g., self-heating could cause a high decrease in luminance of light emitted by these diodes.

Additionally, in references [21,22], it is shown that an increase in the semiconductor device internal temperature causes a visible decrease of its lifetime. Besides, the parameters of mounting process [17,23] and the area of soldering pads [14] also influence the device thermal. Manufacturers of power LEDs are continuously improving the quality of packages for these devices, which are characterized by lower and lower values of junction-to-case thermal resistance $R_{thj\text{-}c}$. The value of this parameter depends, e.g., on physical and chemical processes used during the packaging of these devices [23]. In the assembly of power LEDs, a soldering process is typically used. As previously shown [17,24,25], the composition of the soldering alloy, the type of reflow oven, and the soldering temperature profile in time can influence thermal resistance of a soldering joint between the case of a power LED and the PCB.

Górecki et al in the paper [7] describe a problem of multipath heat transfer between the device package and the ambient, whereas Górecki and Zarębski in the paper [14] the influence of selected factors characterizing, e.g., the assembly process of the tested devices on its thermal resistance are analyzed. In order to improve the efficiency of removal of heat generated in power LEDs, a special terminal of these devices is used that makes it possible to conduct only heat between the junction of this device and the PCB. Of course, in order to use such a terminal a special pad (thermal pad) must be situated on the PCB. Górecki and Ptak in the paper [26] we showed the measurements results illustrating the influence of the area of a thermal pad on thermal resistances of selected power LEDs. In these investigations, we used a custom PCB designed by the authors. Górecki et al in the paper [18], some results of measurements thermal parameters of power LEDs mounted on the MCPCB with soldered and not soldered thermal pad are presented. These measurements were performed for the tested devices operating with free convection cooling.

This paper, which is an extended version of our paper [27], illustrates the influence of the use of a thermal pad on thermal, electric and optical properties of selected power LEDs assembled in different types of packages and situated on typical metal core PCBs (MCPCBs) offered by manufacturers of the tested LEDs. The measurement results presented in this paper were performed for power LEDs situated on the cold plate. The temperature of this cold plate was regulated in a wide range of its value. This paper consists of the following parts. First, the employed measurement method is introduced in detail. Next, the tested devices are described. Finally, the obtained results of investigations are presented and discussed.

2. Measurement Method

This Section describes the measurement method and the set-up used to obtain characteristics of the tested diodes illustrating an influence of the operating conditions on electric, optical and thermal properties of these devices. All the considered parameters are measured simultaneously with the use of the set-up described below. A view of the considered set-up is shown in Figure 1a. This set-up consists of a transient thermal tester, a water-cooling system with a cold plate, a light-tight chamber, a luxmeter, and a radiometer. The interior of the light-tight chamber with the cold plate are shown in Figure 1b [28]. The cold plate with a tested power LED is placed inside the light-tight chamber. The thermocouple is used to measure the temperature of the cold plate. The block diagram showing the main components of the measurement system is presented in Figure 2.

Tested devices are heated by the transient thermal tester, which is also used to record their dynamic temperature responses, i.e., the variations of the voltage drop across LED junctions in time. When isothermal characteristics are measured, the tested devices, soldered to an MCPCB, are placed on a cold plate. The temperature of the cold plate is stabilized at a preset temperature value by a thermostat forcing liquid flow through the plate. Taking into account that, in the case of power devices, the temperature regulation time might be unacceptably long, the control system response might be accelerated, as demonstrated in [29], owing to the use of Peltier thermoelectric modules inserted between the plate and the PCB. However, if non-isothermal measurements are to be taken in the natural convection cooling conditions, the MCPCB is placed horizontally in thermally insulating clamps. For optical measurements, devices on the cold plate or in the clamp are placed inside the light-tight box and the flux density of emitted light is measured by a radiometer. The operation of the entire measurement system is controlled by a PC where all measurement data are stored.

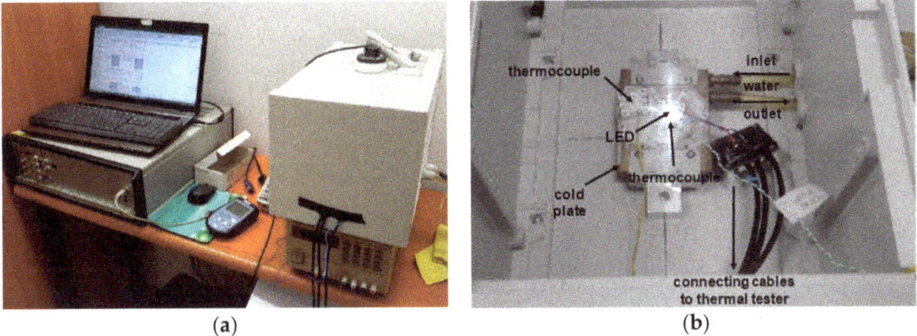

Figure 1. (**a**) View of the measurement equipment; (**b**) the interior of the light-tight chamber with the cold plate and a power LED [29].

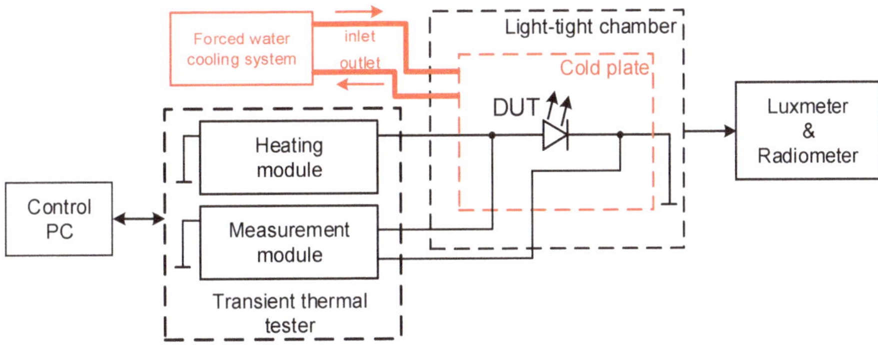

Figure 2. Block diagram of the measurement set-up.

Thermal resistance was measured with the commercial T3Ster $^{®}$ equipment manufactured by the MicReD division of Mentor Graphics (Budapest, Hungary) [30]. This equipment is now the industrial standard for device thermal characterization [15] and it realizes the classical pulse measurement method described e.g. in the papers [9]. In this method the tested device is excited by rectangular current pulses. The frequency of this signal is very low, and the duty cycle is close to 1. The user can select the values of low measurement current I_L and high heating current I_H. The voltage drop values of tested diodes are measured at the low (V_L) and high (V_H) forward current values.

The device under test was placed on the cold plate, whose temperature value was stabilized by a thermostat. The system for water forced cooling of electronic devices is described in the paper [31]. In order to assure higher thermal conductance, some silicon thermal paste was applied between the cold plate surface and the metal core printed circuit board (MCPCB) on which the tested diodes were mounted. The cold plate temperature T_a was regulated during the experiment over the range from 10 °C to 90 °C.

The measurement equipment is dedicated to thermal parameters of typical semiconductor devices, e.g., p–n diodes, but it does not include instruments making it possible to measure optical parameters of power LEDs. Therefore, it is possible to measure with this system only electric thermal resistance defined in the JEDEC standard [15]. When only electric thermal resistance is measured, the influence of optical power on the results of measurements is neglected.

In order to measure thermal resistance of the considered type of power LEDs, the measurement equipment shown in Figure 1 was used. The thermal resistance R_{th} is measured using the following formula:

$$R_{th} = \frac{V_{LE} - V_{LB}}{\alpha_T \cdot \left(V_H \cdot I_H - P_{opt}\right)} \tag{1}$$

where α_T denotes the slope of the thermometric characteristic $V_L(T)$ describing the dependence of the diode forward voltage V_L on temperature at the forward current equal to I_M; V_{LB} and V_{LE} denote the values of the diode forward voltage measured at the current I_M when measurements, respectively, start and end; I_H and V_H denote forward current and forward voltage of the tested diodes during heating process at the steady state, whereas P_{opt} is the optical power emitted by tested device.

The measurements of the surface optical power density were performed using the HD2302 radiometer [32] manufactured by Delta Ohm (Caselle di Selvazzano, Italy). The probe of the radiometer was situated at the distance of 17 cm directly above the light source. The optical power emitted by the investigated diodes was measured using the method presented in [18,33]. This method is based on the measurements of surface density of power of the emitted light by means of the radiometer, the data provided by the diode manufacturer, and the application of some classic geometrical dependencies.

The junction temperature values of the tested diodes at the steady state were measured using the standard pulse electric method [34]. Measurements were performed for different values of LED forward current over a wide range of values of cold plate temperature. The values of diode current I_D and voltage V_D as well as their junction temperature T_j and the surface power density of the emitted radiation Φe were registered simultaneously.

The electric current-voltage characteristics of the tested LEDs measured at the thermal steady state. The measurements were carried out by measuring thermal resistance. The coordinates of points lying on these characteristics are (V_H, I_H).

3. Tested Devices

The electric, thermal and optical parameters were measured for two types of power LEDs manufactured by the Cree, Inc. (Durham (North Carolina), USA); XPLAWT-00-0000-000BV50E3 (further on called XPLAWT) and MCE4WT-A2-0000-JE5 (further on called MCE). The values of selected parameters characterising the properties of tested devices are provided in Table 1.

Table 1. The values of the operating parameters of the tested LEDs.

Diode	I_{Dmax} [A]	P_D [W]	V_F [V]	T_{jmax} [°C]	Viewing Angle [°]	$R_{th\,j\text{-}s}$ [K/W]	ϕ_V [lm]
XPLAWT	3	10	2.95@1.05A	150	125	2.2	460@1.05A
MCE	0.7	2.8	3.2@0.35A	150	110	3	100@0.35A

For the diode XPLAWT, the nominal dissipated power of these LEDs amounts to 10 W, the maximum forward current is equal to 3 A, and the luminous flux at the current of 1050 mA is 460 lm [35]. In turn, for the MCE diode the nominal dissipated power amounts to 2.8 W, the maximum forward current is equal to 0.7 A, and the luminous flux at the current of 350 mA is 100 lm [36]. The typical thermal resistance between the junction and the soldering point $R_{thj\text{-}s}$ provided in the datasheets is equal for the considered LEDs to 2.2 K/W and 3 K/W, respectively. The MCE diode contains four independently operating structures, but in our investigations only one of them is powered.

The measurements were taken for the diodes XPLAWT situated on the MCPCBs presented in Figure 3a. These boards have dimensions of 25 mm x 25 mm and the thickness of 2 mm. The package of one diode was soldered to the thermal pad, also shown in the figure, whereas the other one did not use the thermal pad.

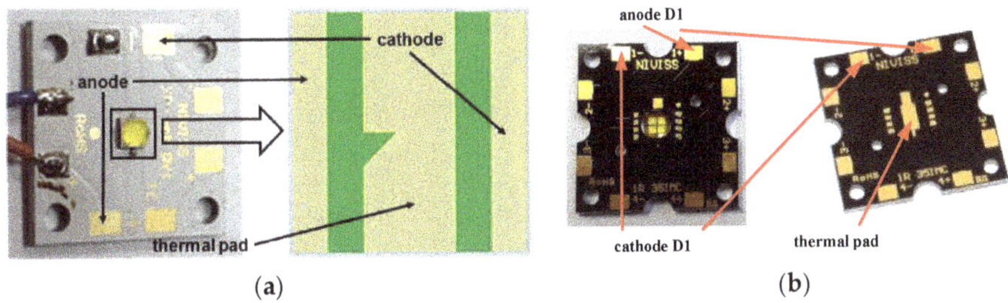

Figure 3. (**a**) View of the metal core printed circuit board (MCPCB) with the XPLAWT LED and the view of this LED electrode layout for this LED; (**b**) view of the MCE LED soldered to the MCPCB and the layout of this MCPCB.

On the other hand, the MCE diode is a compound device containing four LED structures, which can be powered independently. In the current investigations only one structure was used. The view of the MCE diode and the layout of the PCB used to assemble this diode are shown in Figure 3b. The dimensions of the MCPCB are 35 mm x 35 mm, and its thickness is equal to 2 mm. The thickness of the dielectric layer is 60 μm. Comparing the MCPCBs used for the assembly of both tested power LEDs, one can observe that their surface areas differ by almost two times. Besides, the shape and the dimensions of thermal pads are also visibly different.

4. Results

Using the measurement method described in Section 2, selected characteristics illustrating electric, optical, and thermal properties of the tested diodes were obtained for these devices operating with soldered thermal pads and non-soldered thermal pads. The electric properties of the considered devices are described by the non-isothermal current-voltage characteristics. The thermal properties are illustrated by the dependencies of thermal resistance on the diode forward current. Finally, the optical properties of these power LEDs are characterized by dependencies of optical power and luminance on the forward current. Moreover, radiant efficiency of the tested power LEDs was calculated. Selected results of these investigations are shown in Figures 4–13. In these figures, the solid lines denote the measurement results obtained for the diodes with the thermal pad soldered, and the dashed ones - for the diodes operating without the thermal pad soldered.

Figure 4 presents the non-isothermal current voltage characteristics of the XPLAWT diodes obtained at selected values of temperature of the cold plate T_a, which is equal to temperature of the MCPCB. As it is visible in the figure, an increase in temperature T_a shifts the characteristics to the left. This effect is observed as a result of self-heating, which is more visible for the diode operating without the thermal pad. In this case the increase in the diode internal temperature over the cold plate temperature is higher. For both manners of assembly, the differences in the junction temperature and the forward voltage drop increase with current. On the other hand, an increase in junction temperature at a constant value of forward voltage causes approximately exponential increase of the forward current.

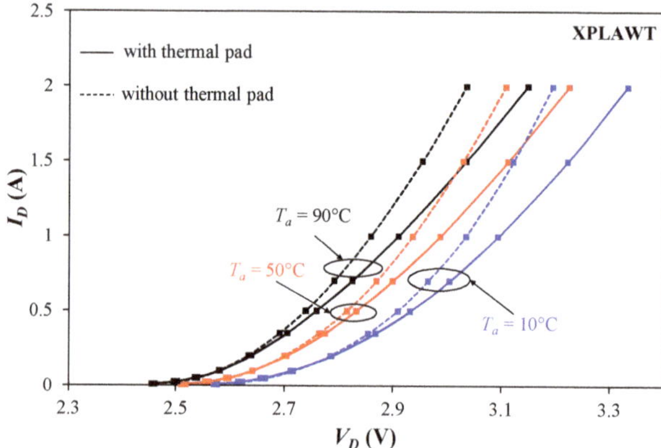

Figure 4. Measured DC I-V characteristics of the XPLAWT diodes at selected values of cold plate temperature.

In Figure 4 it is shown that differences in the forward voltage of the diode XPLAWT due to the variation of the device junction temperature T_j between diodes operating with a thermal pad and without it are equal even to 150 mV at the current of 2 A. In turn, the differences in values of junction temperature at the same value of current exceed even 60 °C. The maximum value of junction temperature of the diode operating without a thermal pad is equal to even 145 °C at $T_a = 90$ °C and $I_D = 2$ A. The temperature coefficient of forward voltage changes with the value of forward current and temperature T_a. At the forward current equal to 2 A and temperature $T_a = 10$ °C, this coefficient is equal to −2.5 mV/K.

Figure 5 presents for different forward current values the measured diode voltage change ΔV_D in response to the variation of the cold plate temperature from 90 °C to 10 °C ($\Delta T_a = -80$ °C) in the case of MCE diodes operating with the thermal pad and without it. The value of ΔV_D was obtained as the difference of values of forward voltage of the tested LEDs measured at the same value of their forward current and at both above mentioned values of cold pate temperature. As observed, the value of ΔV_D is an increasing function of current I_D. Due to the self-heating phenomenon, the value of the considered voltage change is higher for the diode operating without the thermal pad. The observed differences between values of ΔV_D obtained for both considered mounting methods attain even 60 mV at the current equal to 0.7 A.

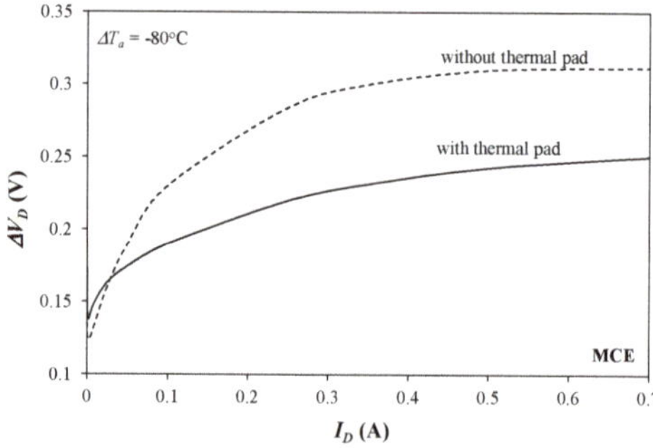

Figure 5. Measured MCE diode forward voltage change ΔV_D in response to the cold plate variation by 80 °C in function of the diode forward current.

Figure 6 illustrates the dependence of the forward voltage of the MCE diode on temperature T_a at two different forward current values. As observed, an increase in temperature of the cold plate causes a decrease in the diode forward voltage. The observed dependence $V_D(T_a)$ is non-linear. The values of V_D voltage obtained for the diode operating with the thermal pad are lower than for the diode operating without the pad, but the differences between these values decrease with the increase of temperature T_a and with the decrease in forward current I_D.

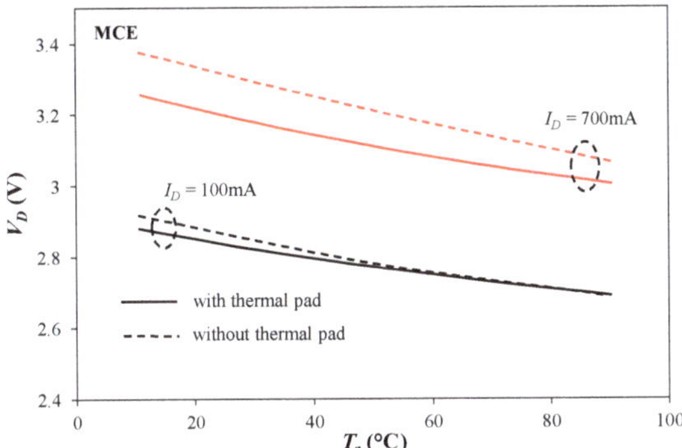

Figure 6. Measured dependences of the forward voltage of MCE diodes on the cold plate temperature for different forward current values.

Figure 7 illustrates for different cold plate temperature values the influence of the forward current on the junction temperature of the MCE diodes operating in both types of mounting manner considered here. Obviously, an increase in forward current and the cold plate temperature causes an increase in the junction temperature. It is worth observing that the diode operating with the thermal path has even a 25 °C lower value of junction temperature T_j than the diode of the same type operating without the thermal pad. The influence of the thermal pad on the junction temperature is the most visible for the lowest value of temperature T_a. It can be also observed that an increase in the value of forward current causes a decrease of the difference between values of temperature T_j obtained at different values of temperature T_a. It is a result of a decreasing thermal resistance value of the tested diodes with an increase in the cold plate temperature.

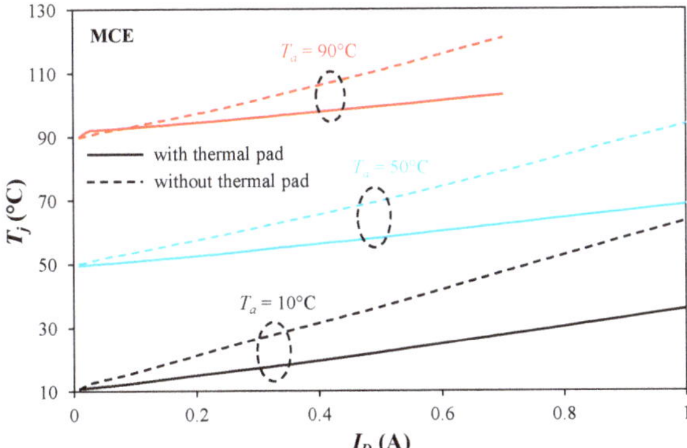

Figure 7. Measured dependences of the junction temperature of the MCE diodes on forward current for different values of the cold plate temperature.

The influence of device mounting manner of the considered diodes on their thermal resistance is illustrated in Figures 8 and 9. In Figure 8, the measured dependences of thermal resistance of the XPLAWT (Figure 8a) and MCE (Figure 8b) diodes on their forward current are shown. The measurements were performed at different values of cold plate temperature. It can be easily noticed that owing to the use of the thermal pad the diode thermal resistance of XPLAWT diode is effectively reduced. Differences in the value of this parameter exceed even 40% for the low value of forward current (100 mA). At higher values of current these differences decrease. For the current equal to 2 A, the value of thermal resistance decreases by less than 30%. For the LED operating without the thermal pad, this decrease is smaller, and it does not exceed 25%.

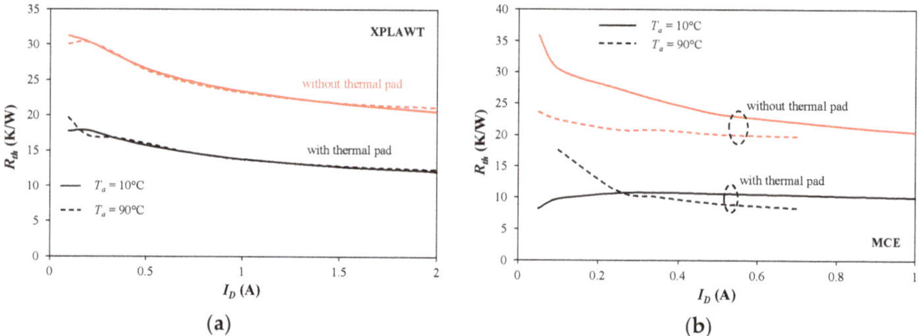

Figure 8. Measured dependences of thermal resistance on forward current for different cold plate temperature values for: (**a**) the XPLAWT; (**b**) MCE diode.

In practice, the results shown in Figure 8a mean that for the forward current equal to 2 A an excess of the device internal temperature above the ambient temperature is equal to about 90 °C for the LED with the thermal pad and above 130 °C for the LED without the thermal pad, thus demonstrating its importance for the thermal performance of the device. It is also worth noticing that there are no visible differences between the results of measurements obtained for different cold plate temperature values. This proves that in the considered case the value of thermal resistance results mainly from efficiency of heat conduction between the diode junction and the cold plate. The influence of convection and radiation on the thermal resistance value R_{th} is negligibly small. The rate of heat removal through the cold plate is very high, and it does not depend on the velocity of cooling liquid.

It is visible in Figure 8b that for the MCE diode a strong influence of the cold plate temperature on thermal resistance is observed. In the considered range of temperature variations, the thermal resistance can change by over 30%, and for higher T_a values, thermal resistance R_{th} is reduced. For the considered diode operating with the thermal pad, an increasing function describes the dependence $R_{th}(I_D)$ for low values of current I_D. The influence of this current I_D is very weak for the diode operating with the thermal pad, whereas it is very strong for the diode operating without the thermal pad.

Figure 9 presents the dependence of thermal resistance of the MCE diode on the the cold plate temperature. Looking at Figure 9, it is visible that there exists a certain minimum of the thermal resistance R_{th} at the cold plate temperature value T_a equal to about 50 °C. It is also easy to observe that for the diode operating without the thermal pad the changes in thermal resistance are much bigger than for the diode operating with the pad.

The last part of the presented experimental results illustrates the influence of the thermal pad on the optical parameters of tested power LEDs. The measured surface power density of light emitted by the XPLAWT diode presented in Figure 10 is an increasing function of the diode forward current I_D. The increase of the cold plate temperature reduces the emitted light power. Moreover, it is also visible that owing to the use of the thermal pad it is possible to obtain higher values of power density. These differences become more apparent with the increased diode current and exceed even 10%.

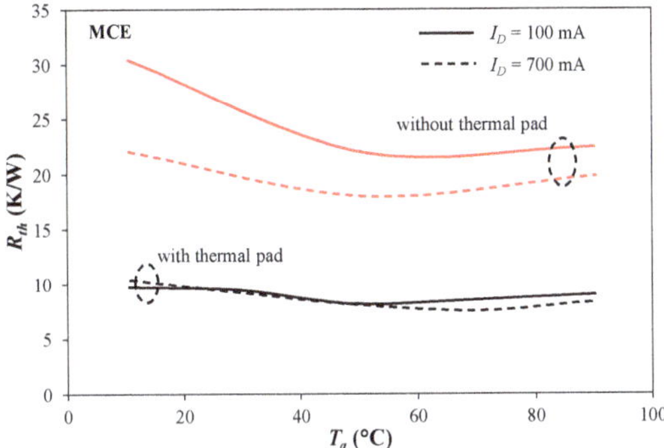

Figure 9. Measured dependences of thermal resistance of the MCE diode on the cold plate temperature for different forward current values.

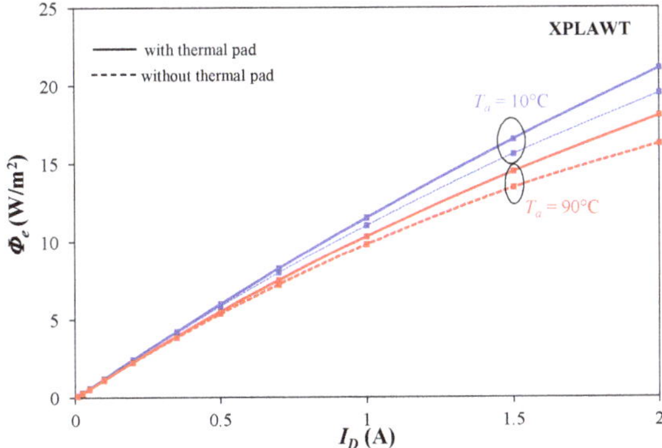

Figure 10. Measured dependences of the surface power density of light emitted by the XPLAWT diode on the LED forward current for different cold plate temperature values.

Figure 11 shows dependences of a change in the surface power density of the emitted light $\Delta\Phi_e$ on forward current for the MCE diode. This change was measured while changing temperature of the cold plate over the range from 10 °C to 90 °C. This change is an increasing function on the diode forward current, and at $I_D = 0.7$ A, it attains even 0.7 W/m^2. The changes in the value of Φ_e obtained for both the considered kinds of mounting the tested power LEDs do not visibly differ between each other, and they are comparable with the measurement error.

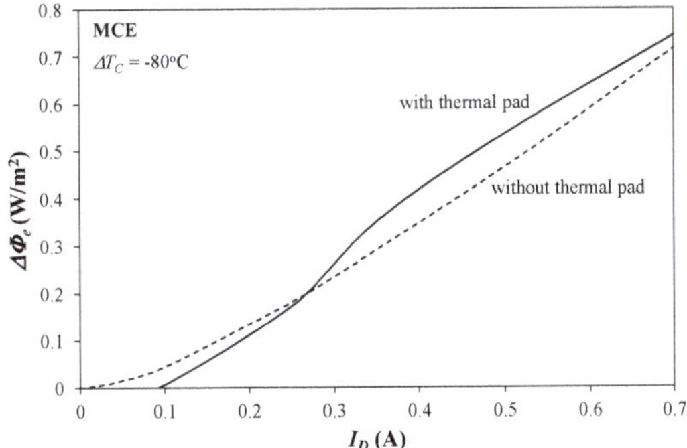

Figure 11. Measured dependences of the surface power density change $\Delta\Phi_e$ of the diodes MCE on forward current.

In turn, Figure 12 illustrates the dependence of the surface power density of the emitted light on temperature of the cold plate at selected values of forward current. The considered dependence $\Phi_e(T_a)$ is a decreasing function. The slope of this dependences is much higher for the higher of the considered values of the diode forward current. This slope is equal to even −0.225 %/K for $I_D = 0.7$ A, whereas for $I_D = 0.1$ A, this slope is practically equal to zero.

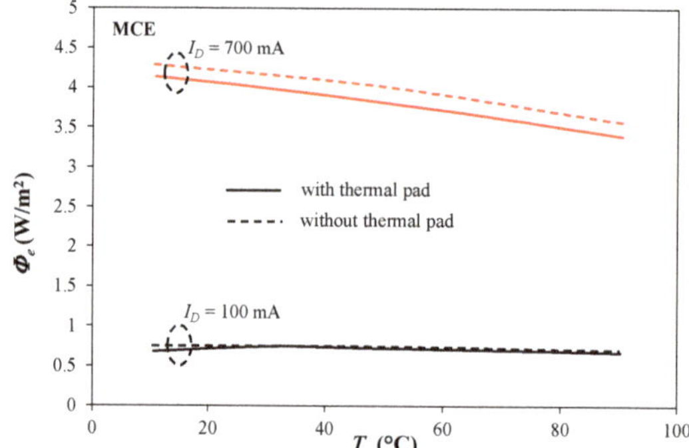

Figure 12. Measured dependences of the surface power density Φ_e of the diodes MCE on temperature of the cold plate.

The computed dependences of the efficiency η_{opt} of the conversion of electrical energy into light in the tested LEDs is shown in Figure 13a for the diode XPLAWT and in Figure 13b for the diode MCE. This efficiency is equal to the quotient of the optical power P_{opt} and the product of the diode current I_D, and the diode forward voltage drop across its junction V_D. The computed efficiency η_{opt}, in all cases has a distinct maximum. This maximum is observed at the current equal to about 30 mA for the diode without the thermal pad and around 100 mA for the diode with the thermal pad. Interestingly enough, at lower current values, a higher efficiency of conversion is observed for the diode without the thermal pad. However, within this range of currents, the power of the emitted radiation is relatively small, and junction temperature of the diode is nearly to the cold plate temperature.

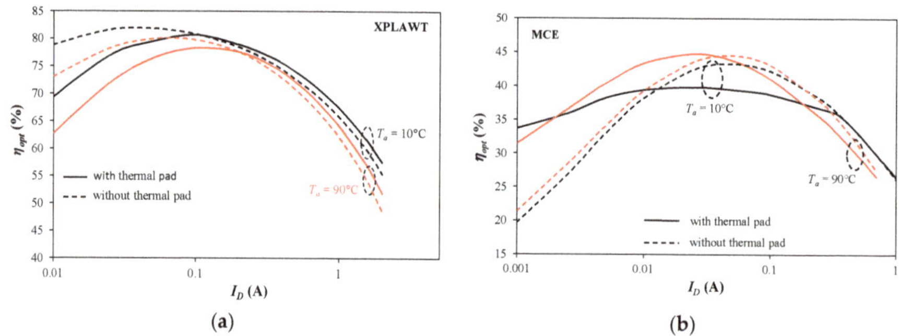

Figure 13. Measured dependences of radiant efficiency on the forward current for different cold plate temperature values for: (**a**) the XPLAWT; (**b**) MCE diodes.

Within the range of forward current values typical for the investigated diodes, the use of the thermal pad makes it possible to achieve higher efficiency η_{opt} and a higher power value of the emitted radiation. The change of the cold plate temperature does not influence in an essential way the relationship between the efficiency values computed for both manners of the diode mounting, since an increase of the cold plate temperature by 80 °C causes the decrease of efficiency by only several percent. It is interesting that the relationship between the characteristics obtained for the MCE diodes operating in different mounting conditions and the cold plate temperature are much different for the XPLAWT diode. Additionally, for the diode with the soldered thermal pad important differences in values of the efficiency η_{opt} obtained for different T_a values are noticeable. These differences exceed even 7%. For both diodes the maximum value of the efficiency is observed at the I_D current in the range from 20 to 50 mA.

In order to compare the influence of thermal pad on properties of the considered power LEDs, some values of parameters of these diodes are collected in Table 2.

Table 2. The values of selected electric, optical and thermal parameters of the tested power LEDs operating with soldered (WTP) or not soldered (NTP) thermal pads.

Diode	Thermal Pad	$\Delta V_D@I_D$	$R_{th}@I_D$	$\Delta T_j@I_D$	$\Delta \Phi_e@I_D$
XPLAWT	NTP	0.159 V@2 A	20.5 K/W@2 A	58.4 K@2 A	3.25 W/m²@2 A
	WTP	0.182 V@2 A	12.3 K/W@2 A	34.2 K@2 A	3.09 W/m²@2 A
MCE	NTP	0.312 V@0.7 A	20.4 K/W@1 A	37.2 K@0.7 A	0.72 W/m²@0.7 A
	WTP	0.251 V@0.7 A	10 K/W@1 A	27.5 K@0.7 A	0.75 W/m²@0.7 A

The change in forward voltage ΔV_D at maximal value of forward current considered here and the change of the cold plate temperature from 10 °C to 90 °C has much smaller values for the diode XPLAWT, but the highest value of ΔV_D voltage is observed for the MCE diode without soldered thermal pads. The value of thermal resistance R_{th} measured for the above mentioned values of forward current I_D is much smaller for the tested devices with soldered thermal pad. Differences in values of this parameter exceed even 50% for MCE diode. Consequently, an excess in junction temperature ΔT_j of the tested diode at selected value of current I_D caused by self-heating phenomenon is much higher for the diode operating without thermal pad.

Finally, the change in surface power density $\Delta \Phi_e$ at selected value of the current I_D at changing temperature of the cold plate by 80 °C is not big and it does not exceed 5% for both the diodes. Analyzing presented results of measurements, it can be stated that the use of thermal pads makes it possible to reduce visibly the values of thermal resistance and junction temperature of the considered devices, but changes in values of optical and electric parameters are not significant. Consistent with the classic theory of reliability of semiconductor devices [21] and obtained results of measurements, the lifetime of the power LEDs operating with soldered thermal pad can be three times longer than for the same devices operating without the thermal pad.

5. Conclusions

This paper presented the results of measurements illustrating the influence of the thermal pad on electric, thermal and optical parameters of selected power LEDs. As demonstrated, these parameters influence one another. For example, an increase in thermal resistance causes a decrease in electrical power consumed by the considered diodes. In turn, an increase in junction temperature causes a decrease in the density of emitted light flux.

Based on these results, it is clearly visible that the thermal pad considerably reduces thermal resistance of the considered diodes. This reduction is more visible for the diode situated on the larger MCPCBs. For the MCE diode, its thermal resistance is a decreasing function of forward current and cold plate temperature. It was demonstrated that due to differences in the thermal resistance of power LEDs mounted in a different manner, the differences in the junction temperature of these diodes at the same value of forward current can attain even 30 °C. Owing to the more effective cooling of the LED mounted with the thermal pad soldered, it is possible to attain higher power values of the emitted light. This power could be even by 10% higher than for the XPLAWT diode operating without the thermal pad. The observed difference in the power value is an increasing function of the forward current.

Furthermore, it was shown that using the thermal pad results in the increase in the value of the efficiency of the electrical energy conversion into light, especially for higher values of the diode forward current. It was also shown that the increased temperature over the ambient causes the decrease of the conversion efficiency. It is worth pointing out that the efficiency of the tested XPLAWT diodes for low values of forward current exceeded even 80%. Moreover, it is worth noticing that an increase in temperature of the cold plate causes a visible decrease in the diode forward voltage and in the power density of the emitted light. This decrease is the most visible for high forward current values.

The results of performed investigations could be useful for designers of solid-state light sources or substrates dedicated for their applications. Taking into account presented results it might be possible to significantly improve the efficiency and lifetime of the power LEDs used in light sources.

Author Contributions: Conceptualization, K.G. and M.J.; methodology, K.G., P.P., and M.J.; investigation, P.P. and T.T.; writing—original draft preparation, K.G. and M.J.; writing—review and editing, K.G., P.P., and M.J.; visualization, K.G. and P.P.; supervision, K.G. and M.J. All authors have read and agreed to the published version of the manuscript.

Funding: This research was funded by the program of the Ministry of Science and Higher Education called "Regionalna Inicjatywa Doskonałości" in the years 2019–2022, project number 006/RID/2018/19, sum of financing 11 870 000 PLN.

Conflicts of Interest: The authors declare no conflict of interest.

References

1. Schubert, E.F. *Light Emitting Diodes*, 3rd ed.; Rensselaer Polytechnic Institute: Troy, NY, USA, 2018.
2. Lasance, C.J.M.; Poppe, A. *Thermal Management for LED Applications*; Springer: Dordrecht, The Netherlands, 2014.
3. Poppe, A. Multi-domain compact modeling of LEDs: An overview of models and experimental data. *Microelectron. J.* **2015**, *46*, 1138–1151. [CrossRef]
4. Górecki, K. Modelling mutual thermal interactions between power LEDs in SPICE. *Microelectron. Reliab.* **2015**, *55*, 389–395. [CrossRef]
5. Górecki, P.; Górecki, K. Modelling dc Characteristics of the IGBT Module with Thermal Phenomena Taken into Account. In Proceedings of the 13th IEEE International Conference on Compatibility, Power Electronics and Power Engineering IEEE CPE POWERENG 2019, Sonderborg, Denmark, 23–25 April 2019. paper SF-001201. [CrossRef]
6. Zarębski, J.; Górecki, K. SPICE-aided modelling of dc characteristics of power bipolar transistors with selfheating taken into account. *Int. J. Numer. Model. Electron. Netw. Devices Fields* **2009**, *22*, 422–433. [CrossRef]
7. Górecki, K.; Zarębski, J.; Górecki, P.; Ptak, P. Compact thermal models of semiconductor devices—A review. *Int. J. Electron. Telecommun.* **2019**, *65*, 151–158. [CrossRef]
8. Bagnoli, P.E.; Casarosa, C.; Ciampi, M.; Dallago, E. Thermal resistance analysis by induced transient (TRAIT) method for power electronic devices thermal characterization—part I. Fundamentals and theory. *IEEE Trans. Power Electron.* **1998**, *13*, 1208–1219. [CrossRef]
9. Zhu, H.; Lu, J.; Wu, T.; Guo, Z.; Zhu, L.; Xiao, J.; Gao, Y.; Lin, Y.; Chen, Z. A Bipolar-Pulse Voltage Method for Junction Temperature Measurement of Alternating Current Light-Emitting Diodes. *IEEE Trans. Electron Devices* **2017**, *64*, 2326–2329. [CrossRef]
10. Farkas, G.; Bein, M.C.; Gaal, L. Multi Domain Modelling of Power LEDs Based on Measured Isothermal and Transient I-V-L Characteristics. In Proceedings of the 22nd International Workshop on Thermal Investigations of ICs and Systems Therminic, Budapest, Hungary, 21–23 September 2016; pp. 181–186.
11. Szekely, V. A new evaluation method of thermal transient measurement results. *Microelectron. J.* **1997**, *28*, 277–292. [CrossRef]
12. Blackburn, D.L. Temperature Measurements of Semiconductor Devices—A Review. In Proceedings of the 20th IEEE Semiconductor Thermal Measurement and Management Symposium SEMI-THERM, San Jose, CA, USA, 11 March 2004; pp. 70–80. [CrossRef]
13. Górecki, K.; Górecki, P.; Zarębski, J. Measurements of parameters of the thermal model of the IGBT module. *IEEE Trans. Instrum. Meas.* **2019**, *68*, 4864–4875. [CrossRef]
14. Górecki, K.; Zarębski, J. Modelling the influence of selected factors on thermal resistance of semiconductor devices. *IEEE Trans. Component. Packag. Manuf. Technol.* **2014**, *4*, 421–428. [CrossRef]

15. JEDEC Standard JESD51-51. Implementation of Electrical Test Method for the Measurement of Light-Emitting Diodes. 2012. Available online: https://www.jedec.org/sites/default/files/docs/JESD51-52.pdf (accessed on 18 July 2020).
16. Górecki, K.; Ptak, P. New method of measurements transient thermal impedance and radial power of power LEDs. *IEEE Trans. Instrum. Meas.* **2020**, *69*, 212–220. [CrossRef]
17. Górecki, K.; Dziurdzia, B.; Ptak, P. The influence of a soldering manner on thermal properties of LED modules. *Solder. Surf. Mt. Technol.* **2018**, *30*, 81–86. [CrossRef]
18. Górecki, K.; Ptak, P.; Janicki, M.; Torzewicz, T. Influence of Cooling Conditions of Power LEDs on Their Electrical, Thermal and Optical Parameters. In Proceedings of the 25th International Conference Mixed Design of Integrated Circuits and Systems MIXDES 2018, Gdynia, Poland, 21–23 June 2018; pp. 237–242.
19. Torzewicz, T.; Ptak, P.; Górecki, K.; Janicki, M. Influence of LED Operating Point and Cooling Conditions on Compact Thermal Model Element Values. In Proceedings of the 24th International Workshop on Thermal Investigastions of ICs and Systems Therminic, Sztokholm, Sweden, 26–28 September 2018. [CrossRef]
20. Górecki, K.; Ptak, P. The influence of the Mounting Manner of the Power LEDs on Its Thermal and Optical Parameters. In Proceedings of the 21st International Conference Mixed Design of Integrated Circuits and Systems MIXDES, Lublin, Poland, 19–21 June 2014; pp. 303–308.
21. Narendran, N.; Gu, Y. Life of LED-based white light sources. *J. Disp. Technol.* **2005**, *1*, 167–171. [CrossRef]
22. Castellazzi, A.; Gerstenmaier, Y.C.; Kraus, R.; Wachutka, G.K.M. Reliability analysis and modeling of power MOSFETs in the 42-V-PowerNet. *IEEE Trans. Power Electron.* **2006**, *21*, 603–612. [CrossRef]
23. Górecki, P.; Górecki, K.; Kisiel, R.; Myśliwiec, M. Thermal parameters of monocrystalline GaN Schottky diodes. *IEEE Trans. Electron. Devices* **2019**, *66*, 2132–2138. [CrossRef]
24. Dziurdzia, B.; Górecki, K.; Ptak, P. Influence of a soldering process on thermal parameters of large power LED modules. *IEEE Trans. Componen. Packag. Manuf. Technol.* **2019**, *9*, 2160–2167. [CrossRef]
25. Skwarek, A.; Ptak, P.; Górecki, K.; Hurtony, T.; Illes, B. Microstructure influence of SACX0307-TiO$_2$ composite solder joints on thermal properties of power LED assemblies. *Materials* **2020**, *13*, 1563. [CrossRef] [PubMed]
26. Górecki, K.; Ptak, P. Modelling LED lamps in SPICE with thermal phenomena taken into account. *Microelectron. Reliab.* **2017**, *79*, 440–447. [CrossRef]
27. Górecki, K.; Ptak, P.; Torzewicz, T.; Janicki, M. Influence of the Use of A Thermal Pad on Electric, Optical and Thermal Parameters of Selected Power LEDs. In Proceedings of the 25th International Workshop on thermal Investigations of ICs and Systems Therminic 2019, Lecco, Italy, 25–27 September 2019. [CrossRef]
28. Janicki, M.; Torzewicz, T.; Ptak, P.; Raszkowski, T.; Samson, A.; Górecki, K. Parametric compact thermal models of power LEDs. *Energies* **2019**, *12*, 1724. [CrossRef]
29. Janicki, M.; Kulesza, Z.; Torzewicz, T.; Napieralski, A. Automated Stand for Thermal Characterization of Electronic Packages. In Proceedings of the 27th Annual IEEE Semiconductor Thermal Measurement and Management Symposium Semi-Therm, San Jose, CA, USA, 20–24 March 2011; pp. 199–202. [CrossRef]
30. Datasheet T3Ster Equipment. Available online: https://corner-stone.com.tw/wp-content/uploads/2017/06/T3ster-technical-information.1.pdf (accessed on 18 June 2020).
31. Datasheet Thermal Scientific Fisher Haake A25. Available online: https://pim-resources.coleparmer.com/instruction-manual/12135-xx.pdf (accessed on 18 June 2020).
32. Datasheet DeltaOhm HD2302. Available online: http://www.otm.sg/files/HD2302_M_uk.pdf (accessed on 20 March 2020).
33. Górecki, K.; Ptak, P. New dynamic electro-thermo-optical model of power LEDs. *Microelectron. Reliab.* **2018**, *91*, 1–7. [CrossRef]
34. Blackburn, D.L.; Oettinger, F.F. Transient thermal response measurements of power transistors. *IEEE Trans. Ind. Electron. Control. Instrum.* **1976**, *IECI-22*, 134–141. [CrossRef]
35. Datasheet Cree XPLAWT-00-0000-000BV50E3. Available online: https://www.cree.com/led-components/media/documents/ds-XPL.pdf (accessed on 18 June 2020).
36. Datasheet Cree MCE4WT-A2-0000-JE5. Available online: https://www.cree.com/led-components/media/documents/XLampMCE.pdf (accessed on 18 June 2020).

MDPI
St. Alban-Anlage 66
4052 Basel
Switzerland
Tel. +41 61 683 77 34
Fax +41 61 302 89 18
www.mdpi.com

Energies Editorial Office
E-mail: energies@mdpi.com
www.mdpi.com/journal/energies